T0298707

Practical Biomedical Signal Analysis Using MATLAB®

Practical Biomedical Signal Analysis Using MATLAB®

Second Edition

Katarzyna J. Blinowska, Jarosław Żygierewicz

CRC Press
Taylor & Francis Group
Boca Raton London New York

CRC Press is an imprint of the
Taylor & Francis Group, an **informa** business

MATLAB® is a trademark of The MathWorks, Inc. and is used with permission. The MathWorks does not warrant the accuracy of the text or exercises in this book. This book's use or discussion of MATLAB® software or related products does not constitute endorsement or sponsorship by The MathWorks of a particular pedagogical approach or particular use of the MATLAB® software.

Second edition published 2022
by CRC Press
6000 Broken Sound Parkway NW, Suite 300, Boca Raton, FL 33487-2742

and by CRC Press
2 Park Square, Milton Park, Abingdon, Oxon, OX14 4RN

© 2022 Taylor & Francis Group, LLC

First edition published by CRC Press 2012

CRC Press is an imprint of Taylor & Francis Group, LLC

ISBN: 978-1-138-36441-7 (hbk)
ISBN: 978-1-032-10552-9 (pbk)
ISBN: 978-0-429-43135-7 (ebk)

Typeset in CMR10
by KnowledgeWorks Global Ltd.

DOI: 10.1201/9780429431357

Access the Support Material: www.routledge.com/9781138364417

Contents

About the Series

The *Series in Medical Physics and Biomedical Engineering* describes the applications of physical sciences, engineering, and mathematics in medicine and clinical research.

The series seeks (but is not restricted to) publications in the following topics:

- Artificial organs

- Assistive technology

- Bioinformatics

- Bioinstrumentation

- Biomaterials

- Biomechanics

- Biomedical engineering

- Clinical engineering

- Imaging

- Implants

- Medical computing and mathematics

- Medical/surgical devices

- Patient monitoring

- Physiological measurement

- Prosthetics

- Radiation protection, health physics, and dosimetry

- Regulatory issues

- Rehabilitation engineering

- Sports medicine

- Systems physiology

- Telemedicine

- Tissue engineering

- Treatment

The *Series in Medical Physics and Biomedical Engineering* is an international series that meets the need for up-to-date texts in this rapidly developing field. Books in the series range in level from introductory graduate textbooks and practical handbooks to more advanced expositions of current research.

The *Series in Medical Physics and Biomedical Engineering* is the official book series of the International Organization for Medical Physics.

The International Organization for Medical Physics

The International Organization for Medical Physics (IOMP) represents over 18,000 medical physicists worldwide and has a membership of 80 national and six regional organizations, together with a number of corporate members. Individual medical physicists of all national member organizations are also automatically members.

The mission of the IOMP is to advance medical physics practice worldwide by disseminating scientific and technical information, fostering the educational and professional development of medical physics and promoting the highest quality medical physics services for patients.

A World Congress on Medical Physics and Biomedical Engineering is held every three years in cooperation with the International Federation for Medical and Biological Engineering (IFMBE) and the International Union for Physical and Engineering Sciences in Medicine (IUPESM). A regionally based international conference, the International Conference on Medical Physics (ICMP) is held between world congresses. The IOMP also sponsors international conferences, workshops and courses.

The IOMP has several programmes to assist medical physicists in developing countries. The joint IOMP Library Programme supports 75 active libraries in 43 developing countries, and the Used Equipment Programme coordinates equipment donations. The Travel Assistance Programme provides a limited number of grants to enable physicists to attend the world congresses.

The IOMP co-sponsors the *Journal of Applied Clinical Medical Physics*. The IOMP publishes, twice a year, an electronic bulletin, *Medical Physics World*. The IOMP also publishes e-Zine, an electronic news letter, about six times a year. The IOMP has an agreement with Taylor & Francis for the publication of the *Series in Medical Physics and Biomedical Engineering* series of textbooks. IOMP members receive a discount.

The IOMP collaborates with international organizations, such as the World Health Organization (WHO), the International Atomic Energy Agency (IAEA) and other international professional bodies such as the International Radiation Protection Association (IRPA) and the International Commission on Radiological Protection (ICRP), to promote the development of medical physics and the safe use of radiation and medical devices.

Guidance on education, training and professional development of medical physicists is issued by the IOMP, which is collaborating with other professional organizations in development of a professional certification system for medical physicists that can be implemented on a global basis.

The IOMP website (www.iomp.org) contains information on all the activities of the IOMP, policy statements 1 and 2 and the 'IOMP: Review and Way Forward' which outlines all the activities of the IOMP and plans for the future.

Preface

This book is intended to guide all those working in the field of biomedical signal analysis and application, particularly for graduate students, researchers at an early stage of their career, industrial researchers, and people interested in the development of signal processing methods. The book is different from other monographs, which are usually collections of papers written by several authors. We tried to present a coherent view of various signal processing methods in the context of their application. Not only do we wish to present the current techniques of biomedical signal processing, but we also want to provide guidance on which methods are appropriate for the given task and given kind of data.

One of the motivations for writing this book was the longstanding experience of the authors in reviewing manuscripts submitted to journals and conference proceedings, which showed how often the signal processing methods are misused. Quite often, sophisticated, but at the same time non-robust and prone to systematic errors, methods are applied for the tasks where more straightforward methods would work better. In this book, we aim to show the advantages and disadvantages of different techniques in the context of their applications.

In the first part of the book, we describe the methods of signal analysis, including the most advanced and new approaches, in an easy and accessible way. We illustrate them with MATLAB® Live Scripts. We omitted proofs of the theorems when necessary, sending the reader to the more specialized mathematical literature. To make the book a practical tool, we refer to MATLAB® routines when available and to the software freely available on the Internet.

In the second part of the book, we describe the application of the methods presented in the first part of the book to the different biomedical signals: electroencephalogram (EEG), electrocorticogram (ECoG), event-related potential (ERP), electrocardiogram (ECG), heart rate variability signal (HRV), electromyograms (EMG), magnetoencephalograms (MEG), magnetocardiograms (MCG), phonocardiograms (PCG), and otoacoustic emissions (OAE).

In this second edition, we also address the analysis of fMRI (BOLD) and functional near-infrared spectroscopy (fNIRS) time series. We discuss new topics that gained interest in recent years, namely, phase-amplitude coupling, wearable devices, multimodal signal analysis, and brain-computer interfaces. Major updates concern multiple channel analysis and connectivity measures. We included a short introduction presenting the basic syntax and functionality relevant to signal processing for those new to MATLAB.

Different approaches to solving particular problems are presented to indicate which methods seem to be most appropriate for a given application. Possible pitfalls, which may be encountered in applying the specific methodology, are pointed out.

We hope that this book will be a practical help to students and researchers in choosing the appropriate methods, designing their own, and adding new value to the growing field of biomedical research.

You can access MATLAB scripts and data in the support material section at: www.routledge.com/9781138364417

List of Abbreviations

AIC	Akaike information criterion	EP	evoked potentials
ApEn	approximate entropy	ERD	event-related
ANN	artificial neural networks		desynchronization
AR	autoregressive model	ERP	event-related potential
ARMA	autoregressive moving	ERS	event-related
	average model		synchronization
BAEP	brain stem auditory evoked	FA	factor analysis
	potentials	FAD	frequency amplitude
BCG	ballistocardiogram		damping (method)
BSPM	body surface potential	fECG	fetal electrocardiogram
	mapping	FDR	false discovery rate
BSR	burst suppression ratio	FFT	fast Fourier transform
BSS	blind source separation	FIR	finite impulse response
CAP	cyclic alternating pattern	fMCG	fetal magnetocardiogram
CFC	cross-frequency coupling	fMRI	functional magnetic
CSD	current source density		resonance imaging
CWT	continuous wavelet	fNIRS	functional near-infrared
	transform		spectroscopy
DFA	detrended fluctuation	FT	Fourier transform
	analysis	FWE	family wise error
DFT	discrete Fourier transform	FWER	family wise error rate
IDFT	inverse discrete Fourier	GAD	Gabor atom density
	transform	GFP	global field power
DWT	discrete wavelet transform	GGC	Granger-Geweke causality
ECG	electrocardiogram	GS	generalized synchronization
ECoG	electrocorticogram	HHT	Hilbert-Huang transform
EEG	electroencephalogram	HMM	hidden Markov model
EF	evoked fields	HSD	honesty significant
EGG	electrogastrogram		difference test
EIG	electrointestinogram	IC	independent component
EMD	empirical mode	ICA	independent component
	decomposition		analysis
EMG	electromyogram	IDFT	inverse discrete Fourier
EEnG	electroenterogram		transform
EOG	electrooculogram	iEEG	intracranial
ERC	event-related causality		electroencephalogram

IIR	infinite impulse response	PLV	phase-locking value
IQ	information quantity	PM	Poincare map
KL	Kullback-Leibler (entropy)	PPG	photoplethysmogram
LDA	linear discriminant analysis	PS	phase synchronization
LDS	linear dynamic system	PSD	power spectral density
LFP	local field potentials	PSP	post-synaptic potential
LTI	linear time invariant	REM	rapid eye movement
MCG	magnetocardiogram	RP	recurrence plot
MCP	multiple comparison problem	SaEn	sample entropy
		SCG	seismocardiogram
MDL	minimum description length (criterium)	sEMG	surface electromyogram
		SIQ	subband information quantity
mECG	maternal electrocardiogram		
MEnG	magnetoenterogram	SnPM	statistical non-parametric mapping
MGG	magnetogastrogram		
MEG	magnetoelectroence-phalogram	SOAE	spontaneous otoacoustic emissions
MI	mutual information	SOBI	second-order blind inference
MLP	multilayer perceptron	SPM	statistical parametric mapping
MMSE	minimum mean square error		
MP	matching pursuit	SSOAE	synchronized spontaneous otoacoustic emissions
MPC	multiple comparison problem		
		STFT	short time Fourier transform
MPF	median power frequency		
MTM	multi taper method	SVD	singular value decomposition
MU	motor unit		
MUAP	motor unit action potential	TE	transfer entropy
OAE	otoacoustic emissions	TWA	T-wave alternans
PAC	phase-amplitude coupling	SWA	slow wave activity
PCA	principal component analysis	WP	wavelet packets
		WT	wavelet transform
PCG	phonocardiogram	WVD	Wigner-Ville distribution
PCI	phase clustering index		

1

A Short Introduction to MATLAB®

1.1 Introduction

MATLAB® is a commercial platform offered by The MathWorks, Inc., the USA which delivers a high-level, matrix-based language and an integrated development environment. Together with specialized libraries, so-called tool-boxes, it is a very efficient system for fast prototyping, and complex scientific and engineering computations. What is also very important many algorithms used in biomedical signal processing were implemented and published as MATLAB functions or toolboxes. In the following sections, we shortly present basic syntax and concepts, needed for the understanding of the examples given in the book and in accompanying Live Script demos. This may be helpful for a novice MATLAB user. A deeper knowledge of MATLAB can be obtained from many MATLAB books and the very useful build-in documentation system of MATLAB.

1.2 Where Is Help?

The documentation can be accessed by the GUI, or from the command prompt. In the latter case just type `doc` (this opens the graphical documentation window) or `help` (if you prefer a more compact textual help). The general set of commands can be displayed by issuing `doc matlab/general`.

1.3 Vectors and Matrixes

In signal processing, it is very convenient to implement the signals as vectors or matrixes. The most basic type of variable in MATLAB is a matrix. In fact, a scalar value is represented as 1×1 matrix. So vectors are $N \times 1$ (a column vector) or $1 \times N$ (a row vector).

DOI: 10.1201/9780429431357-1

The matrixes can be entered either from command line/editor (type and execute the examples in the *command line*, and look into the *workspace window*):

```
A=[1 2 3 4; 5 6 7 8; 8 9 1 2];
disp(A)
```

or produced by a function:

```
B=rand(3,3);
disp(B)
```

or, in practical cases most often, loaded from a file (this we will show later in Sect. 1.8).

Note the semicolon at the end of each line—it prevents MATLAB from displaying its result in the command window. A semicolon inside the matrix forming command finishes a given row and starts another one.

1.4   Matrix Operations

1.4.1   Algebraic Operations

In an intuitive way we can do the matrix algebraic operations:

```
B = [1 2;3 4];
disp(B)
C = B + B;
disp(C)
D = C - 2* B;
disp(D)
M = B/B;
disp(M)
```

An operator preceded by a dot evaluates the operation element-wise. Please, compare results of:

```
G = B*B;
disp(G)
H = B.*B;
disp(H)
```

There is a set of matrix operators. The basic ones are:

transpose: `B_tr = B';`

determinant: `d = det(B);`

trace: `d = trace(B);`

diagonal: `di = diag(B);`

inverse: `B_inv = B^(-1);`

sum: `s = sum(B);` this sums by default along the first dimension, other direction can be indicated in the function call e.g., `s = sum(B,2);`

1.4.2   Matrix Indexing

Matrix addressing starts from 1. We get an element of a matrix like this:

```
B = [1 2;3 4];
disp(B)
      1      2
      3      4
>> B(1,2)
ans =
2
```

We assign its value in this way:

```
>> B(1,2) = 4;
>> B
B =
1 4
3 4
```

Note, that when you modify matrix elements, the matrix size adjusts automatically and it can create elements that were not directly assigned, setting them to 0. Note the element B(2,3) in the example below:

```
>> B(1,3) = 4;
>> B
B =
1 4 4
3 4 0
```

Range control is done only when retrieving matrix elements:

```
>> B(3,1)
??? Index exceeds matrix dimensions.
```

The colon operator ":", is one of the most useful operators in MATLAB. It occurs in several different use cases. The expression 1:10 produces a row vector with elements from 1 to 10.

```
>> 1:10
ans =
1 2 3 4 5 6 7 8 9 10
```

If we give an increment, then we get a vector with the desired difference between elements, for example:

```
>> 10: -2.5: 0
ans =
10.0000 7.5000 5.0000 2.5000 0
```

This operator used in the matrix index gives an easy access to its fragments, e.g.: `A(1:k, j)` gets the first k elements of the jth column of matrix A. Semicolon by itself stands for all the elements of a given row or column e.g.: `sum(A(:, end))` calculates the sum of the elements of the last (keyword **end**) column of A. The operator [] is used to concatenate matrixes. For example:

```
B = ones(2,2);
C = [B B + 1; B + 2 B + 3];
disp(C);
```

We can delete columns or rows in a matrix. This shows how to delete the 2nd row of matrix C (compare content of matrix C before and after the deletion):

```
C = [1 2 3 4; 5 6 7 8; 9 10 11 12; 13 14 15 16]
disp(C);
C(2,:) = [];
disp(C);
```

1.4.3   Logical Indexing

Evaluation of logical expressions on matrixes yields logical arrays, with elements 0 or 1, where 1 indicates the elements for which the expression was true.

```
>> a
a =
1 2 3
4 5 6
0 1 1
>> a>2
ans =
0 0 1
1 1 1
0 0 0
```

We can use such array as a mask to modify the original matrix, as shown below:

```
>> a(ans) = 7
a =
1 2 7
7 7 7
0 0 0
```

1.4.4 Example Exercise

Create a matrix with numbers of the magic square of size 4, then count the sum of the elements in each row, column, and both diagonals. A possible solution[1]:

```
M = magic(4);
% count the sum of the elements in each row,
s_r = sum(M,2);
disp(s_r)
% in each column,
s_c = sum(M,1);
disp(s_c)
% and both diagonals.
s_d1 = sum(diag(M));
disp(s_d1)
s_d2 = sum(diag(fliplr(M)));
disp(s_d2)
```

1.5 Conditionals

Often an algorithm needs to branch based on runtime conditions. MATLAB has syntax for this in the form of `if` statement, try this example for different values of x:

```
x = -1;
if x == 0
    disp('x is zero')
elseif x > 0
    disp('x is positive')
else
    disp('x is negative')
end
```

Note that the block of the statement finishes with the keyword **end**. The indentions are not necessary, but they make the code more readable.

If there are many options, it may be better to use the switch command. For instance:

```
method = 'bilinear';
switch method
    case {'linear','bilinear'}
```

---

[1]The code is delivered in `matlab/c1/Ch1_magic_square.m`

```
            disp('Method is linear')
    case 'cubic'
            disp('Method is cubic')
    case 'nearest'
            disp('Method is nearest')
    otherwise
            disp('Unknown method.')
end
```

1.6   Loops

We would use loops if we need to execute a block of commands for a number of times. There are two types of loops:

while executes a group of statements an indefinite number of times, based on some logical condition; example:

```
i = 10
while i>1
    i = i-1;
    disp(i)
end
```

for executes a group of statements a fixed number of times, example:

```
for k = 1:10
    disp(k*(1:10))
end
```

Additionally, we can modify behavior of loops with:

continue passes control to the next iteration of a for or while loop, skipping any remaining statements in the body of the loop

break terminates execution of a for or while loop.

1.7   Scripts and Functions

A sequence of MATLAB commands saved in a text file (with extension .m) is a script. You can execute them by entering the file name (without extension) at the prompt in the command window. This is equivalent to typing all of the commands from the file. The script has access to all variables found in the workspace, variables created in the script are visible in the workspace. Most MATLAB commands are functions, some are precompiled and work very fast, but a significant number is delivered as text files that are interpreted at runtime (they run slower). The advantage is that we can look into these files and learn a lot.

In MATLAB, you can also create your own functions—built from existing ones. The file containing a function should have the same name as the function with the extension ".m". This is because MATLAB identifies the names of scripts and functions by the names of files on the search path. The first line of a function file defines the syntax of a function call, e.g.:

```
function [mean, stdev] = stat(x)
    % STAT computes mean and standard deviation of x.
    n = length(x);
    mean = sum(x) / n;
    stdev = sqrt( sum((x - mean).^ 2) / n);
```

The above code[2] defines the function stat (should be saved in the file stat.m). This function accepts the vector x as an argument and returns two values: mean and stdev. Variables used inside the function are local, i.e. they are not visible in the workspace outside the scope of the function.

You can call the stat function as:

```
>> x = 1:10;
>> [m, s] = stat(x)
m =
5.5000
s =
2.8723
```

Return from the function occurs after reaching the end of the function body. An earlier conditional return can be achieved with the command **return**.

1.8   Working with Binary Files

1.8.1   Saving to and Loading from Binary Files

The following examples[3] illustrate how to write two sine functions at 2 Hz and 3 Hz to a binary file:

First, prepare the signals.

---

[2]The code is delivered in matlab/c1/Ch1_stat.m
[3]The code is delivered in matlab/c1/Ch1_working_with_binary_files.m together with the exemplary signals.

```
T = 1;
Fs = 128;
t = 0:1/Fs:T-1/Fs;
s1 = sin(2*pi*2*t);
s2 = sin(2*pi*3*t);
signal = zeros(2,T * Fs);
signal(1,:) = s1;
signal(2,:) = s2;
```

We store the consecutive signals in the rows of matrix signal. This is convenient for writing them as a multiplexed file.[4]

Saving the binary signal:

```
f_out = fopen('test_signal1.bin', 'w','l'); %open a file for writing
fwrite(f_out, signal,'double','ieee-le'); % write the signal data
                                           % to the file identified by
                                           % f_out as double precision
                                           % floats with the machine
                                           % format little-endian
fclose(f_out); % close the file
```

The opposite—reading a binary file:

```
ch = 2;                           % specify the number of channels
f_in = fopen('test_signal1.bin', 'r'); % open the file for reading
s = fread(f_in,[ch,Inf],'double','ieee-le' ); % read all data (Inf)
                                               % from the file
fclose(f_in); % close the file
```

Plot and compare the saved and the loaded data:

```
figure()
subplot(221)
plot(t,signal(1,:))
ylabel('channel 1')
title('saved signal')
subplot(223)
plot(t,signal(2,:))
ylabel('channel 2')
subplot(222)
plot(t,s(1,:))
title('loaded signals')
subplot(2,2,4)
plot(t,s(2,:))
```

---

[4]In a multiplexed file containing n samples of k channel signal, samples are ordered in the file in the following way:

$$\underbrace{s_1c_1, s_1c_2, s_1c_3, \ldots, s_1c_k,}_{1^{st}\text{ sample}} \underbrace{s_2c_1, s_2c_2, s_2c_3, \ldots, s_2c_k,}_{2^{nd}\text{ sample}} \ldots, \underbrace{s_nc_1, s_nc_2, s_nc_3, \ldots, s_nc_k,}_{k^{th}\text{ sample}}$$

where s is sample and c is channel.

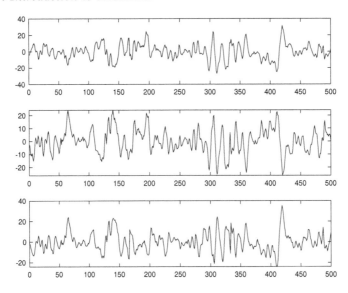

FIGURE 1.1
Correct traces of the signal form the file `3_channel_signal.bin`.

1.8.2   Saving and Loading Signals Using .mat Files

If working only in the MATLAB environment it is convenient to use `save/load` functions. The same operations as above can be performed in the following way:

```
% saving the binary signal:
save('test_signal1.mat', 'signal'); % note, variables
                                     % are given by name (string)
% the opposit - reading a binary file
load('test_signal1.mat');
```

1.8.3   Exercises

1.8.3.1   Unknown Data Type

Suppose, we have a 3-channel binary file with an EEG signal. Unfortunately, we have forgotten what is the type of the recorded data. Load the file `3_channel_signal.bin` and using the plots and Figure 1.1 with signal fragments guess which of the following type of stored variables is true: "float", "uint" or "double". Note, how the traces change if you enter the wrong data type.

1.8.3.2 Unknown Number of Channels

Suppose we have a n-channel binary file with some ECG, EEG, and EMG signals, the type of variables is float, but we don't know the correct number of channels. Load **n_channel_signal.bin** file and try to guess the correct number of channels by plotting the signals. If you guess the correct number of channels you should see the characteristic QRS complex of an ECG signal in the last channel.

2

Introductory Concepts

2.1 Stochastic and Deterministic Signals, Concepts of Stationarity and Ergodicity

A signal is a physical quantity that can carry information. Physical and biological signals may be classified as stochastic or deterministic. A stochastic signal contrary to a deterministic one cannot be described by a mathematical function. An example of a deterministic signal may be the time course of voltage on a discharging capacity or the position of a pendulum during its swing. A typical random process may be the number of particles emitted by the radioactive source in a unit of time or the output of a noise generator. Physiological signals can be qualified as stochastic signals, but they usually consist of a deterministic and a random component. In some signals, the random component is more pronounced while in others the deterministic element prevails. An example of a stochastic signal, where a random component is important is EEG. The other class of signals can be represented by an ECG which has a quite pronounced deterministic component related to propagation of the electrical activity in the heart structures, although some random component coming from biological noise is also present.

A process may be observed in time. A set of observations of quantity x in function of time t forms the time series $x(t)$. In many cases, the biophysical time series can be considered as a realization of a process, in particular, a stochastic process.

If K will be the assembly of k events ($k \in K$) and to each of these events we assign function $x_k(t)$ called realization of the process $\xi(t)$, the stochastic process can be defined as a set of functions:

$$\xi(t) = \{x_1(t), x_2(t), \ldots, x_N(t)\} \tag{2.1}$$

where $x_k(t)$ are the random functions of variable t.

In the framework of the theory of stochastic processes a physical or biophysical process can be described by means of the expected values of the estimators found by the ensemble averaging over realizations. The expected value of a stochastic process is an average over all realizations of the process, weighted by the probabilities of their occurrence. The mean value $\mu_x(t_1)$ of the stochastic process $\xi(t)$ in the time t_1 can be found by summation of the

---

DOI: 10.1201/9780429431357-2

actual values of each realization in time t_1 weighted by the probability of the occurrence of the given realization $p(x_k, t_1)$:

$$\mu_x(t_1) = E[\xi(t_1)] = \lim_{N \to \infty} \sum_{k=1}^{N} x_k(t_1) p(x_k, t_1) \qquad (2.2)$$

$E[.]$ denotes expected value. In general the expected value of the given function $f(\xi)$ may be expressed by:

$$E[f(\xi(t_1))] = \lim_{N \to \infty} \sum_{k=1}^{N} f(x_k(t_1)) p(x_k, t_1) \qquad (2.3)$$

If the probability of occurrence of each realization is the same, which frequently is the case, the equation (2.3) is simplified:

$$E[f(\xi(t_1))] = \lim_{N \to \infty} \frac{1}{N} \sum_{k=1}^{N} f(x_k(t_1)) \qquad (2.4)$$

In particular, function $f(\xi)$ can represent moments or joint moments of the processes $\xi(t)$. Moment of order n is : $f(\xi) = \xi^n$. In these terms mean value (2.2) is the first-order moment and mean square value ψ^2 is the second-order moment of the process:

$$\psi^2(t_1) = E[\xi^2(t_1)] = \lim_{N \to \infty} \sum_{k=1}^{N} x_k^2(t_1) p(x_k, t_1) \qquad (2.5)$$

Central moments m_n about the mean are calculated in respect to the mean value μ_x. The first central moment is zero. The second order central moment is variance:

$$m_2 = \sigma_x^2 = E\left[(\xi - \mu_x)^2\right] \qquad (2.6)$$

where σ is the standard deviation. The third order central moment in an analogous way is defined as:

$$m_3 = E\left[(\xi - \mu_x)^3\right] \qquad (2.7)$$

Parameter β_1 related to m_3:

$$\beta_1 = \frac{m_3}{m_2^{3/2}} = \frac{m_3}{\sigma^3} \qquad (2.8)$$

is called skewness, since it is equal to 0 for symmetric probability distributions of $p(x_k, t_1)$.
Kurtosis:

$$\beta_2 = \frac{m_4}{m_2^2} \qquad (2.9)$$

is a measure of flatness of the distribution. For the normal distribution, kurtosis is equal to 3. A high kurtosis distribution has a sharper peak and longer, fatter tails, in contrast to a low kurtosis distribution which has a more rounded peak and shorter thinner tails. Often instead of kurtosis parameter e—excess of kurtosis: $e = \beta_2 - 3$ is used. The subtraction of 3 at the end of this formula is often explained as a correction to make the kurtosis of the normal distribution equal to zero. For the normally distributed variables (variables whose distribution is described by Gaussian), central odd moments are equal to zero and central even moments take values:

$$m_{2n+1} = 0 \qquad m_{2n} = (2n - 1)m_2^{2n} \tag{2.10}$$

Calculation of skewness and kurtosis can be used to assess if the distribution is roughly normal.

The relation of two processes $\xi(t) = \{x_1(t), \ldots, x_N(t)\}$ and $\eta(t) = \{y_1(t), \ldots, y_N(t)\}$ can be characterized by joint moments. Joint moment of the first order $R_{xy}(t)$ and joined central moment $C_{xy}(t)$ of process $\xi(t)$ are called, respectively, cross-correlation and cross-covariance:

$$R_{xy}(t_1, \tau) = E\left[\xi(t_1)\eta(t_1 + \tau)\right] \tag{2.11}$$

$$C_{xy}(t_1, \tau) = E\left[(\xi(t_1) - \mu_x(t_1))(\eta(t_1 + \tau) - \mu_y(t_1))\right] \tag{2.12}$$

where τ is the time shift between signals x and y.

A special case of the joint moments occurs when they are applied to the same process, that is $\xi(t) = \eta(t)$. Then the first order joint moment $R_x(t)$ is called autocorrelation and joined central moment $C_x(t)$ of process $\xi(t)$ is called autocovariance.

Now we can define:

Stationarity: For the stochastic process $\xi(t)$ the infinite number of moments and joint moments can be calculated. If all moments and joint moments do not depend on time, the process is called stationary in the strict sense. In a case when mean value μ_x and autocorrelation $R_x(\tau)$ do not depend on time the process is called stationary in the broader sense, or weakly stationary. Usually weak stationarity implies stationarity in the strict sense, and for testing stationarity usually only mean value and autocorrelation are calculated.

Ergodicity: The process is called ergodic when its mean value calculated in time (for the infinite time) is equal to the mean value calculated by ensemble averaging (according to equation 2.2). Ergodicity means that one realization is representative of the whole process, namely that it contains the whole information about the process. Stationarity of a process implies its ergodicity. For ergodic processes, we can describe the properties of the process by averaging one realization over time, instead of ensemble averaging.

Under the assumption of ergodicity moment of order n is expressed by:

$$m_n = \lim_{T \to \infty} \int_0^T x^n(t)p(x)\,dt \qquad (2.13)$$

2.2   Discrete Signals

In nature, the biomedical signals are continuous in time and in space. We use computers to store and analyze the data. To adapt the natural continuous data to the digital computer systems we need to digitize them. That is, we have to sample the physical values in certain moments in time or places in space and assign them a numeric value with finite precision. This leads to the notion of two processes: sampling (selecting discrete moments in time) and quantization (assigning a value of finite precision to an amplitude).

2.2.1   The Sampling Theorem

Let's first consider sampling. The most crucial question is how often the signal $f(t)$ must be sampled? The intuitive answer is that, if $f(t)$ contains no frequencies[1] higher than F_N, $f(t)$ cannot change to a substantially new value in a time less than one-half cycle of the highest frequency; that is, $\frac{1}{2F_N}$. This intuition is indeed true. The Nyquist-Shannon sampling theorem [544] states that:

> If a function $f(t)$ contains no frequencies higher than F_N cycles per second, it is completely determined by giving its ordinates at a series of points spaced $\frac{1}{2F_N}$ seconds apart.

The frequency F_N is called the Nyquist frequency and $2F_N$ is the minimal sampling frequency. The "completely determined" phrase means here that we can restore the unmeasured values of the original signal, given the discrete representation sampled according to the Nyquist-Shannon theorem (Figure 2.1).

A reconstruction can be derived via sinc function $f(x) = \frac{\sin \pi x}{\pi x}$. Each sample value is multiplied by the sinc function scaled so that the zero-crossings of the sinc function occur at the sampling instants and that the sinc function's central point is shifted to the time of that sample, nT, where T is the sampling period (Figure 2.1 b). All of these shifted and scaled functions are then added together to recover the original signal (Figure 2.1 c). The scaled and time-shifted sinc functions are continuous, so the sum is also continuous, which makes the result of this operation a continuous signal. This procedure is represented by the Whittaker-Shannon interpolation formula. Let $x[n] := x(nT)$

---

[1] Frequencies are measured in cycles per second—cps, or in Hz—$[\text{Hz}] = \frac{1}{[s]}$.

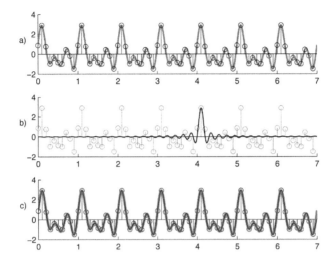

FIGURE 2.1
Illustration of sampling and interpolating of signals. a) The continuous signal
(gray) is sampled at points indicated by circles. b) The impulse response of
the Whittaker-Shannon interpolation formula for a selected point. c) Reconstruction (black) of signal computed according to (2.14).

for $n \in \mathbb{Z}$ be the nth sample. We assume that the highest frequency present
in the sampled signal is F_N and that it is smaller than half of the sampling
frequency $F_N < \frac{1}{2}F_s$. Then the function $f(t)$ is represented by:

$$f(t) = \sum_{n=-\infty}^{\infty} x[n]\mathrm{sinc}\left(\frac{t - nT}{T}\right) = \sum_{n=-\infty}^{\infty} x[n]\frac{\sin \pi(2F_st - n)}{\pi(2F_st - n)} \qquad (2.14)$$

MATLAB example

MATLAB code reproducing Figure 2.1 is given in:
`matlab/c2/Ch2_Whittaker_Shannon_interpolation.m`

2.2.1.1   Aliasing

What happens if the assumption of the sampling theorem is not fulfilled and
the original signal contains frequencies higher than the Nyquist frequency? In
such cases we observe an effect called *aliasing*—different signal components
become indistinguishable (aliased). If the signal of frequency $f_0 \in \left(\frac{1}{2}F_s, F_s\right)$
is sampled with frequency F_s then it has the same samples as the signal with
frequency $f_1 = F_s - f_0$. Note that $|f_1| < \frac{1}{2}F_s$. The sampled signal contains

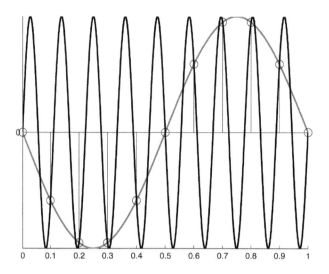

FIGURE 2.2

Illustration of the aliasing effect. Samples taken every 0.1 s from a 9 Hz sinusoid (black) are exactly the same as these taken from the 1 Hz sinusoid (gray).

additional low-frequency components that were not present in the original signal. An illustration of that effect is shown in Figure 2.2.

MATLAB example: Sampling and aliasing

In this exercise, we want to examine the effect of signal sampling in time. *If I sample a signal with a frequency of Fs = 100 Hz, can I reproduce the signal at any frequency?* A possible implementation of the steps below are in the file `matlab/c2/Ch2_sampling.m`, but we encourage you to try to implement the following steps by yourself.

1. Create vector t representing the time of t = 1 s sampled with the frequency Fs =100 Hz.

2. Create the signal s as sine with frequency f = 10 Hz.

3. Plot this signal with points and lines.

4. Plot signals with frequencies of 10, 20, 40, 50, 90 Hz

Something on the previous figure looked strange. Everything was fine up to the frequency of 50 Hz. But the sine with frequency 90 Hz looked very similar to that with frequency 10 Hz. This is the effect of aliasing. Let's have a closer look. For our needs, we will generate signals with a very high frequency, which will be for us an approximation of continuous signals. With their help, we will present the effect of identifying (aliasing).

1. Create a vector representing the "almost continuous" time. It will be 1000 values between [0,1] taken with an interval of 0.001.

2. Now, generate two sine waves: one with frequency −1 and the second with frequency 9.

3. Plot both sine waves.

4. Now, sample the time and our "almost continuous" sine waves with a sampling period of 0.1. (You have to take every 100th element).

5. Draw the points from the sampled signals overlaid (use hold on) on the background of "almost continuous" sine waves. To make the points visible, we suggest using the o and + markers.

You should see something similar to the Figure 2.2. Please observe the mutual position of the points. *Is it possible to distinguish a sinusoid with a frequency of −1 Hz from a sinusoid with a frequency of 9 Hz if both are sampled at a frequency of 10 Hz? How can you generalize this observation?*

2.2.2 Quantization Error

When we measure signal values, we usually want to convert them to numbers for further processing. The numbers in digital systems are represented with a finite precision. The analog to digital converter (ADC) uses a certain number, N, of bits to represent the number. It divides the full range R of measurement values into 2^N levels. Therefore we can estimate the quantization error as not bigger than $\frac{R}{2^N}$ (Figure 2.3).

This error sometimes has to be taken into consideration, especially when the amplitudes of measured signals span across orders of magnitude. An example here can be EEG measurement. Let's assume that we have adjusted the amplification of signal so that $\pm 200\,\mu V$ covers the full range of a 12 bit ADC. This range is divided into bins of $400/2^{12} = 400/4096 \approx 0.1\,\mu V$. It means that we measure the amplitude of the signal with precession $\pm 0.05\,\mu V$.

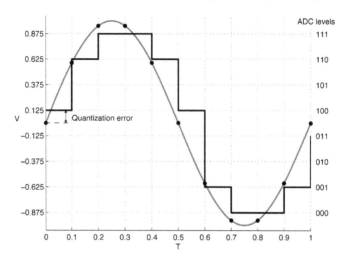

FIGURE 2.3
Illustration of the quantization error for a 3 bit ADC. The digitized representation (black line) of the continuous sinusoid (gray line). The range of 2 [V] is divided into 8 levels. The sampled signal values (black dots) are rounded to the nearest ADC level.

2.3 Linear Time Invariant Systems

In signal processing, there is an important class of systems called linear time-invariant systems—LTI (in case of sampled signals this is sometimes named linear shift-invariant). We can think of the system as a box which modifies the input with a linear operator L to produce the output:

$$input \longrightarrow \boxed{L\{input\}} \longrightarrow output$$

The basic properties of such a system are:

1. Linearity: superposition of inputs produces superposition of outputs, formally: if an input $x_1(t)$ produces output $y_1(t)$:

$$L\{x_1(t)\} = y_1(t)$$

and input $x_2(t)$ produces output $y_2(t)$

$$L\{x_2(t)\} = y_2(t)$$

then the output of the superposition will be:

$$L\{a_1x_1(t)+a_2x_2(t)\} = a_1L\{x_1(t)\}+a_2L\{x_2(t)\} = a_1y_1(t)+a_2y_2(t)$$

2. Time invariance: for a given input the system produces identical outputs no matter when we apply the input. More formally:
   if $L\{x(t)\} = y(t)$ then $L\{x(t+T)\} = y(t+T)$

The important property of LTI systems is that they are completely characterized by their impulse response function. The impulse response function can be understood as the output of the system due to the single impulse[2] at the input. It is so because we can think of the input signal as consisting of such impulses. In the case of a discrete signal, it is very easy to imagine. In case of the continuous signal, we can imagine it as a series of infinitely close, infinitely narrow impulses. For each input impulse, the system reacts in the same way. It generates a response which is proportional (weighted by the amplitude of the impulse) to impulse response function. The responses to consecutive impulses are summed up with the response due to the former inputs (Figure 2.4). Such operation is called *convolution*. For convolution, we shall use symbol: $*$. The process is illustrated in Figure 2.4. Formally the operation of the LTI system can be expressed in the following way. Let's denote the impulse response function as $h[n]$. Next, let us recall the definition of the Kronecker delta:

$$\delta[n] = \begin{cases} 1 & \text{if} \quad n = 0 \\ 0 & \text{if} \quad n \neq 0 \end{cases}, \quad \text{and } n \in \mathbb{Z} \qquad (2.15)$$

Using this function any discrete sequence $x[n]$ can be expressed as:

$$x[n] = \sum_k x[k]\delta[n-k] \qquad (2.16)$$

and the output of the LTI system at time n due to single impulse at moment k as[3]:

$$h_k[n] = L\{\delta[n-k]\} = h[n-k] \qquad (2.17)$$

The output of the LTI system

$$y[n] = L\{x[n]\} \qquad (2.18)$$

can be computed by substituting (2.16) into (2.18):

$$y[n] = L\left\{\sum_k x[k]\delta[n-k]\right\} \qquad (2.19)$$

Due to the linearity of L and property (2.17) we have:

$$y[n] = \sum_k x[k]L\{\delta[n-k]\} = \sum_k x[k]h[n-k] = (x*h)[n] \qquad (2.20)$$

---

[2]In case of continuous time system the impulse is the Dirac's delta—an infinitely sharp peak bounding unit area; in case of discrete systems it is a Kronecker delta—a sample of value 1 at the given moment in time.

[3]In this formula we also use the time invariance property.

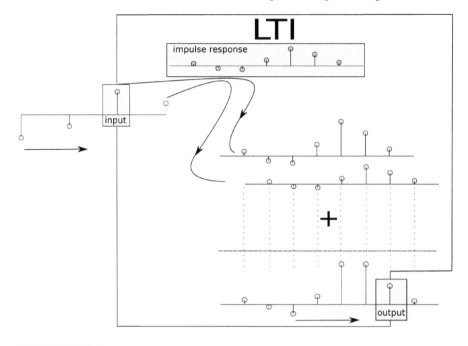

FIGURE 2.4
Idea of output production by an LTI system—convolution of input spikes with the impulse response function. The impulse response of the LTI system is multiplied by the amplitude of the input impulse. This is the response to the current input. The current response is added to the time-shifted responses to the previous inputs.

For any linear operator, there is a class of functions, called *eigenfunctions*, that is not distorted when the operator is applied to them. The only result of the application of the operator to its eigenfunction is multiplication by a number (in general it can be a complex number). Complex exponentials are the eigenfunctions for the LTI system. For an LTI system with signals in the real domain[4] such eigenfunctions are the sinusoids. It follows from the linearity of LTI system and the Euler formulas:

$$\cos x = \frac{1}{2}\left(e^{ix} + e^{-ix}\right)$$
$$\sin x = \frac{1}{2i}\left(e^{ix} - e^{-ix}\right)$$

(2.21)

Please note, that when expressing a real signal in the language of complex exponentials we must introduce a negative valued frequency since any oscillation in a real signal described by f cycles per second is represented by a pair of

---

[4]It means that the signal values are real numbers.

complex exponentials with frequencies f and $-f$. Since complex exponentials are eigenfunctions of LTI system, from equation (2.21) it follows that when a sinusoid is passed through the LTI system it can change the amplitude and phase but not the frequency. This property is very useful when dealing with the LTI systems since sinusoids form a basis in the space of real functions. Hence we can express exactly any real function as a (possibly infinite) sum (or integral) of sinusoids with specific amplitudes, frequencies, and phases. This property is the reason Fourier analysis is used so extensively in signal processing.

MATLAB demo: Signals as vectors

If you would like to see in what sense we can treat signals as vectors please go through Live Script: `matlab/c2/Ch2_demo1.mlx`

2.4 Duality of Time and Frequency Domains

In the previous section, we noticed that a tool which allows translating the input signal to a sum or integral of sinusoids would be very helpful when dealing with LTI systems. Such a tool is the Fourier transform. In fact, there is always a pair of transforms: one, from the time domain to the frequency domain, we shall denote as $\mathcal{F}\{x(t)\}$ and the inverse transform, from the frequency to the time domain, is denoted as $\mathcal{F}^{-1}\{X(\omega)\}$. The scheme below illustrates the operation of the transformation:

$$x(t) \underset{\mathcal{F}^{-1}}{\overset{\mathcal{F}}{\rightleftarrows}} X(\omega)$$

In the time domain, we think of a signal $x(t)$ as a series of values at certain moments in time. In the frequency domain, we think about the same signal $X(\omega)$ as of a specific set of frequencies. Each frequency has its own amplitude and phase. These two representations of a signal are equivalent. That is, we can transform signals without any loss of information from time to frequency representation and vice versa.

The frequency can be expressed in radians per second—in this case, we shall denote it as ω or in cycles per second—in this case, we shall denote it as f. Both quantities are related: $f = 2\pi\omega$.

Different kinds of the Fourier transform are used, depending on the signal. They will be described below.

2.4.1   Continuous Periodic Signal

Let us first consider the simplest case: the signal $x(t)$ is periodic with period T. Such a signal can be expressed as a series:

$$x(t) = \sum_{n=-\infty}^{\infty} c_n e^{i\frac{2\pi t}{T}n} \tag{2.22}$$

where:

$$c_n = \frac{1}{T} \int_0^T x(t) e^{-i\frac{2\pi t}{T}n}\, dt \tag{2.23}$$

This fact can be easily checked by substitution:

$$\int_0^T x(t) e^{-i\frac{2\pi t}{T}k}\, dt \quad =$$

$$= \int_0^T \sum_{n=-\infty}^{\infty} c_n e^{i\frac{2\pi t}{T}n} e^{-i\frac{2\pi t}{T}k}\, dt$$

$$= \sum_{n=-\infty}^{\infty} \int_0^T c_n e^{i\frac{2\pi t}{T}(n-k)}\, dt \tag{2.24}$$

$$= \sum_{n=k} \int_0^T c_n\, dt + \sum_{n \neq k} \underbrace{\int_0^T c_n e^{i\frac{2\pi t}{T}(n-k)}\, dt}_{=0}$$

$$= T c_n$$

We can think of c_n in expression (2.23) as the contribution of the frequency $f_n = \frac{n}{T}$ to the signal $x(t)$

$$c_n = X\left(\frac{n}{T}\right) \tag{2.25}$$

Hence a periodic signal can be expressed by the linear combination of complex exponentials with a discrete set of frequencies.

2.4.2   Infinite Continuous Signal

We can extend the formula (2.23) for aperiodic signals. The trick is that we consider the whole infinite aperiodic signal domain as a single period of an infinite periodic signal. In the limit $T \to \infty$ we obtain:

$$x(t) = \int_{-\infty}^{\infty} X(f) e^{i2\pi ft}\, df \tag{2.26}$$

$$X(f) = \int_{-\infty}^{\infty} x(t) e^{-i2\pi ft}\, dt \tag{2.27}$$

2.4.3 Finite Discrete Signal

In practice, we deal with discrete signals of finite duration. The Fourier transform that operates on this kind of signal is called the discrete Fourier transform (DFT) and the algorithms that implement it are fast Fourier transform (FFT).

The DFT formula can be derived from (2.23). The signal to be transformed is N samples long $x = \{x[0], \ldots, x[n], \ldots x[N-1]\}$ and the samples are taken every T_s seconds. It is assumed that the finite signal x is just one period of the infinite periodic sequence with period $T = N \cdot T_s$. The process of sampling can be written as $x[n] = x(nT_s) = x(t)\delta(t - nT_s)$. Substituting this into (2.23) gives:

$$
\begin{aligned}
X[k] &= \frac{1}{T} \int_0^T x(t)\delta(t - nT_s)e^{-i\frac{2\pi t}{T}k}\, dt \\
&= \frac{1}{NT_s} \sum_{n=0}^{N-1} x[n]e^{-i\frac{2\pi knT_s}{T}}T_s = \frac{1}{N} \sum_{n=0}^{N-1} x[n]e^{-i\frac{2\pi}{N}kn} \quad (2.28)
\end{aligned}
$$

From the above formula, it follows that k in the range $k = 0, \ldots, N-1$ produces different components in the sum. From the Euler formulas (2.21) it follows that for a real signal a pair of conjunct complex exponentials is needed to represent one frequency. Thus for real signals, there are only $N/2$ distinct frequency components. The inverse discrete Fourier transform (IDFT) is given by

$$
x[n] = \sum_{k=0}^{N-1} X[k]e^{i\frac{2\pi}{N}kn} \qquad n = 0, \ldots, N-1. \quad (2.29)
$$

Note, that the signs of the exponents and the normalization factors by which the DFT and IDFT are multiplied (here $1/N$ and 1) are conventions, and may be written differently by other authors. The only requirements of these conventions are that the DFT and IDFT have opposite sign exponents and that the product of their normalization factors is $1/N$.

2.4.4 Basic Properties of Fourier Transform

Given signals $x(t)$, $y(t)$, and $z(t)$ we denote their Fourier transforms by $X(f)$, $Y(f)$, and $Z(f)$, respectively. The Fourier transform has the following basic properties [485]:

Linearity: For any complex numbers a and b:

$$
z(t) = ax(t) + by(t) \quad \Rightarrow \quad Z(f) = a \cdot X(f) + b \cdot Y(f)
$$

Translation: For any real number t_0:

$$
z(t) = x(t - t_0) \quad \Rightarrow \quad Z(f) = e^{-2\pi it_0 f} X(f)
$$

Modulation: For any real number f_0:

$$z(t) = e^{2\pi f_0} x(t) \quad \Rightarrow \quad Z(f) = X(f - f_0)$$

Scaling: For all non-zero real numbers a:

$$z(t) = x(at) \quad \Rightarrow \quad Z(f) = \frac{1}{|a|} X\left(\frac{f}{a}\right)$$

The case $a = -1$ leads to the time-reversal property, which states:

$$z(t) = x(-t) \quad \Rightarrow \quad Z(f) = X(-f)$$

Conjugation:
$$z(t) = x(t)^* \quad \Rightarrow \quad Z(f) = X(-f)^*$$

The $*$ symbol throughout the book denotes the complex conjugation.

Convolution theorem:

$$y(t) = (x * z)(t) \quad \Leftrightarrow \quad Y(f) = X(f) \cdot Z(f) \tag{2.30}$$

This theorem works also in the opposite direction:

$$Y(f) = (X * H)(f) \quad \Leftrightarrow \quad y(t) = x(t) \cdot z(t) \tag{2.31}$$

This theorem has many applications. It allows changing the convolution operation in one of the dual (time or frequency) spaces into the multiplication in the other space. Combined with the FFT algorithm the convolution theorem allows for fast computations of convolution. It also provides insight into the consequences of windowing the signals, or applications of filters.

MATLAB demo: Fourier transform

If you would like to practice basic properties of Fourier transform please go through Live Script: `matlab/c2/Ch2_demo2.mlx`

2.4.5 Power Spectrum: The Plancherel Theorem and Parseval's Theorem

Let's consider the $x[n]$ as samples of a voltage drop across a resistor with the unit resistance $R = 1$. Then the $P = x[n]^2/R$ is the power dissipated by that resistor. By analogy in the signal processing language a square absolute value of a sample is called *instantaneous signal power*.

If $X[k]$ and $Y[k]$ are the DFTs of $x[n]$ and $y[n]$, respectively, then the Plancherel theorem states:

$$\sum_{n=0}^{N-1} x[n]y^*[n] = \frac{1}{N} \sum_{k=0}^{N-1} X[k]Y^*[k] \qquad (2.32)$$

where the star denotes complex conjugation. Parseval's theorem is a special case of the Plancherel theorem and states:

$$\sum_{n=0}^{N-1} |x[n]|^2 = \frac{1}{N} \sum_{k=0}^{N-1} |X[k]|^2 . \qquad (2.33)$$

Because of the Parseval's theorem (2.33) we can think of $\frac{1}{N}|X[k]|^2$ as the portion of signal power carried by the complex exponential component of frequency indexed by k. If we process real signals, then the complex exponential in Fourier series come in conjugate pairs indexed by k and $N - k$ for $k \in 1,\ldots,N/2$. Each of the components of the pair carries half of the power related to the oscillation with frequency $f_k = \frac{k}{N}F_s$ (F_s is the sampling frequency). To recover the total power of oscillations at frequency f_k we need to sum the two parts. That is:

$$P(f_k) = \frac{1}{N}\left(|X[k]|^2 + |X[N-k]|^2\right) \qquad (2.34)$$

For real signals, the power spectrum is often displayed by plotting only the power (2.34) for the positive frequencies.

2.4.6 Z-Transform

A more general case of the discrete signal transformation is the Z-transform. It is especially useful when considering parametric models or filters.

In a more general context, the Z-transform is a discrete version of the Laplace transform. For a discrete signal $x[n]$ the Z-transform is given by:

$$X(z) = Z\{x[n]\} = \sum_{n=0}^{\infty} x[n]z^{-n} \qquad (2.35)$$

where $z = Ae^{i\phi}$ is a complex number. The discrete Fourier transform is a special case of the Z-transform.

Properties of the Z-transform:

Linearity:

$$Z\{a_1 x_1[n] + a_2 x_2[n]\} = a_1 X_1(z) + a_2 X_2(z)$$

Time translation:

$$Z\{x[n-k]\} = z^{-k}X(z)$$

Transform of an impulse:

$$Z\{\delta[n]\} = 1$$

The transform of an impulse together with the time translation yields:

$$Z\{\delta[n - n_0]\} = z^{-n_0}$$

Taking into account the linearity of Z we can compute the transform of a linear combination of the p signal samples:

$$Z\{x[n] + a_1 x[n-1] + \cdots + a_p x[n-p]\} = (1 + a_1 z^{-1} + \cdots + a_p z^{-p})X(z)$$
$$= A(z)X(z)$$

$$(2.36)$$

This result will be very useful for discussion of AR model and filtering.

2.4.7 Uncertainty Principle

In the previous sections, we considered signals represented in either time or frequency domain. For stationary signals, it is enough to know one of the representations. Real-life signals are often a mixture of stationary and non-stationary elements, e.g., alpha spindles, sleep spindles, K-complexes, or other graphoelements together with ongoing background EEG activity. We are often interested in characterizing the non-stationary transients. It is natural to think about their frequency f_0 and localization in time t_0. It is also obvious that such a transient has a certain duration in time—that is, it is not localized in a single time point but it has a span in time. The time span can be characterized by σ_t. The localization of the transient in the frequency domain also has a finite resolution characterized by frequency span σ_f. Those two spans are bounded by the uncertainty principle. Before we continue with the formal notation, let's try to understand the principle heuristically. Let's think about a fragment of a sinusoid observed over time $T = (t - \sigma_t, t + \sigma_t)$, as shown in the Figure 2.5. We can calculate its frequency dividing the number of periods observed during time T by the length of T. As we shrink the time T the localization in time of the fragment of the sinusoid becomes more precise, but less and less precise is the estimation of the number of cycles, and hence the frequency.

Now let's put it formally. We treat the signal energy representation in time $|x(t)|^2$ and representaion in frequency $|X(f)|^2$ as probability distributions with normalizing factor $E_x = \int_{-\infty}^{\infty} |x(t)|^2 dt$. Then we can define mean time:

$$t_0 = \frac{1}{E_x} \int_{-\infty}^{\infty} t\,|x(t)|^2\,dt \qquad (2.37)$$

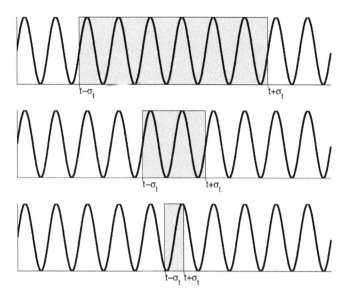

FIGURE 2.5
Illustration of the uncertainty principle. As the observation time shrinks, the time localization of the observation improves, but the estimate of the frequency deteriorates.

Mean frequency:

$$f_0 = \frac{1}{E_x} \int_{-\infty}^{\infty} f \, |X(f)|^2 \, df \tag{2.38}$$

Time span:

$$\sigma_t^2 = \frac{1}{E_x} \int_{-\infty}^{\infty} (t - t_0)^2 |x(t)|^2 \, dt \tag{2.39}$$

Frequency span:

$$\sigma_f^2 = \frac{1}{E_x} \int_{-\infty}^{\infty} (f - f_0)^2 |X(f)|^2 \, df \tag{2.40}$$

It turns out that the product of the time and frequency span is bounded [161]:

$$\sigma_t^2 \sigma_f^2 \geq \frac{1}{16\pi^2} \tag{2.41}$$

The equality is reached by Gabor functions (Gaussian envelope modulated by cosine). It is important to realize this property, especially when working with time-frequency representations of a signal. Many different methods of time-frequency representations make various trade-offs between time and frequency span but in each of them the inequality (2.41) holds.

2.5   Hypotheses Testing

2.5.1   The Null and Alternative Hypothesis

The key to success in statistical hypothesis testing is the correct formulation of the problem. We need to specify it as two options: the null hypothesis H_0 and the alternative one H_1. The two hypotheses must be disjoint and complementary. We usually try to set the option which we would like to reject as the null hypothesis since we can control the probability of erroneous rejection of the null hypothesis.

To actually perform a test, a function called *statistic* $S(x)$ is needed. We assume that the data are random variables and represent a sample taken from some population characterized by properties which are in accordance with the null hypothesis. A statistic is a function of a random variable. The main thing that must be known about the statistic is the probability with which it takes different values for the random variables conforming to the null hypothesis.

If for a given data the computed statistic is $S_x = S(x)$ then the p value returned by the test is the probability of observing statistics with values equal or more extreme than S_x for data conforming to H_0. If the p value is high, then we assume that the data conform to the null hypothesis. But if the probability of observing such a value of the statistic is low then we can doubt that the data agree with the null hypothesis. Consequently, we reject that hypothesis and accept the alternative one. The critical level of probability used to make the decision is called the significance level α. It expresses how low the p value must be to doubt the null hypothesis.

2.5.2   Types of Tests

To select the type of test, we need to answer the question: do we know the probability distribution from which the data were sampled?

Yes, we know or can assume, or can transform, e.g., with the Box-Cox transform [66], the data to one of the known probability distributions. In this case, we select appropriate classical parametric tests based on the normal, t, F, χ^2 or some other known statistics. In MATLAB Statistics and Machine Learning Toolbox many such tests are available. To test the normality assumption we can use Lilliefors' test [339], implemented as `lillietest`, or use a qualitative graphical test implemented as `normplot`.

No, we do not know the probability distribution. In this case, we have two possibilities:

- Use a classical non-parametric test e.g.:

  Wilcoxon rank sum test tests, if two independent samples come from identical continuous distributions with equal medians, against the alternative that they do not have equal medians. In MAT-LAB Statistics and Machine Learning Toolbox, it is implemented as `ranksum`.

Wilcoxon signed rank test one-sample or paired-sample Wilcoxon signed rank test. It tests, if a sample comes from a continuous distribution symmetric about a specified median, against the alternative that it does not have that median. In MATLAB Statistics and Machine Learning Toolbox x it is implemented as `signrank`.

Sign test one-sample or paired-sample sign test. It tests, if a sample comes from an arbitrary continuous distribution with a specified median, against the alternative that it does not have that median. In MATLAB Statistics and Machine Learning Toolbox, it is implemented as `signtest`.

- Use a resampling (bootstrap or permutation) test [138]. In this type of test in principle, any function of the data can be used as a statistic. We need to formulate a statistical model of the process that generates the data. The model is often very simple and relies on appropriate resampling of the original dataset, e.g., we draw a random sample from the original dataset with replacement. It is crucial that the model (the resampling process) conforms to the null hypothesis. The model is simulated many times and for each realization, the statistic is computed. In this way, the empirical distribution of the statistic under the null hypothesis is formed. The empirical distribution of the statistic is used then to evaluate the probability of observing the value of statistic equal to or more extreme than the value for the original dataset.

2.5.3 Multiple Comparisons Problem

Let us assume that we perform a t test with the significance level α. This means that under the true null hypothesis H_0 we can observe with the probability α value greater than the critical:

$$P(t \geq t_\alpha) = \alpha$$
$$P(t < t_\alpha) = 1 - \alpha \qquad (2.42)$$

For n independent tests, the probability that none of them will not give t value greater than the critical is $(1 - \alpha)^n$. Thus the probability of observing at least one of the n values exceeding the critical one (probability of the family wise error FWER) is:

$$P(t \geq t_\alpha; \text{in n tests}) = P^{FWER} = 1 - (1 - \alpha)^n \qquad (2.43)$$

Let us consider tests performed at significance level $\alpha = 0.05$, The probability computed from (2.43) gives the chance of observing extreme values of statistic for data conforming to the null hypothesis just due to fluctuations. For a single test the above formula gives $P(t \geq t_\alpha; \text{in 1 test}) = 0.05$, for $n = 10$, $P(t \geq t_\alpha; \text{in 10 test}) \approx 0.4$, for $n = 100$, $P(t \geq t_\alpha; \text{in 100 test}) \approx 0.994$.

In case of some dependence (e.g., correlation) among the tests, the above-described problem is less severe but also present. There is clearly a need to control the error of false rejections of the null hypothesis due to the multiple comparisons problem (MCP). Table 2.1 summarizes the possible outcomes of m hypothesis tests. As a result of the application of a test, we obtain one of

TABLE 2.1

Outcomes of m hypothesis tests

	# Declared non-significant (H_0 accepted)	# Declared significant (H_0 rejected)	Total
# True null hypotheses	U	V	m_0
# False null hypotheses	T	S	$m - m_0$
Total	$m - R$	R	m

two true statements: a true null hypothesis is accepted (number of such cases is denoted U) or false null hypothesis is rejected (number of such cases is denoted S). There is a possibility to commit one of two types of errors. Type I error is when a true null hypothesis is rejected (the number of such cases is denoted as V); Type II error—the alternative hypothesis is true, but the null hypothesis is accepted (denoted as T). The total number of rejected null hypotheses is denoted by R.

The total number of tested hypotheses m is known. The number of true null hypotheses m_0 and the number of false null hypotheses $m_1 = m - m_0$ are unknown. The number of cases V, T, U, S, and R are treated as random variables, but only R can be observed. The family-wise error rate ($FWER$) is the probability of falsely rejecting one or more true null hypotheses among all the hypotheses when performing multiple tests:

$$FWER = P(V \geq 1) = 1 - P(V = 0) \qquad (2.44)$$

In studies where one specifies a finite number of a priori inferences, families of hypotheses are defined for which conclusions need to be jointly accurate or by which hypotheses are similar in content or purpose. If these inferences are unrelated in terms of their content or intended use (although they may be statistically dependent), then they should be treated separately and not jointly [225].

Sometimes it is not necessary to control the $FWER$ and it is sufficient to control the number of falsely rejected null hypotheses—the false discoveries. The false discovery rate (FDR) is defined as expectedproportion of incorrectly rejected null hypotheses:

$$FDR = E\left[\frac{V}{V+S}\right] = E\left[\frac{V}{R}\right] \tag{2.45}$$

Below we briefly describe three approaches to MCP: correction of the significance level α, statistical maps, and FDR.

2.5.3.1 Correcting the Significance Level

The most straightforward approach is known as Bonferroni correction, which states that if one performs n hypotheses tests on a set of data, then the statistical significance level that should be used for each hypothesis separately should be reduced n times in respect to the value that would be used if only one hypothesis were tested. For example, when testing two hypotheses, instead of an α value of 0.05, one should use the α value of 0.025. The Bonferroni correction is a safeguard against multiple tests of statistical significance on the same data. On the other hand, this correction is conservative in the case of correlated tests, which means that the significance level gets lower than necessary to protect against the rejections of the null hypothesis due to fluctuations.

In special, but very common cases of comparison of a set of mean values, it is suggested to consider Tukey's HSD (honestly significant difference) test. Tukey's method considers the pairwise differences. Scheffé's method applies to the set of estimates of all possible contrasts among the factor level means. An arbitrary contrast is a linear combination of two or more means of factor levels whose coefficients add up to zero. If only pairwise comparisons are to be made, the Tukey method will result in a narrower confidence limit, which is preferable. In the general case when many or all contrasts might be of interest, the Scheffé method tends to give narrower confidence limits and is, therefore, the preferred method [11].

2.5.3.2 Parametric and Non-Parametric Statistical Maps

The multiple comparisons problem is critical for comparison of images or volume data collected under different experimental conditions. Usually, such images or volumes consist of a huge number of elements: pixels, voxels, or resels (resolution elements). For each element statistical models (parametric or non-parametric) are assumed. Hypotheses expressed in terms of the model parameters are assessed with univariate statistics.

In case of the parametric approach (statistical parametric mapping), the general linear models are applied to describe the variability in the data in terms of experimental and confounding effects, and residual variability. In order to control the $FWER$ adjustments are made, based on the number of resels in the image and the theory of continuous random fields in order to set a new criterion for statistical significance that adjusts for the problem of multiple comparisons [171]. This methodology, with application to neuroimaging and MEG/EEG data, is implemented in SPM—a MATLAB software package, written by members and collaborators of the Wellcome Trust Centre for Neuroimaging. SPM is free but copyright software, distributed under the terms of the GNU General Public Licence. SPM homepage: `http://www.fil.ion.ucl.ac.uk/spm/`.

In case of the non-parametric approach (statistical non-parametric mapping, SnPM) the idea is simple: if the different experimental conditions do not make any different effect on the measured quantity, then the assignment of labels to the conditions is arbitrary. Any reallocation of the labels to the data would lead to an equally plausible statistic image. So, considering the statistic images associated with all possible re-labelings of the data, we can derive the distribution of statistic images possible for this data. Then, we can test the null hypothesis of no experimental effect by comparing the statistic for the actual labeling of the experiment with this re-labeled distribution. If out of N possible relabelings the actual labeling gives the rth most extreme statistic, then the probability of observing that value under the null hypothesis is r/N. The details are worked out in [Holmes et al., 1996, and Nichols and Holmes, 2002]. SnPM is implemented as a free MATLAB toolbox: `http://www.nisox.org/Software/SnPM13/`.

2.5.3.3 False Discovery Rate

In case of independent tests, Benjamini and Hochberg [43] showed that the Simes' procedure [551], described below, ensures that its expected fraction of false rejection of null hypothesis, is less than a given q. This procedure is valid when the m tests are independent. Let $H_1 \ldots H_m$ be the null hypotheses and $P_1 \ldots P_m$ their corresponding p-values. Order these values in increasing order and denote them by $P_{(1)} \ldots P_{(m)}$. For a given q, find the largest k such that:

$$P_{(k)} \leq \frac{k}{m} q \qquad (2.46)$$

Then reject all $H_{(i)}$ for $i = 1, \ldots, k$.

In case of dependent tests Benjamini and Yekutieli [44] proposed correction of the threshold (2.46) such that:

$$P_{(k)} \leq \frac{k}{m \cdot c(m)} q \qquad (2.47)$$

where:

- $c(m) = 1$ if the tests are independent, or are positively correlated

- $c(m) = \sum_{i=1}^{m} \frac{1}{i}$ if the tests are negatively correlated

2.6 Surrogate Data Techniques

The surrogate data concept was introduced in [594] and further developed in [138, 8] primarily to distinguish non-linear from stochastic time series. However, the method could be used as well to test for the consistent dependencies between multivariate time series.

The method of surrogate data is basically an application of the *bootstrap* method to infer the properties of the time series. In a surrogate data test, the null hypothesis concerns the statistical model that generates the signal. Any function of data can be used as statistics. In general, the test consists of the following steps:

1. select a function of data to be the statistics;

2. compute the statistics for the original signal;

3. generate a new signal according to the model assumed in the null hypothesis which shares with the original signal as many properties as possible (e.g., mean, variance, and Fourier spectrum);

4. compute the statistics for the new signal;

5. repeat steps 3 and 4 many times to construct the distribution of the statistics;

6. use that distribution to compute the probability of observing the original or more extreme value of statistics for processes conforming to the null hypothesis.

Two approaches to surrogate data construction will be described below. The simplest question one would like to answer is whether there is evidence for any dynamics at all in the time series, i.e., if there are any relations between the consecutive samples. The null hypothesis, in this case, is that the observed data is fully described by a series in which each sample is an independent random variable taken from identical distribution. The surrogate data can be generated by shuffling the time-order of the original time series. The surrogate data will have the same amplitude distribution as the original data, but any temporal correlations that may have been in the original data are destroyed.

The test for linearity corresponds to the null hypothesis that all the structure in the time series is given by the autocorrelation function, or equivalently, by the Fourier power spectrum. The test may be performed by fitting an autoregressive model to the series and examination of the residuals, or by randomizing the phases of the Fourier transform. The second approach is recommended since it is more stable numerically. The main steps of the procedure are the following:

1. compute the Fourier transform $X(f)$ of the original data $x(t)$;

2. generate a randomized version of the frequency representation of the

signal $Y(f) = X(f)e^{i\phi(f)}$ by multiplying each complex amplitude $X(f)$ by $e^{i\phi(f)}$, where $\phi(f)$ is a random value, independently drawn for each frequency f, from the interval $[0, 2\pi)$;

3. in order to get real components from the inverse Fourier transform symmetrize the phases, so that $\phi(f) = -\phi(-f)$;

4. perform the inverse transform.

In this way, we obtain surrogate data $y(t)$ characterized by the same power spectrum as the original time series $x(t)$, but with the correlation structure destroyed. Then the statistical test is performed comparing the estimators (i.e., statistics) obtained from the original data with the distribution of that estimator obtained from surrogate data.

In case of multivariate time series, the surrogates with randomized phases may be used to test for the presence of the phase dependencies between the signals. The rejection of the null hypothesis assuming independence between the time series does not mean the presence of the non-linearities. In biological time series analysis, the Occam's razor approach should be used: we should seek the simplest model consistent with the data and the surrogate data tests are helpful in this respect.

MATLAB demo: Surrogate signal

If you would like to practice surrogate signal generation please go through Live Script: `matlab/c2/Ch2_demo3.mlx`

3

Single Channel (Univariate) Signal

3.1 Filters

Signals may contain many frequency components. Some of them may be of interest, like frequency bands of EEG rhythms; others, like power line artifacts, may be undesired. To remove the unwanted part of the spectral content of the signal one usually uses filters. In order to be implementable the filter must be *causal.* That is, the filter response must only depend on the current and past inputs. If the filter has to depend also on the past outputs it still can be causal by introducing the feedback with appropriate delays.

We shall discuss here only the digital filters implemented in software (not the hardware electronic devices used, e.g., as anti-aliasing filters before sampling). Most of the digital filters perform the following action: they compute a linear combination of a number n_b of past inputs x and a number of n_a of past outputs y to evaluate current output $y[n]$:

$$\begin{aligned} y[n] = b(1) * x[n] + b(2) * x[n-1] + \cdots + b(n_b + 1) * x[n - n_b] \\ - a(2) * y[n-1] - \cdots - a(n_a + 1) * y[n - n_a] \end{aligned} \tag{3.1}$$

The maximum delay, in samples, used in creating each output sample is called the *order* of the filter. In the difference-equation representation (3.1), the order is the larger of numbers n_b and n_a. There are three possibilities:

- If $n_a = 0$ and $n_b \neq 0$ then the filter is called finite impulse response (FIR), which means that once the filter is fed with one non-zero pulse the response will have a finite duration. After n_b time intervals the output sets back to zero. This filter sometimes is called moving average, MA.

- If $n_b = 0$ and $n_a \neq 0$ the filter has an infinite impulse response (IIR). After a single non-zero sample the filter could produce some output forever. It is also called recursive or autoregressive (AR) filter.

- If $n_b \neq 0$ and $n_a \neq 0$ this is the most general filter type. It is also called IIR or ARMA—autoregressive moving average.

The IIR filters usually have lower orders than FIR filters with the same attenuation properties.

DOI: 10.1201/9780429431357-3

The operation of the filter is much easier to understand in the frequency domain. First let's reorder the equation (3.1):

$$y[n] + a(2) * y[n-1] + \cdots + a(n_a + 1) * y[n - n_a]$$
$$= b(1) * x[n] + b(2) * x[n-1] + \cdots + b(n_b + 1) * x[n - n_b] \quad (3.2)$$

Application of Z-transform (Sect. 2.4.6) to both sides of (3.2) yields:

$$A(z)Y(z) = B(z)X(z) \quad (3.3)$$

From this we get:

$$Y(z) = A(z)^{-1}B(z)X(z) = H(z)X(z) \quad (3.4)$$

Function H in (3.4) is called the frequency response function, and it has the form:

$$H(z) = \frac{b(1) + b(2)z^{-1} + \cdots + b(n_b + 1)z^{-n_b}}{a(1) + a(2)z^{-1} + \cdots + a(n_a + 1)z^{-n_a}} \quad (3.5)$$

This function is a ratio of two polynomials. We can factor the numerator and denominator to obtain:

$$H(z) = g\frac{(1 - q_1 z^{-1})(1 - q_2 z^{-1})\ldots(1 - q_{n_b} z^{-1})}{(1 - p_1 z^{-1})(1 - p_2 z^{-1})\ldots(1 - p_{n_a} z^{-1})} \quad (3.6)$$

The numbers $\{q_1, q_2, \ldots, q_{n_b}\}$ are zeros of the numerator and are called zeros of the transfer function. The numbers $\{p_1, p_2, \ldots, p_{n_a}\}$ are zeros of the denominator and are called poles of the transfer function. The filter order equals the number of poles or zeros, whichever is greater. We can obtain frequency dependent transfer function $H(f)$ substituting $z = e^{i2\pi f}$. The function assigns to each frequency f a complex number with magnitude M and phase ϕ:

$$H(f) = M(f)e^{i\phi(f)} \quad (3.7)$$

From equation (3.4) we see that the operation of a filter is a multiplication of each of the Fourier components of the signal by the complex number $H(f)$; that is, filter changes the magnitude and phase of the component. A very useful measure can be derived from the phase of the frequency response function— the group delay. The group delay, defined as:

$$\tau_g(f) = -\frac{d\phi(f)}{df} \quad (3.8)$$

is a measure of the time delay introduced by the filter to the signal component of frequency f. From the formula (3.8) we see that if the phase ϕ depends linearly on the frequency then $\tau_g(f) = const$. This means that all frequency components are equally delayed. The phase structure of the signal is not distorted, which is essential if the filtering is used as a preprocessing step for more advanced methods of signal processing that are phase sensitive. Such a linear phase delay is the property of the FIR filters. In the case of off-line applications the delays introduced by the filter, in any form (not only the linear ones), can be corrected. The technique relies on applying the same filter twice. After filtering the signal in the forward direction, the filtered sequence is reversed in time and run again through the filter. Due to the time reversal property of the Fourier transform (see Sect. 2.4.4) the component $X(f)$ of the original signal is multiplied effectively by

$$H_{eff}(f) = M(f)e^{i\phi(f)} \cdot M(f)e^{-i\phi(f)} = M(f)^2 \tag{3.9}$$

yielding the zero-phase double order filter. In MATLAB this technique is implemented as `filtfilt` function. The standard filtering corresponding to equation 3.1 is implemented as a `filter` function.

MATLAB demo: filtering with zero phase shift

If you would like to see a comparison of standard and zero phase shift filtering please go through Live Script: `matlab/c3/Ch3_01_filtfilt.mlx`

3.1.1   Designing Filters

Practical applications of filters need that they meet certain requirements. Often the characteristics of a filter is expressed as properties of the magnitude responses (the absolute value of the transfer function) such as: the cut-off frequency or frequency band to be attenuated or passed, the amount of attenuation of the unwanted spectral components, steepness of the filter, or the order of the transfer function.

In specifying filter characteristics a unit called decibel [dB] is commonly used. Two levels of signal power P and P_0 differ by n decibels, if

$$n = 10 \log_{10} \frac{P}{P_0}$$

Terms used in specifying filter characteristics are:

Wp —pass band, frequency band that is to be passed through filter without alternation

Ws —stop band, frequency band which has to be attenuated

Rp —allowed peak-to-peak pass band ripples, specified as a positive scalar, expressed in [dB]. If your specification, l is in linear units, you can convert it to decibels using Rp= $40 \log_{10}((1 + l)/(1 - l))$.

Rs —required minimum attenuation in the stop band, expressed in [dB]. If the signal amplitude in linear units is to be attenuated l times, then Rs $= -20 \log_{10} l$.

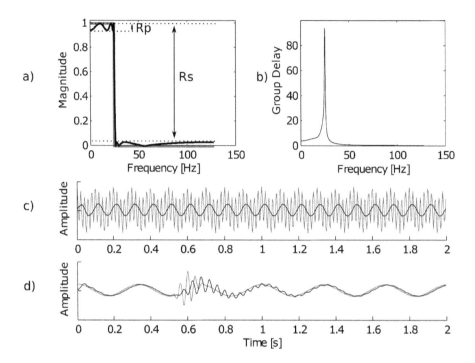

FIGURE 3.1

Illustration of filter properties. a) The magnitude of frequency response function. Gray rectangular outline—the ideal response, black—the magnitude of the designed elliptic 5th order filter. Arrows indicate allowed passband ripples (Rp) and required minimum attenuation in the stop band (Rs). b) Group delay function. c) Application of the designed filter to a signal composed of a 10 Hz and 50 Hz sinusoid: input—gray, output—black line. d) Illustration of the delay and the edge effects of the filter: the input signal (gray) is a 3 Hz sinusoid with a 25 Hz transient at 0.6 s; the output black line shows (i) in the first 0.1 s an edge distortion of the filter (ii) delays: the output 3 Hz sinusoid is delayed slightly compared to the input; the 25 Hz transient is more delayed and spread, which corresponds to the group delay function (b).

These terms are demonstrated in Figure 3.1. Live script reproducing the figure is in: `matlab/c3/Ch3_fig_3_1.mlx`

The requirements concerning filter characteristics are usually contradictory, e.g., for a given filter order the increase of steepness of the filter results in bigger ripples. Therefore some optimization process has to be applied and the resulting filter is a compromise between the set of the conditions. A decision must be made which of the possible filter designs is best for the planned use.

Below we describe briefly the functions from the MATLAB Signal Processing Toolbox that facilitate the design and the validation of the filters. In the filter design function in MATLAB, the frequencies are expressed as a fraction of the Nyquist frequency F_N (namely are scaled so that $F_N = 1$). FIR filters can be designed by means of the following functions:

fir1-allows for design of classical lowpass, bandpass, highpass, bandstop filters. It implements an algorithm based on Fourier transform. If the ideal transfer function is $H_{id}(f_n)$, its inverse Fourier transform is $h_{id}(n)$ and $w(n)$ is a window (by default a Hamming window) then the filter coefficients are $b(n) = w(n)h_{id}(n)$ for $n \in [1, N]$, N is the filter order. This filter has the group delay $\tau_g = N/2$.

fir2-this function allows to specify arbitrary piecewise linear magnitude response. The algorithm first interpolates the desired magnitude response onto a dense evenly spaced grid of points, computes the inverse Fourier transform of the result, and multiplies it by the specified window.

firls-allows to specify arbitrary piecewise characteristics of the magnitude response and minimizes the sum of squared errors between the desired and the actual magnitude response based on algorithm given in [458, pp. 54–83].

firpm-function implements the Parks-McClellan algorithm [458, p. 83]. It designs filters that minimize the maximum error between the desired frequency response and the actual frequency response. Filters designed in this way exhibit an equiripple behavior in their frequency response. The Parks-McClellan FIR filter design algorithm is perhaps the most popular and widely used FIR filter design methodology.

MATLAB demo: FIR filters

If you would like to practice the design of FIR filters please go through Live Script: `matlab/c3/Ch3_02_fir_filters.mlx`

Functions for designing IIR filters:

butter-Butterworth filter gives smooth and monotonic magnitude response function.

cheby1-designs Chebyshev Type I filter. These filters are equiripple in the passband and monotonic in the stopband.

cheby2-designs Chebyshev Type II filter. Filters are monotonic in the passband and equiripple in the stopband. Type II filters do not roll off as fast as type I filters but are free of passband ripple.

ellip-designs elliptic filter. Elliptic filters offer steeper rolloff characteristics than Butterworth or Chebyshev filters but are equiripple in both the pass- and stopbands. In general, elliptic filters, compared to other filters, meet given performance specifications with the lowest order.

Once the filter coefficients are computed it is necessary to analyze the properties of the resulting filter. The transfer function for given filter coefficients $\{a_n\}$ and $\{b_n\}$ can be computed according to equation (3.5); in MATLAB it can be evaluated with `freqz` function. The absolute value of this function gives the frequency magnitude response. A plot of overlaid desired and actual magnitude response allows checking whether the actual magnitude response function is close enough to the desired one (Figure 3.1 a). One should pay attention to the steepness of the magnitude response edges or in other words to the width of the transition between the stop-, and pass-band. Increasing the filter order should make the transition steeper. One should also consider the magnitude of ripples. They can be made smaller by decreasing the filter order or broadening the transition width.

If the filter is to be used in standard, one direction mode, it is also advisable to observe the group delay function (3.8) for delays of different frequency components. This function can be evaluated by means of `grpdelay` from Signal Processing Toolbox (Figure 3.1 b).

If we are interested in the time course of the filtered signal, like in the case of evoked potentials, we should examine also two other characteristics of the filter. These are the impulse and the step responses. We shall consider them more thoroughly while discussing the practical issues arising during the analysis of evoked response potential in Sect. 5.1.7.1.

The process of designing the filters and examining the characteristics can be efficiently accomplished with the tools: `designfilt`[1] and `fvtool`[2]. With `designfilt` you can either specify the requirements as the arguments to the function call and obtain an object representing the designed filter, or you can use it as a design assistant, where the graphical interface will guide you through the design process. The Filter Design Assistant requires Java software and the MATLAB desktop to run. It is not supported if you run MATLAB with the -nojvm, -nodisplay, or -nodesktop options. The `fvtool` helps in the graphical examination of the filter properties. By default, `fvtool(d)` opens FVTool and displays the magnitude response of the digital filter defined by the input d. With FVTool you can also display the phase response, group delay, impulse response, step response, pole-zero plot, and coefficients of the filter. This tool can also export the characteristics.

MATLAB demos

- If you would like to practice the design of IIR filters please go through Live Script: `matlab/c3/Ch3_03_iir_filters.mlxs`

- If you would like to get the intuition about removing power

---

[1]Introduced in R2014a.
[2]Introduced before R2006a.

line noise from the data please go through Live Script:
`matlab/c3/Ch3_04_notch_filter.mlx`

- If you would like to practice filtering the ECG signal please go through Live Script: `matlab/c3/Ch3_05_ecg_filters.mlx`

- If you would like to practice bandpass filtering please go through Live Script: `matlab/c3/Ch3_06_sleep_spindle.mlx`

3.1.2 Changing the Sampling Frequency

In previous sections, we described the main field of filter applications—the selection of relevant frequency bands from the signals and suppression of unwanted frequency bands. Here, we would like to mention one more application where the filters are indispensable: the process of resampling the signal at another sampling frequency. Let's imagine that we need to reduce the sampling frequency (*downsample*) of the signal by half. The simplest idea could be skipping every other sample in the original signal. However, in most cases this would destroy the signal due to the aliasing (see Sect. 2.2.1.1). In order to do it properly, one needs to take care that the assumption of the sampling theorem (Sect. 2.2.1) is fulfilled; that is, the signal contains only frequencies below the Nyquist frequency of the downsampled signal. This is usually achieved by lowpass filtering the original signal before downsampling. If the original signal s_0 is sampled at frequency F_{s_0} and the desired frequency is $F_{s_1} = F_{s_0}/q$ for an integer q, then the frequency band of the signal s_0 needs to be reduced to the range $(0, \frac{1}{2q} F_{s_0})$. In practice, one needs to set the cut-off frequency of the lowpass filter less than $\frac{1}{2q} F_{s_0}$, because the filter roll-off is finite. In MATLAB Signal Processing Toolbox the downsampling with anti-aliasing filtering is implemented in function `decimate`.

MATLAB demo: downsample vs decimate

If you would like to see the difference between downsampling and decimating signal consider the Live Script:
`matlab/c3/Ch3_07_downsampling.mlx`

The opposite process of increasing the sampling rate (*interpolation*) can be accomplished in many ways, without the explicit use of filters. However, the interpolation can also be carried out in the following way. If the sampling frequency has to be increased by integer factor p then between every two samples of the original signal s_0 one needs to insert $p-1$ zeros, and then filter the resulting signal with a lowpass filter. The cut-off frequency of the filter should be set below the Nyquist frequency of the original signal, to remove all the high-frequency components introduced to the signal by the discontinuities at the inserted samples. In MATLAB Signal Processing Toolbox the filter interpolation is implemented as function `interp`.

Finally, using a combination of the above procedures it is possible to change the sampling rate by the rational factor $\frac{p}{q}$. First, the signal is interpolated to p times higher sampling frequency and then downsampled by q. In MATLAB Signal Processing Toolbox the algorithm is implemented as function `resample`.

3.1.3   Matched Filters

In many applications one would like to detect in the signal some specific structures, matching the template s and treating the other signal components as noise. Formally, the signal may be described as composed of the template— s and additive noise v:

$$x = s + v \qquad (3.10)$$

The matched filter is the linear filter that maximizes the output signal-to-noise ratio. The solution is the filter with transfer function h which is complex conjugate time reversal of $\hat{h}$ given by [608]:

$$\hat{h} = Ks \qquad (3.11)$$

where K is a normalizing factor.

MATLAB demo: improving the SNR of QRS complex in ECG

If you would like to see how to construct and utilize a matched filter in a real signal consider the Live Script: `matlab/c3/Ch3_08_ecg_matched_filters.mlx`

3.1.4   Wiener Filter

Typical filters described at the beginning of Sect. 3.1 are designed to have a specific frequency response. It is not the case for a Wiener filter, introduced by Norbert Wiener [642].

Let's assume that a certain system generates signal u. We record the signal with an apparatus which has a known impulse response r. We assume that the measurement is corrupted by the additive Gaussian white noise v. This can be expressed as:

$$x[n] = y[n] + v[n] = \sum_i r[n-i]u[i] + v[n] \qquad (3.12)$$

In such a system u cannot be accessed directly. However, minimizing the square error we can find $\hat{u}$ which estimates u :

$$\hat{u}[n] = \arg\min E\left[\sum_n |u[n] - \hat{u}[n]|^2\right] \tag{3.13}$$

From the Parseval theorem it follows, that the above relation will also hold for the Fourier transforms of the respective signals:

$$\hat{U}[f] = \arg\min E\left[\sum_f |U[f] - \hat{U}[f]|^2\right] \tag{3.14}$$

Taking advantage of the convolution theorem (Sect. 2.4.4) for the signal $y[n] = \sum_i r[n-i]u[i]$ we obtain:

$$Y[f] = U[f]R[f] \tag{3.15}$$

Assuming that the estimator has the form:

$$\hat{U}[f] = \frac{X[f]\Phi[f]}{R[f]} \tag{3.16}$$

It can be shown that condition (3.14) is satisfied for:

$$\Phi[f] = \frac{|Y[f]|^2}{|Y[f]|^2 + \langle|V[f]|^2\rangle} \tag{3.17}$$

$\Phi[f]$ is called a Wiener filter. Thus:

$$H[f] = \frac{\Phi[f]}{R[f]} = \frac{1}{R[f]} \frac{|Y[f]|^2}{|Y[f]|^2 + \langle|V[f]|^2\rangle} \tag{3.18}$$

gives the frequency domain representation of the transfer function of the optimal filter.

In practice neither $Y[f]$ nor $V[f]$ is known. The empirical estimate of $\Phi[f]$ may be computed as:

$$\hat{\Phi}[f] = \begin{cases} \frac{S_x[f] - \hat{S}_v[f]}{S_x[f]} & for \quad S_x[f] > \hat{S}_v[f] \\ 0 & for \quad S_x[f] \le \hat{S}_v[f] \end{cases} \tag{3.19}$$

where $S_x[f]$ is the power spectrum of x and $\hat{S}_v[f]$ is the approximation of the noise power spectrum.

3.2 Probabilistic Models

3.2.1 Hidden Markov Model

An excellent introduction to the Hidden Markov models (HMM) concept, estimation, and illustrative applications can be found in [496]. MATLAB Statistics and Machine Learning Toolbox provides functions and tutorial on the application of HMM (Documentation > Statistics and Machine Learning Toolbox > Cluster Analysis > Hidden Markov Models). Here we review the basic facts only.

Let us consider a system which can be in a number of *states*. In discrete time moments, the state of the system may change. The transitions between states are probabilistic. The probability distribution of the next state depends only on the current state, and it does not depend on the way the system reached the current state. If the states of the system can be observed directly, then it can be mathematically described as a regular Markov model. The Markov model is specified by a set of states and the probabilities of transitions between each pair of states. An extension of the Markov model allowing for much broader applications is a Hidden Markov model (HMM). It is hidden in the sense that the states of the system are not observed directly, but the observations of the model depend in a probabilistic way on the states of the model. Although the states are not observed directly, often some physical sense can be attached to them. As the system evolves in time the state transitions occur and the system passes through a sequence of states $S(t)$. This sequence is reflected in a sequence of observations $Y(t)$. An essential property of HMM is that the conditional probability distribution of the hidden state $S(t)$ at time t, depends only on the hidden state $S(t-1)$. Similarly, the observation $Y(t)$ depends only on the hidden state $S(t)$—both occurring at the same time t. A simple example of such a model is shown in Figure 3.2.

The HMM is specified by the pair of parameters: the number of states (N) and the number of possible observations per state (M), and additionally by three probability distributions governing: state transitions (A), observations (B), initial conditions (π). The HMMs can be used to generate possible sequences of observations. However, more interesting applications of HMMs rely on:

- Estimating the model parameters (N, M, A, B, π) given the sequence of observations. This can be done with maximum likelihood parameter estimation using expectation-maximization Baum-Welch algorithm [36].

- Given a sequence of observations one can evaluate the probability that the sequence was generated by a specific model (N, M, A, B, π).

Imagine that we have a set of observations known to be generated by a number of different models. We can estimate the parameters of each of the models. Later, when a new sequence of observations is obtained, we could identify which model most probably generated that sequence. In this sense, HMMs can be used as classifiers with supervised learning.

Another problem that can be addressed with HMM is the question of the most probable sequence of states given the model and the sequence of observations. The solution to this problem is usually obtained with the Viterbi algorithm [619].

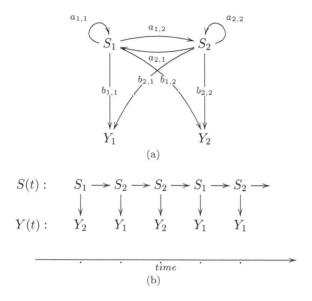

(a)

$S(t):$ $S_1 \rightarrow S_2 \rightarrow S_2 \rightarrow S_1 \rightarrow S_2 \rightarrow$

$Y(t):$ Y_2 Y_1 Y_2 Y_1 Y_1

time

(b)

FIGURE 3.2
(a) An example of two state Hidden Markov model. S_1 and S_2 are states of the model, $a_{i,j}$ are probabilities of transition from state i to j. Y_1 and Y_2 are observations, $b_{i,j}$ are probabilities of observation Y_j if the system is in state S_i. (b) A possible sequence of states and observations. Arrows indicate the dependence.

Besides the HMM related functionality delivered in MATLAB Statistics and Machine Learning Toolbox a toolkit PMTK might be of interest. It supports a large variety of probabilistic models, including HMM. The toolbox can be downloaded from `https://github.com/probml/pmtk3`.

3.2.2 Kalman Filters

A Kalman filter is a method for estimation of the state of a linear dynamic system (LDS) discretized in the time domain. At time k the system is in a *hidden* state x_k. It is hidden since it cannot be directly observed. The observer's knowledge about the state of the system comes from measurements z_k which are distorted by noise w_k. Formally it can be expressed as:

$$z_k = \mathbf{M}_k x_k + w_k \tag{3.20}$$

where $\mathbf{M}_k$ is the matrix describing the linear operation of taking the observation. The measurement noise w_k is assumed to come from a zero mean normal distribution with covariance matrix $\mathbf{R}_k$.

The current state of the system x_k is assumed to depend only on the previous state x_{k-1}, on the current value of a control vector u_k, and current value of a random perturbation v_k. Expressing it as an equation we get:

$$x_k = \mathbf{A}_k x_{k-1} + \mathbf{B}_k u_k + v_k \tag{3.21}$$

where $\mathbf{A}_k$ is the state transition matrix, $\mathbf{B}_k$ is the matrix transforming the control input, v_k is the process noise coming from the zero mean normal distribution with covariance matrix $\mathbf{Q}_k$. The matrixes $\mathbf{M}_k$, $\mathbf{R}_k$, $\mathbf{A}_k$, $\mathbf{B}_k$, and $\mathbf{Q}_k$ may vary in time. The initial state and the noise vectors at each step $\{x_0, w_1, \ldots, w_k, v_1, \ldots, v_k\}$ are all assumed to be mutually independent.

In the estimation of the current state of the system all the above mentioned matrixes are assumed to be known. Two types of error can be defined: *a priori* estimate error:

$$e_k^- = x_k - \hat{x}_k^- \tag{3.22}$$

where $\hat{x}_k^-$ is the estimate of state x_k based only on the knowledge of previous state of the system, with covariance matrix given by:

$$\mathbf{P}_k^- = E[e_k^- e_k^{-T}] \tag{3.23}$$

and *a posteriori* estimate error, with the result z_k of observation known:

$$e_k = x_k - \hat{x}_k \tag{3.24}$$

with covariance matrix given by:

$$\mathbf{P}_k = E[e_k e_k^T] \tag{3.25}$$

Posterior estimate of the system state in current step can be obtained as a linear mixture of the *a priori* estimate of the system state and the error between the actual and estimated result of the observation:

$$\hat{x}_k = \hat{x}_k^- + \mathbf{K}(z_k - \mathbf{M}\hat{x}_k^-) \tag{3.26}$$

The matrix $\mathbf{K}$ is chosen such that the Frobenius norm of the *a posteriori* covariance matrix is minimized. This minimization can be accomplished by first substituting (3.26) into the above definition for e_k (3.24), substituting that into (3.25), performing the indicated expectations, taking the derivative of the trace of the result with respect to $\mathbf{K}$, setting that result equal to zero, and then solving for $\mathbf{K}$. The result can be written as:

$$\mathbf{K}_k = \mathbf{P}_k^- \mathbf{M}^T \left(\mathbf{M}\mathbf{P}_k^- \mathbf{M}^T + \mathbf{R}\right)^{-1} \tag{3.27}$$

The Kalman filter estimates the state of LDS by using a form of feedback control: the filter estimates the process state at some time and then obtains feedback in the form of (noisy) measurements. Two phases of the computation may be distinguished: predict and update. The predict phase uses the state estimate from the previous time step to produce an estimate of the current time step. In the update phase, the current *a priori* prediction is combined with the current observation to refine the *a posteriori* estimate. This algorithm is presented in Figure 3.3. Excellent introduction to

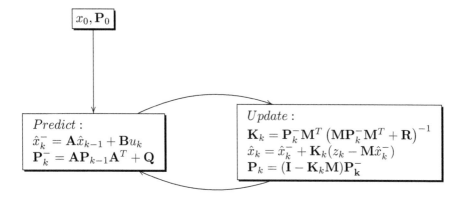

FIGURE 3.3
The computation of successive system states with Kalman filter.

the Kalman filter may be found in [635]. The MATLAB Control System Toolbox supports design and simulation of both a steady state and a time-varying Kalman filter. Demos of usage are given in the official MATLAB documentation and Video and Webinar Series: https://www.mathworks.com/videos/series/understanding-kalman-filters.html.

A free alternative is MATLAB toolbox by Kevin Murphy available at http://people.cs.ubc.ca/~murphyk/Software/Kalman/kalman.html provides Kalman filter formalism for filtering, smoothing, and parameter estimation for linear dynamical systems.

Many real dynamical systems do not exactly fit the model assumed by Kalman filter design. The not-modeled dynamics contributes to the component w_k in equation 3.20. In practice, we cannot distinguish between the uncertainty of the measurement and the not modeled dynamics. We have to assume a certain value of $\mathbf{R}$. Too big $\mathbf{R}$ results in the slow adaptation of the filter, and too small $\mathbf{R}$ leads to instability of the results. $\mathbf{R}$ controls the speed of adaptation, and it has to be chosen optimally.

3.3   Stationary Signals

3.3.1   Analytic Tools in the Time Domain

3.3.1.1   Mean Value, Amplitude Distributions

In Sect. 2.1 formulas for calculation of moments such as mean, variance, standard deviation, correlation used ensemble averaging paradigm. Quite often we have only one realization of a process. Under the assumption of ergodicity one can calculate the estimators by averaging over time instead of averaging over an ensemble.

We have to bear in mind that in case of application of ergodicity assumption, the strict value of the estimator is obtained for an infinitely long time. In practice the shorter the time epochs, the higher the error of estimate is.

3.3.1.2   Entropy and Information Measure

The concept of information measure, in statistics called entropy, was introduced by Shannon [545]. It is connected with the probability of occurrence of a given effect. Let us assume that an event has M possibilities, and i^{th} possibility occurs with probability p_i, then the information connected with the occurrence of ith possibility will be:

$$I_i = -\log p_i \tag{3.28}$$

The expected value of the information is entropy:

$$En = -\sum_{i=1}^{M} p_i \log p_i \tag{3.29}$$

Entropy is a measure of uncertainty associated with the outcome of an event. The higher the entropy, the higher the uncertainty as to which possibility will occur. The highest value of entropy is obtained when all possibilities are equally likely.

In practice for time series, the entropy is calculated from the amplitude distributions. The amplitude range A of a raw sampled signal is divided into K disjointed intervals I_i, for $i = 1, \ldots, K$. The probability distribution can be obtained from the ratio of the frequency of the samples N_i falling into each bin I_i and the total sample number N:

$$p_i = N_i/N \tag{3.30}$$

The distribution $\{p_i\}$ of the sampled signal amplitude is then used to calculate entropy measure according to (3.29).

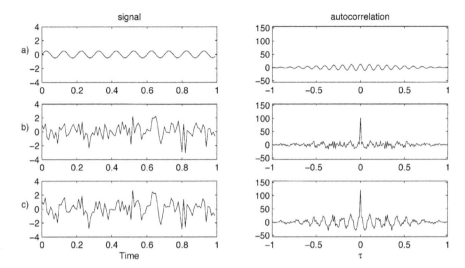

FIGURE 3.4

Illustration of autocorrelation function properties. Left column signals, right column autocorrelation functions of the corresponding signals. a) Periodic signal, b) zero mean Gaussian noise, c) sum of a) and b). The code reproducing this figure, and demonstrating some properties of autocorrelation function is in the Live Script: `matlab/c3/Ch3_fig_3_4.mlx`.

3.3.1.3 Autocorrelation Function

Correlation function was introduced in Sect. 2.1 by equation (2.11) in terms of ensemble averaging. Under the assumption of ergodicity, autocorrelation $R_x(\tau)$ and autocovariance functions $C_x(\tau)$ are defined by:

$$R_x(\tau) = \int_{-\infty}^{+\infty} x(t)x(t+\tau)dt \qquad (3.31)$$

and

$$C_x(\tau) = \int_{-\infty}^{+\infty} (x(t) - \mu_x)(x(t+\tau) - \mu_x)dt \qquad (3.32)$$

where τ is the time lag, μ_x is mean of the signal (equation 2.2). Autocorrelation function $R_x(\tau)$ is always real and symmetric: $R_x(\tau) = R_x(-\tau)$. It takes maximal value for $\tau = 0$. It follows from equations (2.5) and (2.11) that $R_x(\tau = 0) = \psi^2$ (the mean square value of the signal). Variance σ_x^2 is equal to the autocovariance function for lag 0: $C_x(\tau = 0) = \sigma_x^2$. Autocorrelation of a periodic function is also periodic. Autocorrelation of noise decreases rapidly with the time lag τ (Figure 3.4). We can see from Figure 3.4 c that

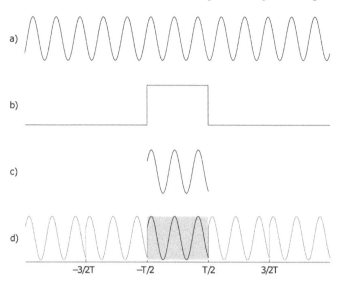

FIGURE 3.5
The illustration of assumptions of DFT. a) The assumed signal. b) Observation
window (equation (3.34)) c) The observed fragment of signal. d) The signal
from observation window (marked with gray rectangle) is periodically repeated
to infinity. The discontinuities of the signal at the window edge introduce
additional components to the frequency spectrum.

autocorrelation can help in the extraction of the periodic signal from noise,
even when the noise amplitude is higher than the signal.

3.3.2 Analytic Tools in the Frequency Domain

3.3.2.1 Estimators of Spectral Power Density Based on Fourier Transform

In physiological signals quite often important information is connected with
the occurrence of specific rhythms; therefore the analysis in the frequency
domain is of particular interest. A commonly used method of estimation of
the power spectrum $S(f)$ is based on the Fourier transform:

$$S(f) = X(f)X^*(f) \tag{3.33}$$

$X(f)$ is the Fourier transform of the signal $s(t)$ defined by equation (2.27),
where Fourier transform is defined for times $\in (-\infty, +\infty)$. Fourier transform
for the sampled signal and for finite time of observation is expressed by formula (2.28), derived under the assumption that the signal given in the time
window, outside it, is periodically repeated to infinity (Figure 3.5). In practice, taking the finite segment of the signal for $t \in (-T/2, T/2)$ is equivalent

to multiplication of the signal by a function w:

$$w_{T/2}(t) = \begin{cases} 0 & \text{for} & t < -T/2 \\ w & \text{for} & -T/2 \le & t \le T/2 \\ 0 & \text{for} & & t > T/2 \end{cases} \quad (3.34)$$

where $w = 1$. This procedure introduces discontinuities at the edges of the window which disturb the spectrum. The discontinuities contribute to the spectrum at many frequencies. Multiplication in the time domain corresponds to convolution in the frequency domain (see equation 2.31). Therefore the transform we obtain can be considered as the convolution of the transform of idealized infinite signal $X(f)$ with the transform of the window function $W_{T/2}(f)$, causing an effect known as spectral leakage. This property deteriorates the spectral estimate; especially disturbing are the side lobes of function $W(f)$.

3.3.2.2 Choice of Windowing Function

A commonly used technique to reduce the spectral leakage is the choice of the proper function $w_{T/2}(t)$. Usually one wants the window w to approach zero at both edges smoothly. In order to improve the estimation of spectral power, windows of different shapes were introduced. Good windows are characterized by a highly concentrated central lobe with very low or quickly diminishing side lobes of their transforms. In MATLAB Signal Processing Toolbox there is a very convenient tool for studying windows properties: `windowDesigner`. It displays the time and frequency representation of the selected window and evaluates its important characteristics:

- Leakage factor—ratio of power in the side lobes to the total window power

- Relative side lobe attenuation—difference in height from the main lobe peak to the highest side lobe peak

- Width of the main lobe at 3 dB below the main lobe peak

The available windows are: `barthannwin`, `bartlett`, `blackman`, `blackmanharris`, `bohmanwin`, `chebwin`, `flattopwin`, `gausswin`, `hamming`, `hann`, `kaiser`, `nuttallwin`, `parzenwin`, `rectwin`, `triang`, `tukeywin`.
     For a signal consisting of multiple frequencies the applied window has considerable influence on the detectability of individual spectral components [213]. Examples of windows and their properties are shown in Figure 3.6.

Errors of Fourier Spectral Estimate

Fourier transform $X(f)$ is a complex number whose real $X_R(f)$ and imaginary $X_I(f)$ parts can be considered as uncorrelated random values of zero mean and equal variance. Since Fourier transform is a linear operation, components $X_R(f)$ and $X_I(f)$ have a normal distribution, if $x(t)$ has normal distribution.

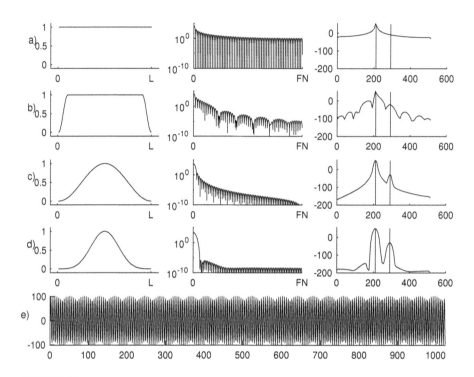

FIGURE 3.6

First column—window in the time domain, second column—window in the frequency domain, third column—spectra of the signal e) obtained by application of respective window function and Fourier transform. e) signal composed of two sinusoids with frequencies $f_1 = 210$ Hz and $f_2 = 290$Hz. The first sinusoid has 100 times greater amplitude than the second one. The windows are: a) rectangular b) Tukey with 10% round off c) Hann window d) Blackmann-Harris. Note the spectrum leakage effects especially pronounced in case of rectangular and Tukey windows. They lead to problems in identifying the peak corresponding to the weaker sinusoid. The code reproducing this figure, and demonstrating the effect of spectral leakage is in the Live Script: `matlab/c3/Ch3_fig_3_6.mlx`.

Therefore value:

$$|X(f)|^2 = X_R^2(f) + X_I^2(f) \qquad (3.35)$$

is a sum of squares of two independent variables of normal distribution; hence each frequency component of estimator $\hat{S}(f)$ has a distribution given by:

$$\frac{\hat{S}(f)}{S(f)} = \frac{\chi_2^2}{2} \qquad (3.36)$$

where χ_2^2 is a chi-square statistics of two degrees of freedom (corresponding to the real and imaginary parts of Fourier transform) [42]. Please note that the above expression is independent of the observation time T, which means that increasing the length of epoch T does not change the distribution function describing the error of the estimator. Extension of the time of observation T causes only the increase of the number of the frequency components in the spectrum. It means that the estimator of the spectral density obtained through Fourier transform is biased; also the error of the estimator is high. Namely for a given frequency f_1 the relative error of $S(f_1)$ is: $\epsilon_r = \sigma_{S_{f_1}}/\mu_{S_{f_1}}$. For the distribution χ_n^2: $\sigma^2 = 2n$ and $\mu = n$, where n is the number of degrees of freedom. For $n = 2$ we get $\epsilon_r = 1$, which means that each single frequency component of $S(f)$ has a relative error 100%. In consequence spectral power calculated through Fourier transform is strongly fluctuating.

In order to improve statistical properties of the estimate two techniques were introduced. The first one relies on averaging over neighboring frequency estimates. Namely we calculate smoothed estimator $\hat{S}_k$ of the form:

$$\hat{S}_k = \frac{1}{l}[S_k + S_{k+1} + \cdots + S_{k+l-1}] \qquad (3.37)$$

Assuming that the frequency components S_i are independent, estimator S_k is characterized by the χ^2 distribution of number degrees of freedom equal $n = 2l$. The relative standard error of frequency estimate will be therefore: $\epsilon_r = \sqrt{\frac{1}{l}}$.

Another way of improving power spectral estimate is averaging the estimates for successive time epochs:

$$\hat{S}_k = \frac{1}{q}[S_{k,1} + S_{k,2} + \cdots + S_{k,j} + \cdots + S_{k,q}] \qquad (3.38)$$

where $S_{k,j}$ is the estimate of the frequency component k based on the time interval j. The number of degrees of freedom, in this case, equals q; therefore the relative error of single frequency estimate will be: $\epsilon_r = \sqrt{\frac{1}{q}}$. This approach is known as Welch's method and is implemented in MATLAB Signal Processing Toolbox as `pwelch`.

Both of these methods require stationarity of the signal in the long enough epoch to either perform averaging over frequency estimates, without excessive decreasing of frequency resolution or divide the epoch into segments long enough to provide necessary frequency resolution.

An alternative method, which can be useful in case of relatively short time epochs, is the multitaper method (MTM) described in [597]. This method uses a sequence of windows that are orthogonal to each other (discrete prolate spheroidal sequences). Each window is used to compute the windowed periodogram of the signal. Subsequently, the periodograms are averaged. This method is implemented in MATLAB Signal Processing Toolbox as pmtm.

MATLAB demo: Welch's method for estimation of power spectral density

If you would like to get experience with the Welch method, consider the Live Script: matlab/c3/Ch3_09_pwelch.mlx

---

MATLAB demo: estimation of power spectrum using Thomson's multitaper method

If you would like to get experience with the multitaper method, consider the Live Script: matlab/c3/Ch3_10_pmtm.mlx

---

Relation of Spectral Density and the Autocorrelation Function

According to the Wiener-Chinchyn formula spectral density function $S(f)$ is a Fourier transform of the autocorrelation function $R(\tau)$:

$$S(f) = \int_{-\infty}^{\infty} R(\tau)e^{i2\pi f\tau}\, d\tau \tag{3.39}$$

Assuming that $R(\tau)$ exists and

$$\int_{-\infty}^{\infty} |R(\tau)|\, d\tau < \infty \tag{3.40}$$

$R(\tau)$ is connected to $S(f)$ by inverse Fourier transform.

$$R(\tau) = \int_{-\infty}^{\infty} S(f)e^{-i2\pi f\tau}\, df \tag{3.41}$$

These relations are visualized in Figure 3.7. From the above formula, it follows that the integral of the spectral density is equal to $R(0)$. Usually, one-sided spectral density estimator $S(f)$ is calculated for $f \in (0, \infty)$, since for real signals $S(f)$ is a symmetric function.

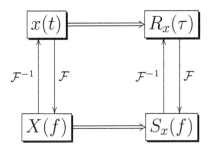

FIGURE 3.7

Illustration of relations between signal $x(t)$, its Fourier transform $X(f)$, its autocorrelation function $R_x(\tau)$, and its spectrum $S_x(f)$.

Bispectrum and Bicoherence

Polyspectra or higher order spectra provide supplementary information to the power spectrum. Third order polyspectrum is called bispectrum. The prefix "bi" refers not to two time series, but to the two frequencies of a single signal. Power spectrum according to equation 3.39 is a Fourier transform of auto-correlation function, which is also called second order cumulant. Bispectrum is a Fourier transform of third order cumulant $R_3(\tau_1, \tau_2)$. For discrete case (sampled signal) correlation is expressed by:

$$R_2(\tau) = \sum_{t=-\infty}^{\infty} x(t)x(t+\tau) \qquad (3.42)$$

and third order cumulant by:

$$R_3(\tau_1, \tau_2) = \sum_{t=-\infty}^{\infty} x(t)x(t+\tau_1)x(t+\tau_2) \qquad (3.43)$$

Bispectrum (BS) is defined as:

$$BS(f_1, f_2) = X(f_1)X(f_2)X^*(f_1 + f_2) \qquad (3.44)$$

where X denotes Fourier transform of signal x. Bispectrum quantifies the relationship between the sinusoids at two primary frequencies f_1 and f_2 and the modulation component at frequency $f_1 + f_2$. Bispectrum is a function of the triplet of frequencies $(f_1, f_2, f_1 + f_2)$ incorporating both power and phase information. Bicoherence, which takes values from the range $[0-1]$, is defined as a squared normalized version of the bispectrum:

$$B^2(f_1, f_2) = \frac{|BS(f_1, f_2)|^2}{S(f_1)S(f_2)S(f_1 + f_2)} \qquad (3.45)$$

Bicoherence is a function which gives the information on non-linear interactions. It is a measure for quantifying the extent of phase coupling between different frequencies in the signal. Namely bicoherence measures the proportion of the signal energy at any bifrequency that is quadratically phase coupled.

In practical applications, the computation of bicoherence is limited by the quality of the data. The statistical error of the spectral estimate is high as was mentioned already in Sect. 3.3.2.1. For bicoherence, these errors cumulate due to the multiplication of spectral terms. Therefore reliable estimation of bicoherence is possible only for a high signal to noise ratio and requires long enough data allowing to compute an average estimate based on several equivalent segments of the signal.

3.3.2.3   Parametric Models: AR, ARMA

A time series model that approximates many discrete stochastic and deterministic processes encountered in practice is represented by the ARMA filter difference equation:

$$x_t = \sum_{i=1}^{p} a_i x_{t-i} - \sum_{k=0}^{r} b_k y_{n-k} \tag{3.46}$$

Where x_t is the output sequence of a causal filter and y_t is an input driving sequence. In the ARMA model, it is assumed that the driving sequence is a zero-mean white noise process.

AR model is defined by the equation:

$$x_t = \sum_{i=1}^{p} a_i x_{t-i} + \epsilon_t \tag{3.47}$$

where ϵ_t is a white noise. It has been shown that the AR model of sufficiently high order can approximate the ARMA model well [380]. The determination of the model order and the coefficients for the AR model is much simpler than for the ARMA model, which requires non-linear algorithms for the estimation of parameters. Many efficient algorithms of AR model fitting exist; therefore, in the following, we shall concentrate on the AR model. More details on ARMA may be found in [380]. Moreover, in transfer function $H(z)$ (equation 3.6), which describes the spectral properties of the system, AR model coefficients determine the poles of $H(z)$, which correspond to the spectral peaks. MA coefficients determine the zeros of $H(z)$, which correspond to the dips of the spectrum. In the practice of biomedical signals processing, we are usually more interested in the spectral peaks, since they correspond to the rhythms present in the time series.

AR Model Parameter Estimation

There is a number of algorithms for estimation of the AR model parameters. Here we present the Yule-Walker algorithm. First, equation (3.47) is multiplied by the sample at time $t - m$:

$$x_t x_{t-m} = \sum_{i=1}^{p} a_i x_{t-i} x_{t-m} + \epsilon_t x_{t-m} \qquad (3.48)$$

then we calculate expected values of the left and right sides of the equation taking advantage of the linearity of the expected value operator:

$$R(m) = E\{x_t x_{t-m}\} = \sum_{i=1}^{p} E\{a_i x_{t-i} x_{t-m}\} + E\{\epsilon_t x_{t-m}\} \qquad (3.49)$$

It is easy to see that the expected value $E\{x_t x_{t-m}\}$ is the autocorrelation function $R(m)$, and $E\{\epsilon_t x_{t-m}\}$ is non-zero only for $t = m$ so:

$$R(m) = \sum_{i=1}^{p} a_i R(m - i) + \sigma_\epsilon^2 \delta(m) \qquad (3.50)$$

where $m = 0, \ldots, p$.

For $m > 0$ we can write a set of equations:

$$\begin{bmatrix} R(0) & R(-1) & \cdots & R(1-p) \\ R(1) & R(0) & R(-1) & \\ \vdots & & & \vdots \\ R(p-1) & \cdots & \cdots & R(0) \end{bmatrix} \begin{bmatrix} a_1 \\ a_2 \\ \vdots \\ a_p \end{bmatrix} = \begin{bmatrix} R(1) \\ R(2) \\ \vdots \\ R(p) \end{bmatrix} \qquad (3.51)$$

and compute the coefficients a.

For $m = 0$ we have:

$$R(0) = \sum_{i=1}^{p} a_i R(-i) + \sigma_\epsilon^2 \qquad (3.52)$$

which allows us to compute σ_ϵ^2.

Methods for estimation of AR model parameters implemented in MATLAB Signal Processing Toolbox [447, 282] are:

- `arburg` Estimate AR model parameters using Burg method

- `arcov` Estimate AR model parameters using covariance method

- `armcov` Estimate AR model parameters using modified covariance method

- `aryule` Estimate AR model parameters using Yule-Walker method

Choice of the AR Model Order

When fitting the AR model to the signal, we have to assume that the autoregressive process can describe the signal. The correct order of the process can be assessed by finding the minimum of the Akaike information criterion (AIC), which is a function of the model order p:

$$AIC(p) = \frac{2p}{N} + \log V \qquad (3.53)$$

where p is the model order (the number of free parameters in the model), N-number of signal samples used for model parameters estimation, V-residual noise variance. The higher the model order, the less variance remains unaccounted; however, an excessive number of model parameters increases the statistical uncertainty of their estimates. The AIC is a sort of cost function that we seek to minimize. The first element of that function expresses punishment for using a high order model, and the second element expresses a reward for reducing the unexplained variance.

AIC is the one mostly applied in practice, but other criteria may be commonly found in the literature, e.g., minimum description length MDL [380]:

$$MDL(p) = \frac{p \log N}{N} + \log V \qquad (3.54)$$

MDL is said to be statistically consistent. The criteria for model order determination work well and give similar results when the AR process reasonably well models the data. An example of an AIC function is shown in Figure 3.8.

MATLAB demo: estimation of model order

If you would like to test both criteria for model selection, try the Live Script: `matlab/c3/Ch3_11_ar_order.mlx`

AR Model Power Spectrum

Once the model is properly fitted to the signal all the signal's properties are contained in the model coefficients. Notably, we can derive an analytical formula for the AR model power spectrum. First we rewrite the model equation (3.47):

$$\sum_{i=0}^{p} a_i x_{t-i} = \epsilon_t \qquad (3.55)$$

Applying Z-transform to both sides of the above equation and taking into account its properties (Sect. 2.4.6) we get:

$$A(f)X(f) = E(f) \qquad (3.56)$$

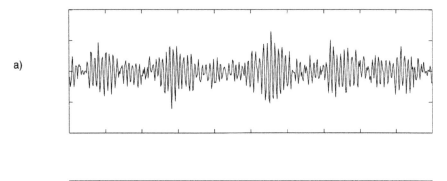

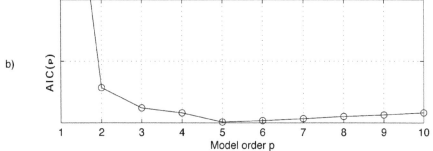

FIGURE 3.8

Example of *AIC* function. a) Signal generated with AR model order 5 ($a_1 = 0.2, a_2 = -0.5, a_3 = -0.3, a_4 = 0.1, a_5 = 0.2$), b) *AIC* for models of order p estimated for that signal. Note a minimum at $p = 5$.

multiplying equation (3.56) by A^{-1} leads to:

$$X(f) = A^{-1}(f)E(f) = H(f)E(f) \tag{3.57}$$

where we defined $H(f) = A^{-1}(f)$. From the definition of power spectrum we obtain:

$$S(f) = X(f)X^*(f) = H(f)VH^*(f) \tag{3.58}$$

Expressing H in terms of model coefficients:

$$H(z) = \frac{1}{\sum_{k=0}^{p} a_k z^{-k}} \tag{3.59}$$

we get:

$$S(f) = \frac{V}{2\pi} \frac{1}{|\sum_{k=0}^{p} a_k z^{-k}|^2}, \qquad z = e^{i2\pi f} \tag{3.60}$$

The AR spectral estimate is maximum entropy method. The entropy of information (strictly speaking the entropy rate for infinite process) is connected with power spectral density function by the formula [560]:

$$En = \frac{1}{4F_N} \int_{-F_N}^{+F_N} \log S(f)df \tag{3.61}$$

It was pointed out by [609] that AR spectral estimate is equivalent to the estimate fulfilling maximum entropy condition. It means that the AR estimate expresses maximum uncertainty for the unknown information but is consistent with the known information.

An illustrative comparison of parametric and non-parametric spectra estimates for EEG signal is shown in Figure 3.9. Estimates obtained with the AR model and with Welch's method show a significant reduction of variation compared with modified periodogram. AR power spectrum is unbiased, free of windowing effects, and it has better statistical properties than the Fourier

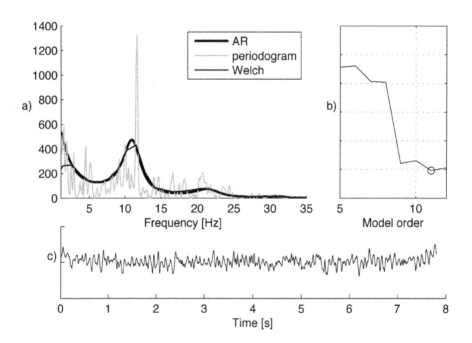

FIGURE 3.9
Comparison of AR and Fourier spectra. a) Spectra: thick line—spectrum of AR model with the lowest AIC, thin black line—power spectral density estimate via Welch's method window length is 1/10 signal length and the overlap is half the window length, gray line–modified periodogram with Blackmann-Harris window. b) *AIC* for different model orders, estimated for the signal c) (minimum is marked with a circle). c) An epoch of EEG signal; sampling frequency 128 Hz.

spectrum because of a greater number of the degrees of freedom in case of AR. Namely, it is equal to N/p, where N is the number of points of the data window and p is a model order. Correct estimation of the model order is important, but small deviations of p do not drastically change the power spectrum. In the case of the AR model fitting, the signal should not be oversampled. Sampling should be just slightly above the double of Nyquist frequency. Oversampling does not bring any new information, and it causes redundancy. In case of the autoregressive model to account for all frequencies, especially low frequencies, we have to increase the number of steps backward in autoregressive process, which means that we have to increase the model order. Estimating more parameters (from the same amount of data) increases the uncertainty of the estimates.

MATLAB demo: estimation of AR model spectra

If you would like to compare the parametric and non-parametric methods of spectrum estimation on some real EEG signal, try the Live Script: `matlab/c3/Ch3_12_AR_FFT_spectra_eeg.mlx`

Parametric Description of the Rhythms by AR Model, FAD Method

The transfer function $H(z)$ has maxima for z values corresponding to the zeroes of the denominator of the expression (3.59). These values of z are the poles z_j lying inside the unit circle $|z| = 1$ in the complex plane. The frequency axis $0 < f < f_N$ in polar coordinates (Figure 3.10) corresponds to the counterclockwise traverse of the upper half of this unit circle. The angular position of z_j lying closest to the unit circle determines the frequency of the corresponding peak, and the radial distance depends on the damping of the relevant frequency component. From the coefficients of the AR model we can derive parameters which characterize oscillatory properties of the underlying process. Namely the transfer function $H(z)$ equation (3.59) for the corresponding continuous system takes a form:

$$H(z) = \sum_{j=1}^{p} C_j \frac{z}{z - z_j} \tag{3.62}$$

It can be interpreted as a system of parallel filters each of them corresponding to a characteristic frequency (Figure 3.11). For $z = z_j$ the resonance for filter j occurs. We can introduce parameters characterizing these filters [165]:

$$\alpha_j = \frac{1}{\Delta t} \log z_j, \quad \beta_j = -Re(\alpha_j), \quad \omega_j = Im(\alpha_j), \quad \phi_j = Arg(C_j), \quad B_j = 2|C_j| \tag{3.63}$$

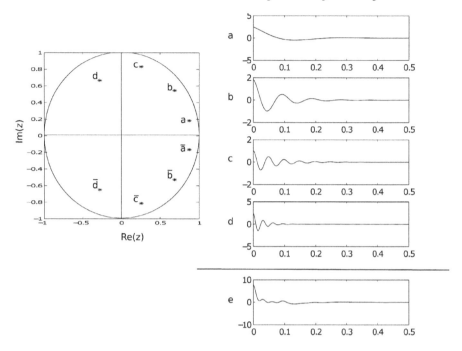

FIGURE 3.10

Left: poles of illustrative AR transfer function in the complex plane. For real signals the transfer function has pairs of complex conjugate poles (here: a and $\bar{a}$, b and $\bar{b}$, c and $\bar{c}$, d and $\bar{d}$). These poles z_j lie inside the unit circle $|z| = 1$ in the complex plane. The frequency f related to the pole ($0 < f < f_N$) corresponds to the angular position of the pole. Right: traces a-d are impulse response functions corresponding to the poles a–d; each of them is in the form of an exponentially damped sinusoid. Trace e—total impulse response of the AR model is the sum of the components a, b, c, and d.

where Δt is the sampling interval. The imaginary part of α_j corresponds to the resonance frequency ω_j and the real part to the damping factor β_j of the oscillator generating this frequency. By rewriting $H(z)$ in terms of parameters from (3.63) and performing inverse Laplace transform we find the impulse response function in the form of damped sinusoids (Figure 3.11):

$$h(t) = \sum_{j=1}^{p} B_j e^{-\beta_j t} cos(\omega_j t + \phi_j) \tag{3.64}$$

In this way, we can find directly (not from the spectral peak positions) the frequencies of the rhythms present in the signal, which can be quite useful, especially for weaker components. We can also find the amplitudes and damping factors of the oscillations. The method of time-series description using parameters: frequency (ω_j), amplitude (B_j), damping (β_j) was named FAD. The

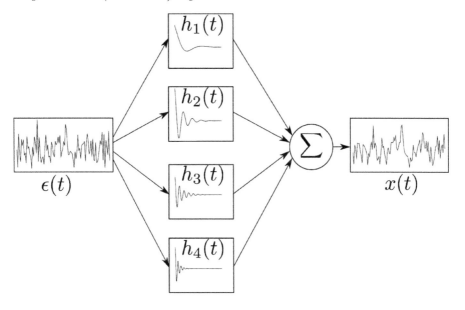

FIGURE 3.11
Transfer function of the AR model with four rhythms as a system of parallel filters each of them corresponding to a characteristic frequency (see equation 3.64).

number of identified oscillatory components depends on the model order; for even model order, it is $p/2$. Model of odd order contains the non-oscillatory component $e^{-\beta}$, which accounts for the usually observed form of the power spectrum decaying with frequency. The correct estimation of the model order is important for the description of signals in terms of FAD parameters. For a too low model order, not all components would be accounted for. For a too high model order, too many poles in the complex plane would appear. They would lie close to the main frequency components and would give the effect of "sharpening" the spectral peaks.

MATLAB demo: parametrization of the signal by FAD technique

If you would like to get some intuition about the parameters estimated by FAD, try the Live Script: `matlab/c3/aCh3_13_FAD.mlx`

3.4   Non-Stationary Signals

3.4.1   Instantaneous Amplitude and Instantaneous Frequency

In case of stationary signals the concept of amplitude and frequency spectra is intuitive. For a non-stationary process, it is more challenging to determine signal amplitude or frequency, since these quantities can vary in time, so it is useful to introduce instantaneous amplitude and frequency. To define the instantaneous amplitude and instantaneous frequency, first, we need to introduce the concept of the analytic signal.

Analytic signal is a complex signal $x_a(t)$ related to a real signal $x(t)$ by formula:

$$x_a(t) = x(t) + ix_h(t) \tag{3.65}$$

where $x_h(t)$ is the Hilbert transform of $x(t)$. The Hilbert transform is a linear operator that can be thought of as the convolution of $x(t)$ with the function $h(t) = \frac{1}{\pi t}$. The spectral power for $x_a(t)$ is non-zero only for positive f.

Thus approximation of the analytic signal can be obtained from the following algorithm:

1. Calculate the FFT of the input sequence x consisting of n samples:
   $X = \mathtt{fft}(x)$

2. Create a vector h:

$$h(j) = \begin{cases} 1 & \text{for} \quad j = 1, (n/2)+1 \\ 2 & \text{for} \quad j = 2, 3, \ldots, (n/2) \\ 0 & \text{for} \quad j = (n/2)+2, \ldots, n \end{cases} \tag{3.66}$$

3. Calculate the element-wise product of X and h: $Y = X \cdot h$

4. Calculate the inverse FFT of Y.

This algorithm is implemented in MATLAB Signal Processing Toolbox as `hilbert`. The analytic signal can be presented in the form:

$$x_a(t) = A(t)\cos(\phi(t)) \tag{3.67}$$

where:

- $A(t) = |x_a(t)|$ is called the instantaneous amplitude,

- $\phi(t) = \arg(x_a(t))$—the instantaneous phase,

- $f(t) = \frac{1}{2\pi} \frac{d\arg(x_a(t))}{dt}$—the instantaneous frequency[3].

---

[3]Note, that in case of a stationary signal, e.g., $x = B\cos(2\pi ft)$ these definitions retrieve

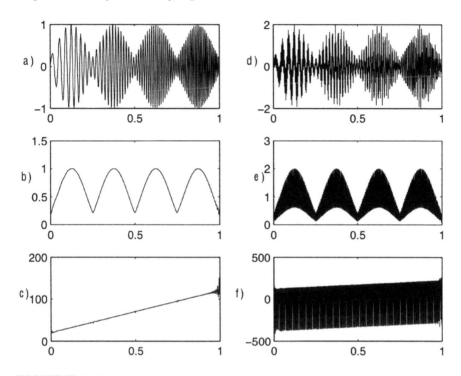

FIGURE 3.12
Instantaneous amplitude and frequency. a) Signal: modulated linear chirp; b)
instantaneous amplitude of signal in panel (a); c) instantaneous frequency of
signal in panel (a); d) signal composed of two modulated linear chirps with
the same slope as in (a) but different intercept frequency; e) instantaneous
amplitude of signal in panel (d); f) instantaneous frequency of signal (d).

The concepts of instantaneous amplitude and frequency are useful in case
of simple signals without overlapping frequency components, e.g., in Figure
3.12 a–c; the instantaneous amplitude recovers the modulation of the chirp,
and instantaneous frequency recovers its linearly changing frequency. However,
the instantaneous measures turn out to be meaningless, if the signal is more
complex, e.g., if there are two structures with different frequencies present at
the same moment. Figure 3.12 d–f illustrates such a case. The signal in panel
d) is composed of two signals analogous to that in panel a), but shifted in
frequency by a constant. Its instantaneous amplitude (panel e) and frequency
(panel f) display fast oscillatory behavior.

---

amplitude $A(t) = B$ and frequency $\phi(t) = f$ of the signal.

MATLAB demo: instantaneous amplitude and frequency

If you would like to get some intuition about the concepts of the instantaneous amplitude and frequency, try the Live Script: `matlab/c3/Ch3_14_Inst_Amp_Freq.mlx`

3.4.2 Analytic Tools in the Time-Frequency Domain

3.4.2.1 Time-Frequency Energy Distributions

Signal energy can be computed in time or in frequency domain as:

$$E_x = \int_{-\infty}^{\infty} |x(t)|^2 dt = \int_{-\infty}^{\infty} |X(f)|^2 df \tag{3.68}$$

and the quantities $|x(t)|^2$ or $|X(f)|^2$ are interpreted as energy density. The idea of signal energy density can be extended to the time-frequency space. The time-frequency energy density $\rho_x(t, f)$ should represent the amount of signal energy assigned to a given time-frequency point (t, f), and fulfill the conditions:

1. The total energy is conserved:

$$E_x = \int_{-\infty}^{\infty} \int_{-\infty}^{\infty} \rho_x(t, f) dt df \tag{3.69}$$

2. The marginal distributions are conserved:

$$\int_{-\infty}^{\infty} \rho_x(t, f) dt = |X(f)|^2 \tag{3.70}$$

$$\int_{-\infty}^{\infty} \rho_x(t, f) df = |x(t)|^2 \tag{3.71}$$

Convenient MATLAB toolbox for analysis of signals in time-frequency domain is the `tftb` available under the terms of the GNU Public License at `http://tftb.nongnu.org/`.

Wigner-Ville Distribution

The basic time-frequency energy density distribution is the Wigner-Ville distribution (WVD) defined as (for signals represented in time domain):

$$W_x(t, f) = \int_{-\infty}^{\infty} x(t + \tau/2) x^*(t - \tau/2) e^{-i2\pi f \tau} d\tau \tag{3.72}$$

or equvialently (for signals represented in the frequency domain):

$$W_x(t, f) = \int_{-\infty}^{\infty} X(f + \xi/2)X^*(f - \xi/2)e^{i2\pi\xi t}d\xi \qquad (3.73)$$

The $W_x(t, f)$ is implemented as `tfrwv` function in the MATLAB toolbox `tftb`.

WVD has the following properties:

- Energy conservation equation (3.69)

- Conservation of marginal distributions equations: (3.70) and (3.71)

- Conservation of time and frequency shifts:

$$y(t) = x(t - t_0) \Rightarrow W_y(t, f) = W_x(t - t_0, f) \qquad (3.74)$$

$$y(t) = x(t)e^{i2\pi f_0 t} \Rightarrow W_y(t, f) = W_x(t, f - f_0) \qquad (3.75)$$

- Conservation of scaling

$$y(t) = \sqrt{k}x(kt) \Rightarrow W_y(t, f) = W_x(kt, f/k) \qquad (3.76)$$

WVD is a quadratic representation, so an intrinsic problem occurs when the signal contains more than one time-frequency component. Let's assume that we analyze a signal y composed of two structures x_1 and x_2:

$$y(t) = x_1(t) + x_2(t) \qquad (3.77)$$

Then, the WVD can be expressed as:

$$W_y(t, f) = W_{x_1}(t, f) + W_{x_2}(t, f) + 2Re\{W_{x_1,x_2}(t, f)\} \qquad (3.78)$$

where $W_{x_1,x_2}(t, f) = \int_{-\infty}^{\infty} x_1(t + \tau/2)x_2^*(t - \tau/2)e^{-i2\pi f\tau}d\tau$ is called a cross-term.

WVD has many desired properties mentioned above and optimal time-frequency resolution, but the presence of cross-terms in real applications can make the interpretation of results difficult. The problem is illustrated in Figure 3.13.

MATLAB demo: properties of Wigner-Ville distribution

If you would like to get some intuition about the basic properties of WVD, try the Live Script: `matlab/c3/Ch3_15_WV.mlx`

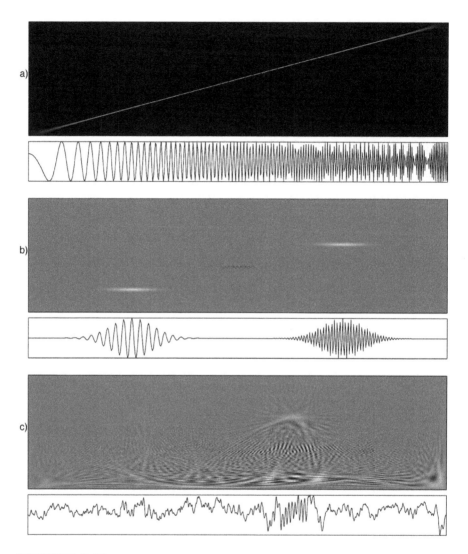

FIGURE 3.13

Illustration of cross-terms in WVD. a) Linear chirp—its time-frequency representation is a straight line, b) two Gabor functions—note the cross-term structure in the time-frequency representation just in the middle of the line connecting the representations of the individual components. c) A real world example: 10 sec. of EEG recorded during sleep stage 2, with a clear sleep spindle visible as a transient oscillatory structure.

Cohen class

The cross-terms oscillate in the time-frequency space with relatively high frequency, as can be observed in Figure 3.13. The property can be used to suppress the influence of the cross-terms simply by applying a low-pass spatial filter on the WVD. A family of distributions obtained by filtering WVD is called Cohen's class:

$$C_x(t, f, \Pi) = \int_{-\infty}^{\infty} \int_{-\infty}^{\infty} \Pi(s - t, \xi - f) W_x(s, \xi) ds d\xi \qquad (3.79)$$

where

$$\Pi(t, f) = \int_{-\infty}^{\infty} \int_{-\infty}^{\infty} f(\xi, \tau) e^{-i2\pi(f\tau + \xi t)} d\tau d\xi \qquad (3.80)$$

and $f(\xi, \tau)$ is the filter kernel. WVD is a member of Cohen's class with $f(\xi, \tau) = 1$. Another commonly used member of the class is the Choi-Williams distribution with the filtering kernel given by two-dimensional Gaussian function:

$$f(\xi, \tau) = \exp\left[-\alpha (\xi\tau)^2\right]. \qquad (3.81)$$

A proper selection of the kernel parameter can significantly reduce cross-terms, as illustrated in Figure 3.14.

MATLAB demo: smoothing the Wigner-Ville distribution

If you would like to get some intuition about effects of smoothing of WVD, try the Live Script: `matlab/c3/Ch3_16_smooth_WV.mlx`

3.4.2.2 Time-Frequency Signal Decompositions

Short Time Fourier Transform and Spectrogram

Time-frequency representation of the signal can also be obtained in a different manner. In this approach one extracts successive short pieces of signal with a window function and computes its frequency representation with a short time Fourier transform (STFT):

$$F_x(t, f; h) = \int_{-\infty}^{\infty} x(u) h^*(u - t) e^{-i2\pi uf} du \qquad (3.82)$$

where h is the window. If the window has a finite energy then the STFT can be reversed:

$$x(t) = \frac{1}{E_h} \int_{-\infty}^{\infty} \int_{-\infty}^{\infty} F_x(u, f; h) h(t - u) e^{i2\pi tf} du \, df \qquad (3.83)$$

where $E_h = \int_{-\infty}^{\infty} |h(t)|^2 dt$. Thus we can view the STFT as a tool that decomposes the signal into waveforms, *time-frequency atoms*, of the form:

$$h_{t,f}(u) = h(u - t) e^{i2\pi fu} \qquad (3.84)$$

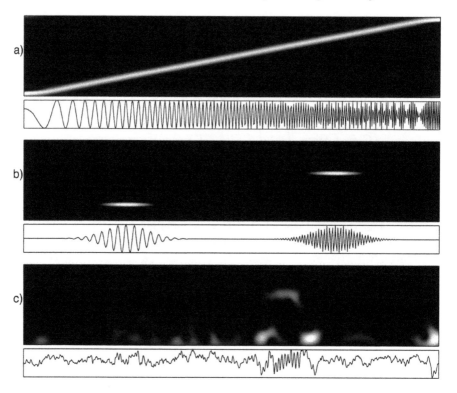

FIGURE 3.14
Ilustrative application of Choi-Williams distribution to the same set of signals as in Figure 3.13. a) Linear chirp—its time-frequency representation is a straight line, b) two Gabor functions—note the cross-term structure in the time-frequency representation is highly reduced. c) A real world example: 10 sec. of EEG recorded during sleep stage 2, with a clear sleep spindle visible as a transient oscillatory structure.

This concept is illustrated in Figure 3.15. Each atom is obtained by translation of a single window h and its modulation with frequency f.

STFT can be used to obtain the distribution of energy in the time-frequency space. If the window function has a unit energy (i.e., $E_h = 1$) then a squared modulus of STFT, a *spectrogram*, is an estimator of energy density (Figure 3.16):

$$S_x(t, f) = \left| \int_{-\infty}^{\infty} x(u)h^*(u - t)e^{-i2\pi fu} du \right|^2 \qquad (3.85)$$

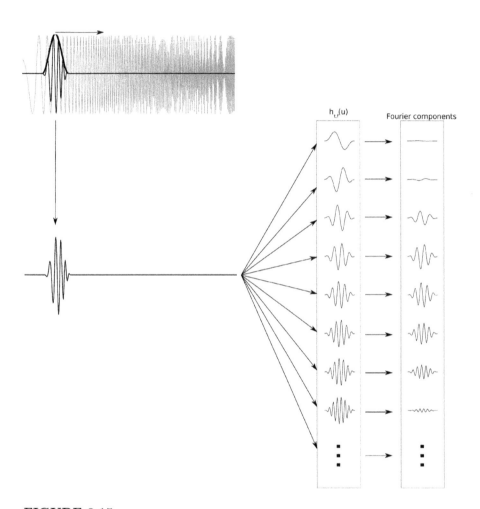

FIGURE 3.15
STFT as atomic decomposition. A sliding window (thick line) extracts a portion of signal roughly localized in time. This windowed signal is "compared" with each of the time frequency atoms of the form $h_{t,f}(u) = h(u - t)e^{i2\pi fu}$, giving the Fourier components ($h_{t,f}(u)$ weighted by intensities).

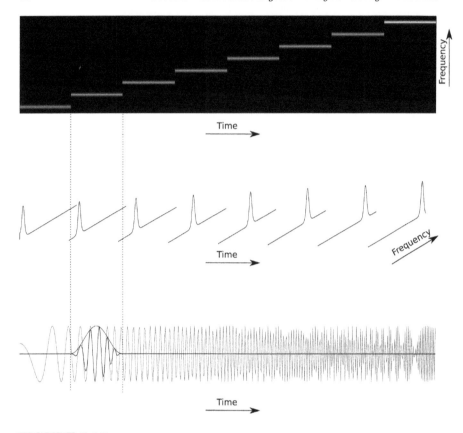

FIGURE 3.16

Illustration of a spectrogram construction. STFT is used to compute the spectrum of a short windowed fragment of the signal. The consecutive spectra are presented as a gray-scale time-frequency map. From the bottom: signal of increasing frequency, spectra in perspective, time-frequency representation.

MATLAB demo: construction of spectrogram

If you would like to see the construction of the spectrogram "from scratch", try the Live Script: `matlab/c3/Ch3_17_spectrogram.mlx`

Spectrogram conserves the translations in time:

$$y(t) = x(t - t_0) \Rightarrow S_y(t, f; h) = S_x(t - t_0, f; h) \qquad (3.86)$$

and in frequency:

$$y(t) = x(t)e^{i2\pi f_0 t} \Rightarrow S_y(t, f; h) = S_x(t, f - f_0; h) \qquad (3.87)$$

Spectrogram is a quadratic form; thus in the representation of multi-component signals the cross-terms are also present:

$$y(t) = x_1(t) + x_2(t) \Rightarrow S_y(t, f) = S_{x_1}(t, f) + S_{x_2}(t, f) + 2Re\{S_{x_1,x_2}(t, f)\} \tag{3.88}$$

where

$$S_{x_1,x_2}(t, f) = F_{x_1}(t, f)F^*_{x_2}(t, f) \tag{3.89}$$

The formula (3.89) shows that the cross-terms are present only if the signal components are close enough in the time-frequency space, i.e., when their individual STFTs overlap. The time and frequency resolution are determined by the properties of the window h. The time resolution can be observed as the width of the spectrogram of the Dirac's delta:

$$x(t) = \delta(t - t_0) \Rightarrow S_x(t, f; h) = e^{-i2\pi t_0 f}h(t - t_0). \tag{3.90}$$

The frequency resolution can be observed as the width of the spectrogram of the complex sinusoid.

$$x(t) = e^{i2\pi f_0 t} \Rightarrow S_x(t, f; h) = e^{-i2\pi t f_0}H(f - f_0) \tag{3.91}$$

It is clear from (3.90) that the shorter duration of time window h, the better is the time resolution. But due to the uncertainty principle (2.41) the shorter the window in time, the broader is its frequency band; thus from (3.91) the poorer the frequency resolution.

MATLAB demo: basic properties of spectrogram

If you would like to get some intuition about the time and frequency shifts, and the cross-terms structure, try the Live Script: `matlab/c3/Ch3_18_spectrogram_properties.mlx`

Continuous Wavelet Transform and Scalogram

The continuous wavelet transform (CWT) is defined as

$$T_x(t, a; \Psi) = \int_{-\infty}^{\infty} x(s)\Psi^*_{t,a}(s)ds \tag{3.92}$$

where $\Psi_{t,a}(s)$ is obtained from the mother wavelet Ψ by time translation t and dilation by scale a:

$$\Psi_{t,a}(s) = \frac{1}{\sqrt{|a|}}\Psi\left(\frac{s - t}{a}\right) \tag{3.93}$$

Please note that $\Psi_{t,a}$ conserves the overall shape of Ψ, in the sense of number of zero-crossings, however it is dilated or contracted. The wavelet Ψ should have the zero mean value. If for scale $a_0 = 1$ the wavelet is concentrated in frequency around f_0, we can bind the scale and frequency by:

$$f_a = \frac{f_0}{a} \tag{3.94}$$

The CWT can be interpreted as decomposition of the signal into the time-frequency atoms, obtained through a projection of the signal on the set of waveforms derived from a single wavelet Ψ by time translations and scaling. In this sense, the CWT is very similar to STFT (3.82), the main difference being that the window length changes with the scale (or due to (3.94) with the frequency). This property results in a different compromise between the time and frequency resolutions in different frequency bands. The frequency resolution is fine at low frequencies at the expense of reduced time resolution. The frequency resolution gradually deteriorates for high-frequency bands but the time resolution for these bands improves.

The CWT can be used to form a time-frequency representation of the energy distribution—scalogram, in a way analogous to the way in which a spectrogram is obtained from STFT. The time-scale representation is:

$$S_x(t, a; \Psi) = |T_x(t, a; \Psi)|^2 \tag{3.95}$$

Due to equation (3.94) it can be understood as time-frequency representation:

$$S_x(t, f; \Psi) = |T_x(t, f_0/f; \Psi)|^2 \tag{3.96}$$

Analogously to the spectrogram, the scalogram is disturbed by cross-terms of multicomponent signals in the regions of time-frequency space where the CWT of individual components overlap.

MATLAB demo: basic properties of scalogram

If you would like to get some intuition about the time and frequency shifts, and the cross-terms structure, try the Live Script: `matlab/c3/Ch3_19_scalogram_properties.mlx`

Discrete Wavelet Transform

The CWT in its theoretical form (3.92) operates on continuous time-scale (time-frequency) space. In any practical application the time and scale dimensions have to be sampled. The sampling, in general, can be performed in any way resulting in different approximations of CWT. The common way is to select $t_0 > 0$ and $a_0 > 0$ and then generate a grid of points $\{t = nt_0a_0{}^m, \quad a = a_0^m\}$ for $m, n \in \mathbb{Z}$

The discrete wavelet transform can be defined then as:

$$T_x(n, m; \Psi) = \int_{-\infty}^{\infty} x(s)\Psi_{n,m}^*(s)ds; \quad m, n \in \mathbb{Z} \tag{3.97}$$

where $\Psi_{n,m}^*$ is a scaled and translated version of Ψ:

$$\Psi_{n,m}(s) = \frac{1}{\sqrt{a_0^m}} \Psi\left(\frac{s - nt_0 a_0^m}{a_0^m}\right) \tag{3.98}$$

Dyadic Wavelet Transform—Multiresolution Signal Decomposition

However, there is one special manner of sampling, which, for some forms of wavelets, creates an orthonormal basis in the time-scale space. This is the *dyadic sampling* obtained for $t_0 = 1$ and $a_0 = 2$. The dyadic sampled wavelets are related to multiresolution analysis, and the transform can be efficiently computed using filter banks [375].

To better understand the multiresolution signal decomposition we should start with considering the so-called *scaling function*. The scaling function is a unique measurable function[4] $\phi(t) \in L^2(\mathbb{R})$ such that, if the dilation of $\phi(t)$ by 2^j is expressed by:

$$\phi_{2^j}(t) = 2^j \phi(2^j t) \tag{3.99}$$

then $\sqrt{2^{-j}}\phi_{2^j}(t - 2^{-j}n)$ for $n, j \in \mathbb{Z}$ is the orthonormal basis in vector space V_{2^j}. The vector space V_{2^j} is the set of all possible approximations of a signal at resolution 2^j. The signal $x(t)$ can be projected on the orthonormal basis:

$$A_{2^j}(t) = 2^{-j} \sum_{n=-\infty}^{\infty} \langle x(u), \phi_{2^j}\left(u - 2^{-j}n\right)\rangle \phi_{2^j}\left(t - 2^{-j}n\right) \tag{3.100}$$

The *discrete approximation of signal* $x(t)$ at resolution 2^j is defined by a set of inner products:

$$A_{2^j}^d x(n) = \langle x(u), \phi_{2^j}(u - 2^{-j}n)\rangle \tag{3.101}$$

The above formula is equivalent to the convolution of signal $x(t)$ with the dilated scaling function $\phi_{2^j}(t)$ which is a low-pass filter. The computation of $A_{2^j}^d x$ for consecutive resolutions 2^j can be interpreted as a low-pass filtering of $x(t)$ followed by uniform sampling at rate 2^j. In practice, the computations of $A_{2^j}^d x$ can be realized iteratively by convolving $A_{2^{j+1}}^d x$ with $\hat{h}(n)$, such that $\hat{h}(n) = h(-n)$ is a mirror filter of:

$$h(n) = \langle \phi_{2^{-1}}(u), \phi(u - n)\rangle \tag{3.102}$$

and downsampling the result by 2. This algorithm can be expressed by the

---

[4]i.e., the integral of squared function is finite.

formula:

$$\begin{cases} A^d_{2^0} = x(n) \\ A^d_{2^j} x(n) = \mathcal{D}_2 \left(\sum_{k=-\infty}^{\infty} \hat{h}(2n-k) A^d_{2^{j+1}} x(n) \right) \end{cases} \qquad (3.103)$$

where $\mathcal{D}_2$ is the operator of downsampling by factor 2.

The low-pass filtering used in the algorithm has the effect that after each iteration the consecutive approximations contain less information. To describe the residual information lost in one step and measure the irregularities of signal at resolution 2^j a function Ψ_{2^j} is constructed, such that:

$$\Psi_{2^j}(t) = 2^j \Psi \left(2^j t \right) \qquad (3.104)$$

and the functions $\Psi_{2^j} \left(t - 2^{-j} n \right)$ are a basis in a vector space O_{2^j}. The vector space O_{2^j} is orthonormal to V_{2^j} and together they span the vector space $V_{2^{j+1}}$:

$$O_{2^j} \perp V_{2^j} \text{ and } O_{2^j} \oplus V_{2^j} = V_{2^{j+1}} \qquad (3.105)$$

The function $\Psi(t)$ is called a wavelet and corresponds to the mother wavelet function in equation 3.93. The projection of a signal $x(t)$ on the space O_{2^j} is called the *discrete detail signal* and is given by the product:

$$D^d_{2^j} x(n) = \langle x(u), \Psi_{2^j} \left(u - 2^j n \right) \rangle \qquad (3.106)$$

These inner products can be computed conveniently from higher resolution discrete approximation by the convolution:

$$D^d_{2^j} x(n) = \sum_{k=-\infty}^{\infty} \hat{g} \left(2n - k \right) A^d_{2^{j+1}}(k) \qquad (3.107)$$

with the mirror filter $\hat{g}(n) = g(-n)$ of the filter $g(n)$ defined by:

$$g(n) = \langle \Psi_{2^{-1}}(u), \phi(u-n) \rangle \qquad (3.108)$$

The filter $g(n)$ can also be computed from filter $h(n)$ with the formula:

$$g(n) = (-1)^{1-n} h(1-n) \qquad (3.109)$$

Filter $g(n)$ is a high-pass filter. The $A^d_{2^j}$ and $D^d_{2^j}$ can be obtained from $A^d_{2^{j+1}}$ by convolving with $h(n)$ and $g(n)$, respectively, and downsampling by factor 2. Computation of discrete approximation by repeating the process for $j < 0$ is called the pyramidal transform. It is illustrated in Figure 3.17.

For a continuous signal, the cascade could be infinite. In practical cases of sampled signals, the process of consecutive downsamplings has a natural stop. The coefficients of the wavelet transform are the outputs of all the detail branches and the last approximation branch.

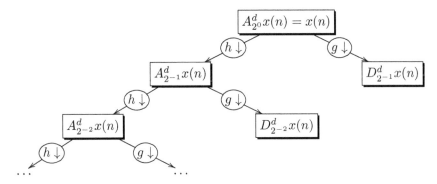

FIGURE 3.17
Cascade of filters producing the discrete wavelet transform coefficients. In this algorithm the signal is passed through the cascade, where at each step the approximation of signal from the previous step is low-pass and high-pass filtered, by convolving with h and g, respectively. Each of the filtered sequences is downsampled by factor 2 (on the scheme it is marked as $\downarrow$). Next the approximation is passed to the following step of the cascade.

For correctly designed filters (quadrature mirror filters), the process of decomposition can be reversed, yielding the inverse discrete wavelet transform. A technical discussion of how to design these filters is beyond the scope of this book and is available, e.g., in [571]. If we implement the wavelet transform as an iterated filter bank, we do not have to specify the wavelet explicitly.

There is one important consequence of discretization of the wavelet transform. The transform is not shift invariant, which means that the wavelet transform of a signal and of the wavelet transform of the same time-shifted signal are not simply shifted versions of each other. An illustration of the structure of the time-frequency space related to the discrete wavelet representation is shown in Figure 3.18.

Wavelet Packets

Wavelet packets (WP) are a modification of the discrete wavelet decomposition. The decomposition into WP is achieved in two steps. First is the modification of the filter/downsampling cascade. In the WP scheme at each level of the cascade, both branches—the approximation and the detail coefficients—are further filtered and downsampled. As a result, a complete tree is obtained, as illustrated in Figure 3.19. The second modification is that from that tree we can select the most suitable decomposition of a given signal, e.g., with respect to an entropy-based criterion [90]. This procedure is called pruning of a decomposition tree. More detailed information on different aspects of wavelet analysis may be found in [376].

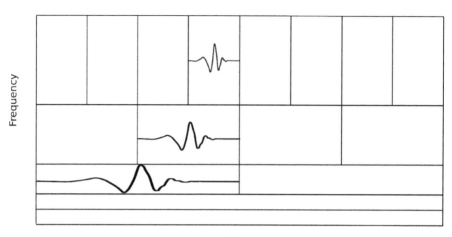

FIGURE 3.18
Structure of the time-frequency plane for discrete wavelets. The wavelet function is dilated and translated.

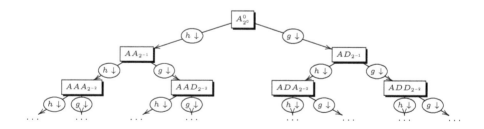

FIGURE 3.19
Three levels of WP decomposition tree.

Wavelets in MATLAB

In MATLAB wavelet analysis can be conveniently performed with the Wavelet Toolbox distributed by MathWorks. The collection of functions delivered by this toolbox can be obtained by command `help wavelet`.

Another MATLAB toolbox for wavelet analysis is WaveLab `http://www-stat.stanford.edu/~wavelab/Wavelab_850/index_wavelab850.html`. WaveLab is a collection of MATLAB functions that implement a variety of algorithms related to wavelet analysis. The techniques made available are: orthogonal and biorthogonal wavelet transforms, translation-invariant wavelets, interpolating wavelet transforms, cosine packets, wavelet packets.

Moreover, this toolbox has good introductory tutorials described here: `http://statweb.stanford.edu/~wavelab/Wavelab_850/AboutWaveLab.pdf`. They cover decomposition, de-noising, shrinkage, and the cartoon guide to wavelets.

Matching Pursuit—MP

The atomic decompositions of multicomponent signals have the desired property of explaining the signal in terms of time-frequency localized structures. The two methods presented above: spectrogram and scalogram are working well, but they are restricted by the a priori set trade-off between the time and frequency resolution in different regions of the time-frequency space. This trade-off does not follow the structures of the signal. In fact, interpretation of the spectrogram or scalogram requires understanding which aspects of the representation are due to the signal and which are due to the properties of the methods.

A time-frequency signal representation that adjusts to the local signal properties is possible in the framework of matching pursuit (MP) [377]. In its basic form MP is an iterative algorithm that in each step finds an element g_{γ_n} (atom) from a set of functions D (dictionary) that best matches the current residue of the decomposition $R^n x$ of signal x; the null residue being the signal:

$$\begin{cases} R^0 x = x \\ R^n x = \langle R^n x, g_{\gamma_n} \rangle g_{\gamma_n} + R^{n+1} x \\ g_{\gamma_n} = \arg\max_{g_{\gamma_i} \in D} |\langle R^n x, g_{\gamma_i} \rangle| \end{cases} \qquad (3.110)$$

where: $\arg\max_{g_{\gamma_i} \in D}$ means the atom g_{γ_i} which gives the highest inner product with the current residue: $R^n x$. Note, that the second equation in (3.110) leads to orthogonality of g_{γ_n} and $R^{n+1} x$, so :

$$\|R^n x\|^2 = |\langle R^n x, g_{\gamma_n} \rangle|^2 + \|R^{n+1} x\|^2 \qquad (3.111)$$

The signal can be expressed as:

$$x = \sum_{n=0}^{k} \langle R^n x, g_{\gamma_n} \rangle g_{\gamma_n} + R^{k+1} x \qquad (3.112)$$

It was proved [105] that the algorithm is convergent, i.e., $\lim_{k \to \infty} \left\| R^k s \right\|^2 = 0$. Thus in this limit we have:

$$x = \sum_{n=0}^{\infty} \langle R^n x, g_{\gamma_n} \rangle g_{\gamma_n} \qquad (3.113)$$

In this way signal x is represented as a weighted sum of atoms (waveforms) from the dictionary D. The iterative procedure is illustrated in Figure 3.20. Taking into account (3.111) we can see that the representation conserves the

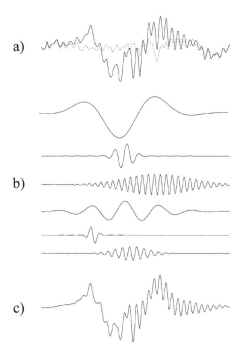

a)

b)

c)

FIGURE 3.20
Illustration of the iterative MP decomposition. a) Signal to be decomposed (black) and the residue after six iterations (dotted line), b) from top to bottom the six consecutive selected atoms, c) the sum of the six atoms.

energy of the signal:

$$\|x\|^2 = \sum_{n=0}^{\infty} |\langle R^n x, g_{\gamma_n} \rangle|^2 \qquad (3.114)$$

The MP decomposition can be considered as an extension of the atomic decompositions offered by STFT or CWT. The main advantage of the MP paradigm is the relaxation of constraint between the frequency band and frequency resolution. The MP algorithm performs decomposition in an extremely redundant set of functions, which results in a very flexible parametrization of the signal structures.

In principal the dictionary D can be any set of functions. In practical implementations (e.g., https://github.com/develancer/empi) the dictionary contains a base of Dirac deltas, a base of sinusoids and a set of Gabor functions:

$$g_\gamma(t) = K(\gamma)e^{-\pi\left(\frac{t-u}{\sigma}\right)^2} \sin\left(2\pi f(t-u) + \phi\right) \tag{3.115}$$

with $K(\gamma)$ normalization factor such that $\|g_\gamma\| = 1$, and $\gamma = \{u, f, \sigma, \phi\}$ are the parameters of functions in the dictionary (u –time translation, f – frequency, σ – time width, ϕ – phase).

The parameters γ can be sampled in various ways. The original idea of Mallat [377] was to follow a dyadic scheme that mimics the oversampled discrete wavelets. For applications where the parameters are used to form statistics of the atoms the dyadic sampling produced estimators that were biased by the structure of the dictionary. The introduction of stochastic dictionaries relied on randomization of the time-frequency coordinates and time width of atoms [128]. This allowed obtaining a bias-free implementation. Further extensions of the MP algorithm allow for analysis of multivariate signals i.e. multichannel and multi-trial decompositions [131, 548] (Sect. 4.6.4).

The atomic decomposition of the signal can be used to produce the time-frequency energy distribution. In this approach, the best properties of energy distributions (WVD) and atomic decompositions can be joined. The WVD of the whole decomposition is:

$$W_x(t,f) = \sum_{n=0}^{\infty} |\langle R^n x, g_{\gamma_n}\rangle|^2 \, W_{g_{\gamma_n}}(t,f)$$

$$+ \sum_{n=0}^{\infty} \sum_{m=0, m\neq n}^{\infty} \langle R^n x, g_{\gamma_n}\rangle \langle R^m x, g_{\gamma_m}\rangle^* \, W_{g_{\gamma_n} g_{\gamma_m}}(t,f) \tag{3.116}$$

where

$$W_{g_{\gamma_n}}(t,f) = \int g_{\gamma_n}\left(t + \frac{\tau}{2}\right) g^*_{\gamma_n}\left(t - \frac{\tau}{2}\right) e^{-i2\pi f \tau} d\tau \tag{3.117}$$

is WVD of individual atoms. The double sum in equation (3.116) corresponds to the crossterms, but since it is given explicitly, it can be omitted yielding the estimator of the energy density distribution in the form:

$$E_x^{MP}(t,f) = \sum_{n=0}^{M} |\langle R^n x, g_{\gamma_n}\rangle|^2 \, W_{g_{\gamma_n}}(t,f) \tag{3.118}$$

This interpretation is valid since normalization of atoms (3.115):

$$\int_{-\infty}^{+\infty} \int_{-\infty}^{+\infty} W_g(t,f) \, dt \, df = \|g\|^2 = 1 \tag{3.119}$$

leads to:

$$\int_{-\infty}^{+\infty} \int_{-\infty}^{+\infty} E_x^{MP}(t, f) \, dt \, df = \|x\|^2 \qquad (3.120)$$

This representation has implicitly no cross-terms and for Gabor atoms offers the highest time-frequency resolution (Sect. 2.4.7).

MP dictionary can be adjusted to be coherent with particular structures in the signal. The dictionary containing asymmetric functions was designed and proved to be useful in the description of components with different time courses of the rising and decaying parts [253] (Sect. 5.4.2).

MATLAB demo: basic properties of matching pursuit

If you would like to get some intuition about the time and frequency properties of MP, try the Live Script: `matlab/c3/Ch3_20_MP.mlx`

MATLAB demo: MP is greedy. It's not a bug, it's a feature.

The Live Script: `matlab/c3/Ch3_21_MP_greedy.mlx` demonstrates the fact that MP is a greedy procedure. It has some specific consequences, which should be taken into account while interpreting the results.

Comparison of Time-frequency Methods

The complex character of biomedical signals and their importance in health research and clinical practice brought a wide variety of signal analysis methods into applications in biomedical research. The most widespread are the spectral methods that make possible the identification of the basic rhythms present in the signal. Conventional methods of the analysis assumed stationarity of the signal, even though interesting processes are often reflected in fast dynamic changes of a signal. This implied the application to the analysis of the signals methods operating in time-frequency space.

The available time-frequency methods can be roughly divided into two categories:

- Those that give directly continuous estimators of energy density in the time-frequency space

- Those that decompose the signal into components localized in the time-frequency space, which can be described by sets of parameters, and at the second step the components can be used to create the estimators of time-frequency energy density distribution

An example of time-frequency energy distribution obtained by means of different methods is shown in Figure 3.21.

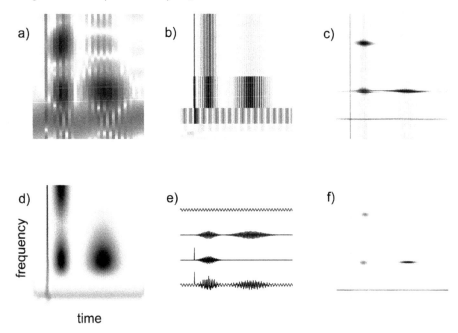

FIGURE 3.21
Comparison of energy density in the time-frequency plane obtained by
different estimators for a signal e): a) spectrogram, b) discrete wavelet
transform, c) Choi-Williams transform, d) continuous wavelets, f) matching
pursuit. Construction of the simulated signal shown in (e), the signal consisting of: a sinusoid, two Gabor functions with the same frequency but different
time positions, a Gabor function with frequency higher than the previous pair,
an impulse. From [54].

In the first category—the Cohen's class of time-frequency distributions—
one obtains the time-frequency estimators of energy density directly without
decomposing the signal into some predefined set of simple elements. This
allows for maximal flexibility in expressing the time-frequency content of the
signal. However, there are two consequences:

- The first consequence is the lack of a parametric description of the signals
  structures.

- The second consequence is that, no matter how much the signal structures
  are separated in time or frequency, they interfere and produce cross-terms.

The problem of compromise between time and frequency resolution manifests
when one selects the proper filter kernel to suppress the cross-terms.

In the second category, the most natural transition from spectral analysis to the analysis in time-frequency space is the use of short-time Fourier

transform (STFT), and a representation of energy density derived from it—the spectrogram. The positive properties of this approach are the speed of computations and the time and frequency shift-invariance, which makes the interpretation of the resulting time-frequency energy density maps easy to interpret. The main drawbacks are: (1) the a priori fixed compromise between time and frequency resolution in the whole time-frequency space, which results in smearing the time-frequency representation, (2) the presence of cross-terms between the neighboring time-frequency structures.

Another common choice in the second category is CWT. From the practical point of view, the main difference from the STFT relies on another compromise between the time and frequency resolution. In the case of CWT, one sacrifices the time resolution for the better frequency resolution of low-frequency components and vice versa for higher-frequency components; also, the change of the frequency of a structure leads to the change of the frequency resolution.

STFT and CWT can be considered as atomic representations of the signal, and as such, give a particular parametric description of the signals. However, the representation in not sparse; in other words, there are too many parameters; hence, they are not very informative.

The sparse representation of the signal is provided by DWT and MP, which leads to the efficient parameterization of the time series. The DWT can decompose the signal into a base of functions, which is a set of waveforms that has no redundancy. There are fast algorithms to compute the DWT. Similar to CWT, the DWT has poor time resolution for low frequencies and poor frequency resolution for high frequencies. The DWT is handy in signal denoising or signals compression applications. The lack of redundancy has a consequence in the loss of time and frequency shift-invariance. DWT may be appropriate for time-locked phenomena, but much less for transients appearing in time at random, since parameters describing a given structure depend on its location inside the considered window.

The decomposition based on the matching pursuit algorithm offers the step-wise adaptive compromise between the time and frequency resolution. The resulting decomposition is time and frequency invariant. The time-frequency energy density estimator derived from the MP decomposition has explicitly no cross-term, which leads to clean and easy-to-interpret time-frequency maps of energy density. The price for the excellent properties of the MP decomposition is the higher computational complexity.

The sparsity of the DWT and MP decompositions has a different character, which affects their applicability. DWT is especially well suited to describing time-locked phenomena since it provides a common base. MP is especially useful for structures appearing in the time series at random. The sparsity of MP stems from the very redundant set of functions, which allows representing the signal structures in a limited number of atoms. The MP decomposition gives a parameterization of the signal structures in terms of the amplitude, frequency, time of occurrence, span in time and frequency, which are close to the intuition of practitioners.

Empirical Mode Decomposition and Hilbert-Huang Transform

The Hilbert-Huang transform (HHT) was proposed by Huang et al. [231]. It consists of two general steps:

- The empirical mode decomposition (EMD) method to decompose a signal into the so-called intrinsic mode function (IMF)

- The Hilbert spectral analysis (HSA) method to obtain instantaneous frequency

The HHT is a non-parametric method and may be applied for analyzing non-stationary and non-linear time-series data.

Empirical mode decomposition (EMD) is a procedure for the decomposition of a signal into so-called intrinsic mode functions (IMF). An IMF is any function with the same number of extrema and zero crossings, with its envelopes being symmetric with respect to zero. The definition of an IMF guarantees a well-behaved Hilbert transform of the IMF. The procedure of extracting an IMF is called sifting. The sifting process is as follows:

1. Between each successive pair of zero crossings, identify a local extremum in the signal.

2. Connect all the local maxima by a cubic spline line as the upper envelope $E_u(t)$.

3. Repeat the procedure for the local minima to produce the lower envelope $E_l(t)$.

4. Compute the mean of the upper and lower envelope: $m_{11}(t) = \frac{1}{2}(E_u(t) + E_l(t))$.

5. A candidate h_{11} for the first IMF component is obtained as the difference between the signal $x(t)$ and $m_{11}(t)$: $h_{11}(t) = x(t) - m_{11}(t)$.

In a general case, the first candidate h_{11}, doesn't satisfy the IMF conditions. In such a case, the sifting is repeated, taking h_{11} as the signal. The sifting is repeated iteratively:

$$h_{1k}(t) = h_{1(k-1)}(t) - m_{1k}(t) \tag{3.121}$$

until the assumed threshold for standard deviation SD computed for the two consecutive siftings is achieved. The SD is defined as:

$$SD = \sum_{t=0}^{T} \frac{|h_{1(k-1)}(t) - h_{1k}(t)|^2}{h_{1(k-1)}^2(t)} \tag{3.122}$$

Authors of the method suggest the SD of 0.2–0.3 [231]. At the end of the sifting process after k iterations the first IMF is obtained:

$$c_1 = h_{1k} \tag{3.123}$$

The c_1 mode should contain the shortest period component of the signal. Subtracting it from the signal gives the first residue:

$$r_1 = x(t) - c_1 \tag{3.124}$$

The procedure of finding consecutive IMFs can be iteratively continued until the variance of the residue is below a predefined threshold, or the residue becomes a monotonic function—the trend (the next IMF cannot be obtained). The signal can be expressed as a sum of the n-empirical modes and a residue:

$$x(t) = \sum_{i=1}^{n} c_i + r_n \tag{3.125}$$

Each of the components can be expressed by means of a Hilbert transform as a product of instantaneous amplitude $a_j(t)$ and an oscillation with instantaneous frequency $\omega_j(t)$ (Sect. 3.4.1): $c_j = a_j(t)e^{i \int \omega_j(t)dt}$. Substituting this to (3.125) gives representation of the signal in the form:

$$x(t) = \sum_{i=1}^{n} a_j(t)e^{i \int \omega_j(t)dt} \tag{3.126}$$

Equation (3.126) makes possible construction of time-frequency representation—the so-called Hilbert spectrum. The weight assigned to each time-frequency coordinate is the local amplitude.

MATLAB demo: basic properties of EMD

If you would like to get some intuition about empirical mode decomposition, try the Live Script: `matlab/c3/Ch3_22_emd.mlx`.

Compare the results obtained by Hilbert spectrum with results obtained by smoothing of WVD in the Live Script `matlab/c3/Ch3_16_smooth_WV.mlx`

3.4.3   Cross-Frequency Coupling

The time-frequency methods, discussed so far, considered the spectral evolution in time. However, qualitatively new information can be gained from the analysis of relations between the different frequencies—this type of analysis is called cross-frequency coupling (CFC). Of special interest is the analysis of relations between the phase of a lower-frequency and the amplitude of the higher-frequency oscillations. This is known as phase-amplitude coupling (PAC). The effects that can be investigated in this way concern modulation, driving, controlling, time-ordering, coordination of activity represented by the higher-frequency oscillation by the slower rhythmic processes. Recently, PAC has been assigned important functional roles in cognition and neural information processing, specifically, in learning and memory [344, 77, 606, 20], spatial navigation [604], sensory signal detection [211], and attentional selection [539]. There is a growing interest in understanding patterns of CFC since they may be relevant for diagnosing and eventually treating various disorders or in designing preventive strategies [665, 46, 109, 302]. Selected applications of PAC analysis are reported in Sect. 5.1.7.3. In the following subsections, we will focus on the methodological issues.

3.4.3.1 Models of Phase-Amplitude Coupling

The phenomenon of PAC can be best understood by considering simple models.

Amplitude Modulation Model

The model used by Tort et al. [605] consists of a low-frequency sine, a high-frequency sine, and white noise. Importantly, the amplitude of the high-frequency sine is modulated accordingly to the phase of the low-frequency oscillation. The modeled signal can be expressed as:

$$s(t) = \overline{A_{f_P}} \sin(2\pi f_P t) + A_{f_A}(t) \sin(2\pi f_A t) + \overline{A_N} W(t) \qquad (3.127)$$

where f_P and f_A are respectively the frequencies of low- and high-frequency oscillation; $\overline{A_{f_P}}$ is the amplitude of low-frequency oscillation, in this model it is a constant; $A_{f_A}(t)$ is the amplitude of the high-frequency oscillation, it depends on the low-frequency oscillation phase; $W(t)$ is Gaussian white noise derived from the standard normal distribution, and $\overline{A_N}$ is a scaling factor for white noise. The high-frequency oscillation amplitude A_{f_A} is modulated by the low-frequency as follows:

$$A_{f_A}(t) = \overline{A_{f_A}} \frac{(1-\chi)\sin(2\pi f_P t) + 1 + \chi}{2} \qquad (3.128)$$

where $\overline{A_{f_A}}$ is a constant determining the maximal amplitude of the high-frequency oscillation and $\chi \in [0, 1]$ controls the strength of the modulation. Low values of χ correspond to stronger coupling, i.e. deeper modulation. An example of signal with this type of PAC is depicted in Figure 3.22.

MATLAB demo: models of phase-amplitude coupling

If you want to get the intuition about the model parameters, consider the LiveScript `matlab/c3/Ch3_23_PAC_model.mlx`

High-frequency Transients

Another physiologically plausible model of PAC relies on the assumption that the low-frequency modulation, in some range of phases, promotes a

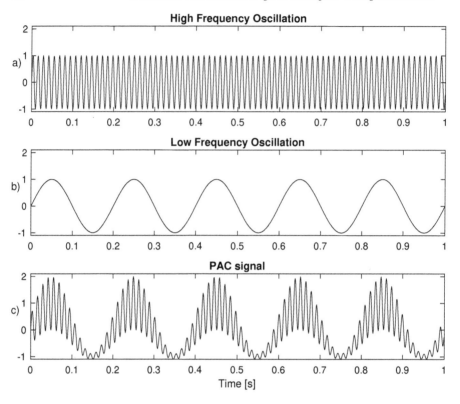

FIGURE 3.22
Example of phase-amplitude coupling according to the model used in [605].
a) The high-freqency oscillation at 77 Hz, b) the low frequency oscillation at
5 Hz, c) PAC signal obtained using formula (3.127 and 3.128) for $\chi = 0.1$.

high-frequency burst. This model can be implemented as a series of gaussian-
shaped high-frequency transients centered at a certain phase of the low-
frequency oscillation.

3.4.3.2   Evaluation of Phase-Amplitude Coupling

Commonly used methods for estimation of PAC rely on selecting two frequency
bands, e.g., by filtering [77, 604], or by wavelet-based approaches, e.g., [421],
followed by evaluation of correlation or dependency between the phase of
the lower-frequency and the amplitude of the higher-frequency oscillation.
The phase and amplitude are usually derived as the instantaneous phase and
amplitude of the filtered components (c.f. Sect.3.4.1). While filtering the high-
frequency component, it is essential to adjust the bandwidth of the filter such
that it encompasses the sidebands related to the modulation of the frequency

f_A. It means that if we consider the PAC between f_P and f_A the bandwidth of the high-frequency filter has to be at least $[f_A - f_P, f_A + f_P]$.

PAC is usually presented in the form of a comodulorgam, i.e., a color-coded map where the color corresponds to the magnitude of coupling, the horizontal axis is the phase-determining frequency, and the vertical axis is the frequency of the component of which amplitude is modulated.

Direct PAC Estimator

Mean Vector Length (MVL) proposed by Canolty [77] relies on computation of a composite signal $s(t)$:

$$s(t) = A_{f_A}(t)e^{i\Phi_{f_P}(t)} \tag{3.129}$$

Its amplitude, $A_{f_A}(t)$, is obtained as the instantaneous amplitude of the high-frequency component. The phase, Φ_{f_P}, is the instantaneous phase of the low-frequency oscillation. These components are derived by band-pass filtering of the investigated signal. Consider samples of this signal as vectors on the complex plane. Summing such vectors is constructive if all of them point in roughly the same direction, or destructive if they point into random directions. In this sense, the mean length of such vectors can measure how much they are clustered around a specific phase of low-frequency oscillations. This idea is outlined in Figure 3.23 d). A drawback of using this simple concept of PAC measure is that it dependents on the absolute amplitude of the high-frequency oscillation [605]. Appropriate normalization can correct this issue. Özkurt et al. [452] proposed the *Direct PAC Estimator* dPAC:

$$\text{dPAC}(f_P, f_A) = \frac{1}{\sqrt{T}} \frac{\left| \sum_{t=0}^{T} A_{f_A}(t) \cdot e^{i\Phi_{f_P}(t)} \right|}{\sqrt{\sum_{t=0}^{T} A_{f_A}(t)^2}} \tag{3.130}$$

To obtain the low and high-frequency components the Authors used a two-way least-squares FIR filter. The order of the filter was equal to the number of samples in three cycles of the corresponding frequency band. To avoid edge effects of filtration, the first and last second of data are excluded from further analysis. The comodulogram is obtained by applying equation (3.130) for each pair of frequencies for phase f_P and amplitude f_A.

Modulation index

Modulation index (MI)[604] measures how much the distribution of the mean high-frequency amplitude across the bins of low-frequency phase deviates from the uniform distribution. The version of MI as used in [605] can be found in the Authors github repository https://github.com/tortlab/phase-amplitude-coupling. The low- and high-frequency oscillations are obtained by filtering the investigated signal $s(t)$ around, respectively, low-frequency f_P and high-frequency f_A using FIR filter.The resulting

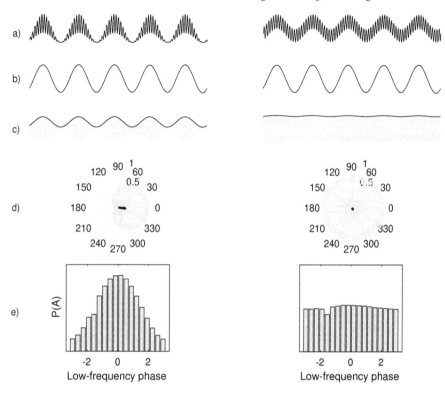

FIGURE 3.23
Outline of the dPAC and MI evaluation. a) signal constructed according
to (eq. 3.128), left for for $\chi = 0.01$, right for $\chi = 0.9$; b) light-gray low-
frequency bandpass filtered component; c) high-frequency bandpass filtered
component, black—instantaneous amplitude of the component; d) representation of: gray—the samples of signal constructed according to (eq. 3.129)
in the complex plain, black arrow is the mean vector; e) distribution of
average amplitude of high-frequency component across the phase bins of the
low-frequency computed according to (eq. 3.131).

narrow-band signals are used to obtain the corresponding analytic signals.
The instantaneous amplitude of the high-frequency analytic signal is denoted
as the $A_{f_A}(t)$, and the instantaneous phase of the low-frequency analytic signal is denoted as $\Phi_{f_P}(t)$. The range of phases $[-\pi, \pi]$ is divided into J equal
bins. Each sample t_i of the amplitude $A_{f_A}(t_i)$ is assigned to the corresponding bin accordingly to the value of $\Phi_{f_P}(t_i)$. The amplitude within each phase
bin is averaged, giving $\langle A_{f_A} \rangle_{\Phi_{f_P}}(j)$, after normalization yielding the empirical

probability distribution:

$$P(j) = \frac{\langle A_{f_A} \rangle_{\Phi_{f_P}}(j)}{\sum_{k=1}^{J} \langle A_{f_A} \rangle_{\Phi_{f_P}}(k)} \qquad (3.131)$$

An illustration of this approach is presented in Figure 3.23 e). The modulation index is evaluated as the Kullback-Leibler distance between the uniform and the empirical distribution of the mean amplitude:

$$\mathrm{MI}(f_P, f_A) = \frac{\log(J) + \sum_{k=1}^{J} P(k) \log[P(k)]}{\log(J)} \qquad (3.132)$$

The comodulogram is obtained by applying equation (3.132) to each pair of frequencies for phase f_P and amplitude f_A.

Extended Modulation Index Analysis

A convenient EEGLAB plugin for performing Modulation index analysis can be obtained from `https://github.com/GabrielaJurkiewicz/ePAC`. It offers the above-described methods and additionally provides an interesting approach based on the analysis of fragments of time-frequency maps aligned to the subsequent maxima of the low-frequency component obtained by band-pass filtering [267]. The maps are computed by scalogram (cf. Sect. 3.4.2.2). The coupling is evaluated according to equations (3.131) and (3.132), except that the power A_{f_A} is obtained from the phase-locked average scalogram. This approach solves the problem of adjusting the high-frequency filter bandwidth. Moreover, the toolbox offers statistical methods appropriate to assess the significance of the effects shown in the comodulograms. The maximal statistic solves the multiple comparisons problem that arises when making the tests for many pairs of frequencies for phase and for amplitude. Besides the standard histogram, the toolbox also provides a phase histogram, which allows identifying the phase of low frequency related to the augmentations of the high-frequency power. To support the interpretation of the coupling in terms of possible epiphenomenal origin, the toolbox computes for each low-frequency component auxiliary plots of the average scalogram phase-locked to the maximum of the component, and the average component itself with the average phase-locked full-band signal overlaid. If the detected coupling extends over a wide frequency range, and the average signal in the instants corresponding to the power augmentation undergoes sharp transitions, it suggests the epiphenomenal origin of the coupling. Exemplary analyses performed with the toolbox are presented in Figure 3.24.

Epiphenomenal Coupling

In physiological settings, we would like to think about PAC as an interplay of two processes, each in itself expressed in the form of oscillation of a given

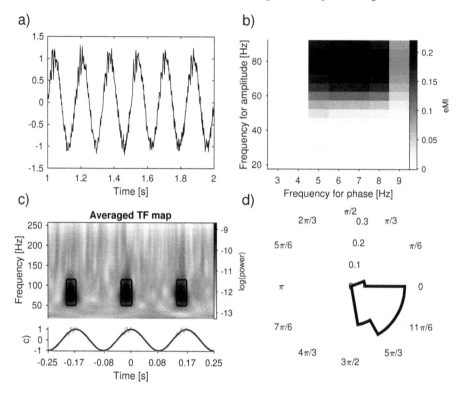

FIGURE 3.24
Example of extended phase-amplitude analysis. a) Fragment of simulated signal with coupling between 6 and 77 Hz; b) comodulogram; c) auxiliary plot showing the average time-frequency map with regions contributing to the coupling marked with black outline, and below the average signal for phase (black) overlaid with average non-filtered signal (gray); d) phase histogram indicates the range of phases at which the coupling occurs.

frequency. However, a similar effect, i.e., a specific relation of different frequencies can also be observed as an epiphenomenon. This potential confound was already pointed out by [304, 615] and [15]. Epiphenomenal PAC can originate from a common source drive, where both low- and high-frequency components are coupled with a certain stimulus, either external or internal. Such coupling may arise, especially in experiments involving event-related effects [622].

Epiphenomenal coupling in comodulograms can also arise due to a cyclic occurrence of broadband transient structures at a rate corresponding to the low-frequency [615]. These structures do not even have to be oscillatory. They cause an inhomogeneous distribution of high-frequency amplitude across the phases of the low-frequency cycle detected by most of the currently used methods for the construction of comodulograms. Examples of this effect were given in [178]. It can be observed in some cases of signals recorded during epileptic seizure—broadband spikes occur at specific phases of the low-frequency waves. The spike-and-wave discharges in epileptic signals should be classified as epiphenomenal coupling. This coupling could be better analyzed and understood in the language of dynamical models. Examples of such effects and heuristic methodology for their automatic classification are discussed in [267].

As reported by Aru et al. [15], and developed in their supplementary literature review, several conditions should be met to indicate meaningful CFC. These conditions are not always met in the literature, resulting in a strong over-interpretation of the effects. Some of the problems of detection of meaningful coupling were recently addressed by the method of time-resolved phase-amplitude coupling (tPAC) [523]. These authors included the condition of coexistence of both low-frequency oscillation for phase and high-frequency amplitude in the signal, and appropriate setting of the bandwidth of the high-frequency bandpass filter. A more elegant mathematical approach for evaluation of the high-frequency power modulations utilizing the generalized Morse wavelets was proposed in [421]. A solution to the analysis of PAC in event-related settings was presented in [622].

MATLAB demo: analysis of phase-amplitude coupling with ePAC

If you want to try the analysis of PAC for simulated genuine and epiphenomenal coupling run the LiveScript `matlab/c3/Ch3_ _24_PAC_demo.mlx`

3.5 Non-Linear Methods of Signal Analysis

Non-linear methods of signal analysis were inspired by the theory of non-linear dynamics—indeed the biomedical signal may be generated by a non-linear process. Dynamical systems are usually defined by a set of first-order ordinary differential equations in a phase space. The phase space is a finite-dimensional vector space $\mathbb{R}^m$, in which a state $x \in \mathbb{R}^m$ is defined. For the deterministic system we can describe the dynamics by an explicit system of m first-order ordinary differential equations :

$$\frac{dx(t)}{dt} = f\left(t, x(t)\right), \qquad x \in \mathbb{R}^m \tag{3.133}$$

If the time is treated as a discrete variable, the representation takes a form of an m-dimensional map:

$$x_{n+1} = F(x_n), \qquad n \in \mathbb{Z} \tag{3.134}$$

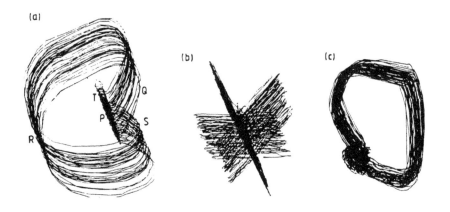

FIGURE 3.25
Phase portraits of human ECG in three-dimensional space. A two-dimensional projection is displayed for two values of the delay τ: (a) 12 ms and (b) 1200 ms. (c) represents the phase portrait constructed from ECG of simultaneously recorded signals from three ECG leads. From [25].

A sequence of points x_n or $x(t)$ solving the above equations is called a trajectory of the dynamical system. Typical trajectories can run away to infinity or can be confined to a particular space, depending on F (or f). An attractor is a geometrical object to which a dynamical system evolves after a long enough time. An attractor can be a point, a curve, a manifold, or a complicated object called a *strange attractor*. An attractor is considered strange if it has a non-integer dimension. Non-linear chaotic systems are described by strange attractors.

Now we have to face the problem that what we observe is not a phase space object but a time series, and we do not know the equations describing the process that generates them. The delay embedding theorem of Takens [584] provides the conditions under which a smooth attractor can be reconstructed from a sequence of observations of the state of a dynamical system. The reconstruction preserves the properties of the dynamical system that do not change under smooth coordinate changes. A reconstruction in d dimensions can be obtained by means of the retarded vectors:

$$\xi(t) = x(t), x(t+\tau), \ldots, x(t+(m-1)\tau) \qquad (3.135)$$

Number m is called the embedding dimension, τ is called delay time or lag. According to the above formula, almost any scalar observation, e.g., time series, is sufficient to learn about the evolution of a complex high-dimensional deterministic evolving system. However, we do not know in advance how long the retarded vector must be, and we do not know the delay τ. Choosing a too small value of τ would give a trivial result, and too large τ would hamper the information about the original system. Usually, the time coordinate of the first minimum of the autocorrelation function is taken as τ. The phase portrait of an ECG obtained by embedding in three-dimensional space is shown in Figure 3.25. Please note (picture b) the distortion of the phase portrait for too large τ. Finding m is even more complex. The method of false neighbors [278] is difficult to apply and doesn't give unequivocal results. Usually, the embedding dimension is found by increasing the dimension step by step.

MATLAB tutorial: embeding ECG signal

If you want to gain some simple intuition about embedding a signal and reconstructing its phase portrait run the LiveScript `matlab/c3/Ch3_25_ECG_phase_portrait.mlx`

3.5.1 Lyapunov Exponent

Lyapunov exponents describe the rates at which nearby trajectories in phase space converge or diverge; they provide estimates of how long the behavior of a mechanical system is predictable before chaotic behavior sets in. The *Lyapunov exponent* or *Lyapunov characteristic exponent* of a dynamical system is a quantity that characterizes the rate of separation of infinitesimally close trajectories. Quantitatively, the separation of two trajectories in phase space with an initial distance ΔZ_0 can be characterized by the formula:

$$|\Delta Z(t)| \approx e^{\lambda t}|Z_0| \tag{3.136}$$

where λ is the Lyapunov exponent. Positive Lyapunov exponent means that the trajectories are diverging, which is usually taken as an indication that the system is chaotic. The number of Lyapunov exponents is equal to the number of dimensions of the phase space.

3.5.2 Correlation Dimension

The concept of generalized dimension (special cases of which are: correlation dimension and Hausdorff dimension) was derived from the notion that geometrical objects have certain dimensions, e.g., a point has a dimension 0, a line—1, a surface—2; in case of chaotic trajectories dimension is not an integer.

The measure called *correlation dimension* was introduced in [196]. It involves definition of the correlation sum $C(r)$ for a collection of points x_n in some vector space to be the fraction of all possible pairs of points which are closer than a given distance r in a particular norm. The basic formula for $C(r)$ is:

$$C(r) = \frac{2}{N(N-1)} \sum_{i=1}^{N-1} \sum_{j=i+1}^{N} \Theta(r - ||x_i - x_j||) \tag{3.137}$$

where Θ is the Heaviside step function,

$$\Theta(x) = \begin{cases} 0 \text{ for } x \leq 0 \\ 1 \text{ for } x > 0 \end{cases} \tag{3.138}$$

The sum counts the pairs (x_i, x_j) whose distance is smaller than r. In the limit $N \to \infty$ and for small r, we expect C to scale like a power law $C(r) \propto r^{D_2}$, so the correlation dimension D_2 is defined by:

$$D_2 = \lim_{r \to 0} \lim_{N \to \infty} \frac{\partial \log C(r, N)}{\partial \log r} \tag{3.139}$$

In practice, from a signal $x(n)$, the embedding vectors are constructed using the Takens theorem for a range of m values. Then one determines the correlation sum $C(r)$ for the range of r and for several embedding dimensions. Then $C(m, r)$ is inspected for the signatures of self-similarity, which is performed by the construction of a double logarithmic plot of $C(r)$ versus r. If the curve does not change its character for successive m we conjecture that the given m is a sufficient embedding dimension and D_2 is found as a slope of a plot of $\log(C(r, N))$ versus $\log(r)$.

The Haussdorf dimension D_H may be defined in the following way: If for the set of points in M dimensions the minimal number of N spheres of diameter l needed to cover the set increases like:

$$N(l) \propto l^{-D_H} \text{ for } l \to 0, \tag{3.140}$$

D_H is a Hausdorff dimension. $D_H \geq D_2$, in most cases $D_H \approx D_2$. We have to bear in mind that the definition (3.139) holds in the limit $N \to \infty$, so in practice, the number of data points of the signal should be large. It has been pointed out by [278] that $C(r)$ can be calculated automatically, whereas a dimension may be assigned only as the result of a careful interpretation of these curves. The correlation dimension is a tool to quantify self-similarity (fractal—non-linear behavior) when it is known to be present. The correlation dimension can be calculated for any signal, also for a purely stochastic time series or colored noise, which does not mean that these series have a non-linear character. The approach which helps in distinguishing nonlinear time series from the stochastic or linear ones is the method of surrogate data (Sect. 2.6).

MATLAB demo: computation of correlation dimension

If you want to see how to estimate the correlation dimension, run the LiveScript `matlab/c3/Ch3_26_corr_dim.mlx`

3.5.3 Detrended Fluctuation Analysis

Detrended fluctuation analysis (DFA) quantifies intrinsic fractal-like correlation properties of dynamic systems. A fundamental feature of a fractal system is scale-invariance or self similarity in different scales. In DFA the variability of the signal is analyzed in respect to local trends in data windows. The method allows to detect long-range correlations embedded in a non-stationary time series. The procedure relies on the conversion of a bounded time series $x_t(t \in \mathbb{N})$ into an unbound process: X_t:

$$X_t = \sum_{i=1}^{t}(x_i - \langle x_i \rangle) \tag{3.141}$$

where X_t is called a cumulative sum and $\langle x_i \rangle$ is the average in the window t. Then the integrated time series is divided into boxes of equal length L, and a local straight line (with slope and intercept parameters a and b) is fitted to the data by the least squares method. Next, the fluctuation—the root-mean-square deviation from the trend is calculated over every window at every time scale:

$$F(L) = \sqrt{\frac{1}{L}\sum_{i=1}^{L}(X_i - a \cdot i - b)^2} \tag{3.142}$$

This detrending procedure followed by the fluctuation measurement process is repeated over the whole signal over all time scales (different box sizes L). Next, a log−log graph of L against $F(L)$ is constructed. A straight line on this graph indicates statistical self-affinity, expressed as $F(L) \propto L^\alpha$. The scaling exponent α is calculated as the slope of a straight line fit to the log−log graph of L against $F(L)$.

The fluctuation exponent α has different values depending on the character of the data. For uncorrelated white noise it has a value close to $\frac{1}{2}$, for correlated process $\alpha > \frac{1}{2}$, for a pink noise $\alpha = 1$ (power decaying as $1/f$), $\alpha = \frac{3}{2}$ corresponds to Brownian noise.

In the case of power-law decaying autocorrelations, the correlation function decays with an exponent γ: $R(L) \sim L^{-\gamma}$ and the power spectrum decays as $S(f) \sim f^{-\beta}$. The three exponents are related by relations: $\gamma = 2 - 2\alpha$, $\beta = 2\alpha - 1$, and $\gamma = 1 - \beta$.

As with most methods that depend upon line fitting, it is always possible to find a number α by the DFA method, but this does not necessarily mean that the time series is self-similar. Self-similarity requires that the points on the $\log - \log$ graph are sufficiently collinear across an extensive range of window sizes. Detrended fluctuation is used in the HRV time series analysis, and its properties will be further discussed in Sect. 5.2.2.3.

MATLAB demo: detrended fluctuation analysis

If you want to get an insight how the DFA is computed, run the LiveScript `matlab/c3/Ch3_27_DFA.mlx`

3.5.4   Recurrence Plots

Usually, the dimension of phase space is higher than two or three, which makes its visualization difficult. Recurrence plot [137] enables us to investigate the m-dimensional phase space trajectory through a two-dimensional representation of its recurrences, namely a recurrence plot (RP) reveals all the times when the phase space trajectory visits roughly the same area in the phase space. In this way, distinct recurrent behavior, e.g., periodicities and also irregular cyclicities, can be detected. The recurrence of states, in the meaning that states become arbitrarily close after some time is a fundamental property of deterministic dynamical systems and is typical for non-linear or chaotic systems.

Recurrence of a state at time i against a different state at time j is marked within a two-dimensional squared matrix with ones and zeros (black and white dots in the plot), where both axes are time axes. More formally RP can be expressed as:

$$R_{i,j} = \Theta(r_i - ||x_i - x_j||), \quad x_i \in R^m, \quad i,j = 1,\ldots,N \qquad (3.143)$$

where N is the number of considered states x_i, r_i is a threshold distance, $||\cdot||$ a norm, and $\Theta(\cdot)$ the Heaviside function. The visual appearance of an RP (Figure 3.26) gives hints about the dynamics of a system. Uncorrelated white noise results in uniformly distributed dots, periodic patterns are connected with cyclicities in the process—time distance between patterns (e.g., lines) corresponds to period, diagonal lines mean that the evolution of states is similar at different times—the process could be deterministic; if these diagonal lines occur beside single isolated points, the process could be chaotic.

MATLAB demo: recurrence plots

If you want to play with the recurrence plots of the logistic map, run the LiveScript `matlab/c3/Ch3_28_recurence_plots.mlx`

The visual interpretation of RPs requires some experience. The RPs are uesd for analysis of specific biomedical processes, e.g., cardiac activity. The main advantage of recurrence plots is that they provide information even for short and non-stationary data, where other non-linear methods fail.

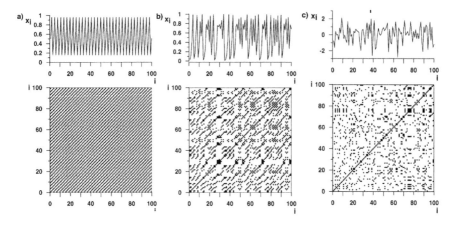

FIGURE 3.26
Examples of the recurrence plots (lower panels) for different signals shown
above: a) a periodic signal obtained from the logistic equation for the control
parameter r = 3.829, b) a chaotic signal obtained from the logistic equation for
r = 3.9999, c) white noise. Note that in the RP for the deterministic chaotic
signal short lines parallel to the diagonal are present. Adapted from [293].

3.5.5   Poincaré Map

Another way to visualize the dynamical system evolution in a phase space
is a Poincaré map. A Poincaré section is the intersection of a trajectory of a
dynamical system in the state space with a certain lower dimensional subspace,
transversal to the flow of the system. The Poincaré map relates two consecutive
intersection points x_{n+1} and x_n, which come from the same side of the plane
(Figure 3.27). A Poincaré map differs from a recurrence plot because it is
defined in a phase space, while a recurrence plot is defined in a time space
(points on this plot depict pairs of time moments when the system visits
roughly the same region of phase space). By means of the Poincaré map it is
possible to reduce the phase space dimensionality, at the same time turning
the continuous time flow into a discrete time map [278].

3.5.6   Approximate, Sample, and Multiscale Entropy

Even for low-dimensional chaotic systems, a huge number of points is required
to achieve convergence of the algorithms estimating dimension or entropy
of the process. To overcome this difficulty and provide a measure capable
of quantifying the changes in process complexity, the modifications of the
entropy measure were proposed. Approximate entropy (ApEn) was introduced
by [481]. ApEn measures the (logarithmic) likelihood that trajectories that
are close to each other remain close upon the next incremental comparison.

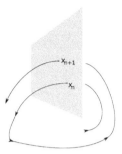

FIGURE 3.27
Construction of the Poincaré map. The Poincaré map relates two consecutive intersection points x_{n+1} and x_n, which come from the same side of the plane.

However, methodological pitfalls in ApEn were pointed out in [503] and [96] and another measure called sample entropy (SaEn) was introduced in [503].

In the SaEn method the vectors (blocks) $\xi(i)$ of differing dimensions k are created by embedding a signal in a way similar to that given in formula (3.135). The reconstructed vectors are the state space (k-dimensional) representation of the dynamics of the system. SaEn is defined as the logarithmic difference between the probability of occurrence of the vector $\xi(i)$ within the chosen distance r in k dimensions and the probability of occurrence of the vector $\xi(i)$ within the same chosen distance in $k + 1$ dimension:

$$\text{SaEn} = \log \frac{\rho^k(r)}{\rho^{k+1}(r)}, \tag{3.144}$$

where $\rho^k(r)$, $\rho^{k+1}(r)$—densities of occurrence in k, $k + 1$ dimensions, respectively. MATLAB code for calculating Sample Entropy is available at: `https://archive.physionet.org/physiotools/sampen/matlab/`. SaEn met some criticism, e.g., that it does not characterize the complexity of the signal completely [192] and soon other modifications appeared.

In the approach called multiscale entropy (MSE) [96], the procedure involves partitioning the signal into disjoined windows, and the data are averaged inside each time window. It is equivalent to downsampling, or low-pass filtering. In the case of a continuous system, the same signal sampled at two different sampling rates would show different behavior in MSE analysis, so the method will quantify the system behavior in different frequency ranges. Building on this idea, a refined composite multivariate multiscale fuzzy entropy was proposed [22]. The authors demonstrated that their extension led to more stable results and less sensitivity to the signals' length in comparison with the other existing multivariate multiscale entropy-based methods.

However, in another approach, SaEn was modified by introducing the time delay δ between the successive components of blocks (or subseries) [192]; δ was chosen as the time point at which autocorrelation function falls below $1/e$. In the density estimation in order not to account for the temporally correlated points, for each center i, the $i + \delta$ surrounding points were discarded.

The above-described entropy measures and detrended fluctuation analysis found applications mainly in heart rate variability analysis, and their utility will be discussed in the section concerning this signal.

MATLAB demo: multiscale entropy

The LiveScript `matlab/c3/Ch3_29_MSE.mlx` gives a short introduction to multiscale entropy.

3.5.7   Limitations of Non-Linear Methods

There are several pitfalls in the application of non-linear methods. One of them is that even for infinite-dimensional, stochastic signals, low-dimensional estimates may be obtained, which was pointed out by [594]. The temporal coherence of the data may be mistaken for the trace of non-linearity. Therefore before applying non-linear methods, one should first check if the signals have traces of non-linearity since the non-linear process needs not produce non-linear time series. The null hypothesis about the stochastic character of the signal may be checked employing surrogate data technique (Sect. 2.6).

The non-linear methods rely to a large extent on the reconstruction of phase space. In order to construct the phase space by embedding, preferably long stationary data epochs are needed. In the case of biomedical signals, long, stationary data segments are often hard to find. Another problem is connected with the fact that the non-linear methods are prone to systematic errors related to the arbitrary choices, e.g., of the bin length in estimating probabilities or lag of embedding procedure.

However, the biggest problem in applying non-linear methods is that they are susceptible to noise. As was pointed out by [278], even for a low-dimensional deterministic signal with a noise component of the order 2-3%, it would be difficult to reasonably estimate correlation dimension, since, in this case, a significant scaling region of the correlation sum could not be found. The reason being that the artifacts of the noise meet the artifacts of the overall shape of the attractor. Since biomedical signals contain a significant noise component, one should approach non-linear methods with caution and apply them only when the non-linearity of a signal is well established and linear methods seem not to work correctly.

4

Multiple Channels (Multivariate) Signals

The technological progress in the biomedical field has led to the construction of recording equipment, allowing registration of activity from multiple sites. Today a typical dataset contains not only two or four but dozens or hundreds of channels. This is especially the case for EEG, MEG, fMRI, and sometimes also for ECG and EMG signals. Analysis of multichannel data can give a better insight into the relations between the investigated signals, but it is a challenging task. Besides many experimental and computational difficulties, the problem quite often lies in the proper application of existing mathematical tools. The techniques capitalizing on the covariance structure of the multichannel (multivariate) data are especially useful in this respect. In this chapter, an introduction to the fundamental aspects of multichannel data processing will be presented.

4.1 Cross-Estimators: Cross-Correlation, Cross-Spectra, Coherence

Joint moments of the order two: cross-correlation R_{xy} and cross-covariance C_{xy}, were defined in Sect. 2.1 (equations 2.11 and 2.12) by means of ensemble averaging formalism. Under the assumption of ergodicity they can be expressed by the formulas:

$$R_{xy}(\tau) = \int_{-\infty}^{\infty} x(t)y(t+\tau)dt \tag{4.1}$$

$$C_{xy}(\tau) = \int_{-\infty}^{\infty} (x(t) - \mu_x)(y(t+\tau) - \mu_y)\, dt \tag{4.2}$$

Cross-correlation and cross-spectrum are bound by means of Fourier transform and inverse Fourier transform (cf. Sect. 3.3.2.2):

$$S_{xy}(f) = \int_{-\infty}^{\infty} R_{xy}(\tau)e^{-i2\pi f\tau}d\tau \tag{4.3}$$

$$R_{xy}(\tau) = \int_{-\infty}^{\infty} S_{xy}(f)e^{i2\pi f\tau}df \tag{4.4}$$

DOI: 10.1201/9780429431357-4

Similarly to the power spectrum, cross-spectrum is usually computed by means of the Fourier transforms X and Y of the signals x and y:

$$S_{xy}(f) = \lim_{T \to \infty} \frac{1}{T} X(f, T) Y^*(f, T) \tag{4.5}$$

The fact that we have only a limited data window of length T has the same consequences as in cases of power spectra estimation (see Sect. 3.3.2.1). Usually, non-rectangular window functions are used, and smoothing is applied. Computation of cross-spectrum is implemented in MATLAB Signal Processing Toolbox as `cpsd` function. It estimates the cross power spectral density of the discrete-time signals using Welch's averaged, modified periodogram method.

MATLAB demo: cross spectra.

The LiveScript `matlab/c4/Ch4_01_cross_spectra.mlx` gives a short introduction to cross spectra.

$S_{xy}(f)$ is a complex value consisting of real and imaginary parts:

$$S_{xy}(f) = Re(S_{xy})(f) + iIm(S_{xy})(f) \tag{4.6}$$

In polar coordinates it can be expressed by the formula:

$$S_{xy}(f) = |S_{xy}(f)| e^{i\Phi_{xy}(f)}, \tag{4.7}$$

where $|S_{xy}(f)| = \sqrt{Re(S_{xy})^2(f) + Im(S_{xy})^2(f)}$ is the modulus and $\Phi_{xy}(f) = \tan^{-1} \frac{Im(S_{xy})}{Re(S_{xy})}$ is the phase.

Coherence is a measure which is often used in biomedical application. It is expressed by the formula:

$$\gamma_{xy}(f) = \frac{S_{xy}(f)}{\sqrt{S_x(f)S_y(f)}} \tag{4.8}$$

where S_x and S_y are spectra of signals x and y. Since $S_{xy}(f) \leq \sqrt{S_x(f)S_y(f)}$, function $|\gamma_{xy}(f)| \leq 1$. Coherence shows the relation between two signals in frequency and phase. The square of the coherence measures the spectral power in a given frequency common to both signals.

MATLAB demo: coherence.

The LiveScript `matlab/c4/Ch4_02_coherence.mlx` gives a short introduction to coherence between a pair of channels.

The above formula defines ordinary (bivariate) coherence. If a data set contains more than two channels, the signals can be related to each other in different ways. Namely, two (or more) signals may simultaneously have a common driving input from the third channel. Depending on the character of relations between channels, some of them may be connected directly with each other, and some connections can be indirect (through other channels). To distinguish between these situations, partial and multiple coherences were introduced.

The construction of partial coherence relies on subtracting influences from all other processes under consideration. For three channels partial coherence is defined as a normalized partial cross spectrum:

$$\kappa_{xy|z}(f) = \frac{|S_{xy|z}(f)|}{\sqrt{S_{xx|z}(f)S_{yy|z}(f)}} \tag{4.9}$$

where partial cross-spectrum is defined as:

$$S_{xy|z}(f) = S_{xy}(f) - S_{xz}(f)S_{zz}^{-1}(f)S_{zy}(f) \tag{4.10}$$

And the partial phase coherence (PPC) is defined as:

$$\phi_{x,y|z}(f) = \arg S_{xy|z}(f) \tag{4.11}$$

For an arbitrary number of channels partial coherence may be defined in terms of the minors of spectral matrix $\mathbf{S}(f)$, which on the diagonal contains spectra and off-diagonal cross-spectra:

$$\kappa_{ij}(f) = \frac{\mathbf{M}_{ij}(f)}{\sqrt{\mathbf{M}_{ii}(f)\mathbf{M}_{jj}(f)}} \tag{4.12}$$

where $\mathbf{M}_{ij}$ is a minor of $\mathbf{S}$ with the ith row and j^{th} column removed. Its properties are similar to ordinary coherence, but it is non-zero only for direct relations between channels. If a signal in a given channel can be explained by a linear combination of some other signals in the set, the partial coherence between them will be low.

Multiple coherence is defined by:

$$G_i(f) = \sqrt{1 - \frac{\det \mathbf{S}(f)}{S_{ii}(f)\mathbf{M}_{ii}(f)}} \tag{4.13}$$

Its value describes the number of common components in the given channel and the other channels in the set. If the value of multiple coherence is close to zero, then the channel has no common components with any other channel of the set. The high value of multiple coherence for a given channel means that a large part of that channel's variance is common to all other signals; it points to the strong relationship between the signals. Partial and multiple coherences can be conveniently found using the autoregressive parametric model MVAR.

4.2 Multivariate Autoregressive Model (MVAR)

4.2.1 Formulation of MVAR Model

MVAR model is an extension of the one channel AR model for an arbitrary number of channels. In the MVAR formalism, sample $x_{i,t}$ in channel i at time t is expressed not only as a linear combination of p previous values of this signal but also through p previous samples of all the other channels of the process. For k channels model of order p may be expressed as:

$$\vec{x}_t = A_1 \vec{x}_{t-1} + A_2 \vec{x}_{t-2} + \cdots A_p \vec{x}_{t-p} + \vec{E}_t \tag{4.14}$$

where : $\vec{x}_t = [x_{1,t}, x_{2,t}, \ldots x_{k,t}]$ is a vector of signal samples at times t in channels $\{1, 2, \ldots k\}$. $\vec{E}_t = [E_{1,t}, E_{2,t}, \ldots E_{k,t}]$ is a vector of noise process samples at time t. The covariance matrix $\mathbf{V}$ of a noise process is expressed as:

$$\vec{E}_t \vec{E}_t^{\mathsf{T}} = \mathbf{V} = \begin{pmatrix} \sigma_1^2 & 0 & \cdots & 0 \\ 0 & \sigma_2^2 & \cdots & 0 \\ \vdots & & & \vdots \\ 0 & \cdots & 0 & \sigma_k^2 \end{pmatrix} \tag{4.15}$$

where T denotes transposition.

$\mathbf{V}$ is a matrix representing residual variances of noises assumed in the model. In the process of model estimation, we obtain an estimate of the variance matrix as $\hat{\mathbf{V}}$, which accounts for the variance not explained by the model coefficients. The matrix $\hat{\mathbf{V}}$ usually contains small non-diagonal elements; their value informs us how well the model fits the data. The MVAR model coefficients for each time lag l are $k \times k$-sized matrices:

$$\mathbf{A}_t = \begin{pmatrix} A_{11}(l) & A_{12}(l) & \cdots & A_{1k}(l) \\ A_{21}(l) & A_{22}(l) & \cdots & A_{2k}(l) \\ \vdots & & & \vdots \\ A_{k1}(l) & \cdots & A_{kk-1}(l) & A_{kk}(l) \end{pmatrix} \tag{4.16}$$

Before starting a fitting procedure, specific preprocessing steps are needed. First, the temporal mean should be subtracted for every channel. Additionally, in most cases, normalization of the data by dividing each channel by its temporal variance is recommended. It is especially useful when data channels have different amplification ratios.

The estimation of model parameters in the case of a multivariate model is similar to the one channel model. The classical technique of AR model parameters estimation is the Yule-Walker algorithm. It requires calculating the correlation matrix $\mathbf{R}$ of the system up to lag p. The model equation (4.14) is multiplied by $\vec{x}_{t+s}^{\mathsf{T}}$, for $s = 0, \ldots, p$ and expectations of both sides of each equation are taken:

$$R_{ij}(s) = \frac{1}{N_s} \sum_{t=1}^{N_s} x_{i,t} x_{j,t+s} \tag{4.17}$$

Assuming that the noise component is not correlated with the signals, we get a set of linear equations to solve (the Yule-Walker equations):

$$\begin{pmatrix} \mathbf{R}(0) & \mathbf{R}(-1) & \cdots & \mathbf{R}(p-1) \\ \mathbf{R}(1) & \mathbf{R}(0) & \cdots & \mathbf{R}(p-2) \\ \vdots & \vdots & & \vdots \\ \mathbf{R}(1-p) & \mathbf{R}(2-p) & \cdots & \mathbf{R}(0) \end{pmatrix} \begin{pmatrix} \mathbf{A}(1) \\ \mathbf{A}(2) \\ \vdots \\ \mathbf{A}(p) \end{pmatrix} = \begin{pmatrix} \mathbf{R}(-1) \\ \mathbf{R}(-2) \\ \vdots \\ \mathbf{R}(-p) \end{pmatrix} \tag{4.18}$$

and

$$\hat{\mathbf{V}} = \sum_{j=0}^{p} \mathbf{A}(j) \mathbf{R}(j) \tag{4.19}$$

This set of equations is similar to the formula (3.51) for one channel model, however the elements $\mathbf{R}(i)$ and $\mathbf{A}(i)$ are $k \times k$ matrices. Other methods of finding MVAR coefficients are the Burg (LWR) recursive algorithm and the covariance algorithm. The Burg algorithm produces high-resolution spectra and is preferred when closely spaced spectral components are to be distinguished. The covariance algorithm or its modification better describes sinusoidal components in spectra. Recently, a Bayesian approach has been proposed for estimating the optimal model order and model parameters. In most cases, however, the spectra produced by different algorithms are very similar to each other. MVAR model and methods of determination of its coefficients are described in [491, 282, 380, 358].

Several criteria of the determination of the MVAR model order were proposed in [358]. Similarly to the one channel case we seek the minimum of the function consisting of two terms: the first one is a reward for minimizing the residual variance, the second one is a punishment for a too high model order. The first term depends on the estimated residual variance $\hat{\mathbf{V}}(p)$ for a given p, the second one is a function of model order, number of channels k, and number of data points N. The criteria presented below differ in respect to the second term:

AIC criterion:

$$AIC(p) = \log[\det(\hat{\mathbf{V}})] + 2\frac{pk^2}{N} \tag{4.20}$$

Hannan-Quin criterion:

$$HQ(p) = \log[\det(\hat{\mathbf{V}})] + 2\log(\log(N))\frac{pk^2}{N} \tag{4.21}$$

Schwartz criterion:

$$SC(p) = \log[\det(\hat{\mathbf{V}})] + \log(N)\frac{pk^2}{N} \tag{4.22}$$

AIC criterion is the one which is mostly used, but some authors [282, 380] claim that it sometimes gives a too high model order.

4.2.2 MVAR in the Frequency Domain

In analogy to the procedure described in Sect. 3.3.2.2 equation (4.14) can be easily transformed to describe relations in the frequency domain:

$$\mathbf{E}(f) = \mathbf{A}(f)\mathbf{X}(f) \tag{4.23}$$

$$\mathbf{X}(f) = \mathbf{A}^{-1}(f)\mathbf{E}(f) = \mathbf{H}(f)\mathbf{E}(f) \tag{4.24}$$

where

$$\mathbf{H}(f) = \left(\sum_{m=0}^{p} \mathbf{A}(m)e^{-2\pi imf\Delta t} \right)^{-1} \tag{4.25}$$

and $\Delta t = \frac{1}{F_s}$; F_s is the sampling frequency.

From the form of that equation we see that the model can be considered as a linear filter with white noises $\mathbf{E}(f)$ on its input (flat dependence on frequency) and the signals $\mathbf{X}(f)$ on its output. The transfer matrix $\mathbf{H}(f)$ contains information about all relations between channels of a process. From the transfer matrix, spectra and cross-spectra may be calculated:

$$\mathbf{S}(f) = \mathbf{X}(f)\mathbf{X}^*(f) = \mathbf{H}(f)\mathbf{E}(f)\mathbf{E}^*(f)\mathbf{H}^*(f) = \mathbf{H}(f)\mathbf{V}\mathbf{H}^*(f) \tag{4.26}$$

where $\mathbf{V}$ is the covariance matrix of the noise process (4.15). The matrix $\mathbf{S}(f)$ contains auto-spectra of each channel on the diagonal and cross -spectra off the diagonal. The simple rule connecting the model order with the number of spectral peaks does not hold in the case of MVAR. Usually, the model order is lower since more coefficients are used to describe spectral properties; their number is equal to k^2p.

4.3 Measures of Directedness

4.3.1 Estimators Based on the Phase Difference

In the study of biomedical signals, a particularly interesting problem is finding the direction of the influence exerted by one channel on the others, i.e., finding the directedness. The most straightforward way to determine the directedness seems the application of cross measures: cross-correlation or coherence. Cross-correlation and coherence are statistically equivalent since they are bound by the Fourier transform, but cross-correlation operates in the time domain and coherence in the frequency domain. Hence these measures emphasize different features of signals.

The amplitude of the cross-correlation describes the similarity of the signals and, if for a given delay τ, the function $R_{xy}(\tau)$ has maximum, we can assume that signal y is delayed by τ in respect to signal x. Thus the direction may be inferred from the delay between signals found for the maximum of the function. It gives information concerning the most pronounced frequency component. When there are different frequency components, the signal has to be filtered first (using procedure not disturbing phases—in MATLAB `filtfilt`).

For signals containing different rhythmical components, coherence is usually a method of choice. Coherence contains phase information (Sect. 4.1), which can be translated into a delay between signals. The delay in samples Δx can be computed from the formula: $\Delta x = \frac{\phi}{\pi} \frac{F_s}{f}$, where f is the frequency for which the phase ϕ is calculated. Since the phase is dependent on frequency, the delays may be found for particular frequencies. However, we have to bear in mind that phase is determined modulo 2π; therefore, the result may be ambiguous.

In the case of directedness measures, the problems connected with driving two channels by a third one are critical and can lead to false propagation patterns. Therefore bivariate correlation and bivariate coherence are hardly applicable for systems with the number of channels higher than two. The application of partial coherences may alleviate this problem; however, the estimation of directedness using phases of partial coherences is prone to high statistical errors. The asymptotic variance for the phase of partial coherence in the case of three channels is:

$$\text{var}\left(\phi_{xy|z}(f)\right) = \frac{1}{n}\left(\frac{1}{\kappa^2_{xy|z}(f)} - 1\right) \tag{4.27}$$

where n is the number of degrees of freedom, and it depends on the method of estimation of cross-spectra.

Therefore, for low values of coherence, the errors will be substantial. Also, in the case of estimation of coherence using Fourier transform, even with smoothing, the number of degrees of freedom is low. Partial coherences found utilizing MVAR are computed under the minimum phase's assumption, and they are not suitable for the determination of directedness.

4.3.2 Causality Measures

4.3.2.1 Granger Causality

In biomedical studies finding causality relations between channels of multivariate processes is of particular interest. The testable definition of causality was introduced by Granger[1] (1969) in the field of economics [194]. Granger defined causality in terms of predictability of time series; namely if some series $y(t)$ contains information in past terms that helps in the prediction of series

---

[1] Nobel Laureate in economics.

$x(t)$, then $y(t)$ is said to cause $x(t)$. More specifically, if we try to predict a value of $x(t)$ using p previous values of the series x only, we get a prediction error ϵ:

$$x(t) = \sum_{j=1}^{p} A'_{11}(j)x(t-j) + \epsilon(t) \tag{4.28}$$

If we try to predict a value of $x(t)$ using p previous values of the series x and p previous values of y we obtain another prediction error ϵ_1:

$$x(t) = \sum_{j=1}^{p} A_{11}(j)x(t-j) + \sum_{j=1}^{p} A_{21}(j)y(t-j) + \epsilon_1(t) \tag{4.29}$$

If the variance ϵ_1 (after including series y to the prediction) is lower than the variance ϵ we say that y causes x in the sense of Granger causality. Similarly we can say that x causes y in the sense of Granger causality when the variance ϵ_2 is reduced after including series x in the prediction of series y:

$$y(t) = \sum_{j=1}^{p} A_{22}(j)y(t-j) + \sum_{j=1}^{p} A_{21}(j)x(t-j) + \epsilon_2 \tag{4.30}$$

Different measures of causality or directionality based on the Granger causality concept were introduced: Granger causality index (GCI) [179], Granger-Geweke causality (GGC), directed transfer function (DTF) [272] and partial directed coherence (PDC) [29]. GCI operates in the time domain, DTF, PDC, GGC in the frequency domain. Initially, Granger causality was defined for a two-channel system. The estimators mentioned above were extended to multichannel systems.

4.3.2.2 Granger Causality Index and Granger-Geweke Causality

For a two channels system, GCI is based directly on principle formulated by Granger. Namely, we check if the second signal's information improves the prediction of the first signal. In order to find out, we have to compare the variance of univariate AR model (equation 4.28) with a variance of the model accounting for the second variable (equation 4.29).

Granger causality index showing the driving of channel x by channel y is defined as the logarithm of the ratio of residual variance for a two channel model to the residual variance of the one channel model:

$$\text{GCI}_{y \to x} = \log \frac{\epsilon_1}{\epsilon} \tag{4.31}$$

This definition can be extended to the multichannel system by considering how the inclusion of the given channel changes the residual variance ratios. To quantify directed influence from a channel x_j to x_i for n-channel autoregressive process in time domain we consider n and $n-1$ dimensional MVAR models. First, the model is fitted to a whole n-channel system, leading to the residual variance $\hat{V}_{i,n}(t) = \text{var}(E_{i,n}(t))$ for signal x_i. Next, a $n-1$ dimensional MVAR model is fitted for $n-1$ channels, excluding channel j, which leads to the residual variance $\hat{V}_{i,n-1}(t) = \text{var}(E_{i,n-1}(t))$. Then Granger causality index is defined as:

$$\text{GCI}_{j\to i}(t) = \log \frac{\hat{V}_{i,n}(t)}{\hat{V}_{i,n-1}(t)} \tag{4.32}$$

GCI is smaller or equal to 1, since the variance of the n-dimensional system is lower than the residual variance of a smaller $n-1$ dimensional system. $\text{GCI}(t)$ is used to estimate causality relations in time domain.

In biomedical applications, frequently, the spectral content of signals and transmission in different frequency bands is of interest, so the frequency-dependent estimators of connectivity were developed. Granger causality concept was extended to the frequency domain by Geweke [179, 180], who introduced an estimator called Granger-Geweke Causality (GGC). GGC describing directed influence from channel j to i is expressed by formula:

$$\text{GGC}_{j\to i}(f) = \log \left(\frac{S_{ii}(f)}{S_{ii}(f) - (\hat{V}_{jj} - \hat{V}_{ij}^2/\hat{V}_{ii})|H_{ij}(f)|^2} \right) \tag{4.33}$$

where $\hat{V}$ is the estimated noise covariance matrix. The idea standing behind this definition is that the causal influence depends on the relative sizes of the spectral power $S_{ii}(f)$ in channel i and the intrinsic power expressed by: $S_{ii}(f) - (\hat{V}_{jj} - \hat{V}_{ij}^2/\hat{V}_{ii})|H_{ij}(f)|^2$. When the intrinsic power equals the total power, the causal power is zero. GGC increases when the causal power rises.

Other estimators operating in the frequency domain are DTF and PDC.

4.3.2.3 Directed Transfer Function

Directed transfer function (DTF) introduced by Kaminski and Blinowska [272] is based on the properties of the transfer matrix $\mathbf{H}(f)$ of MVAR which is asymmetric and contains spectral and phase information concerning relations between channels. DTF describes the causal influence of channel j on channel i at frequency f:

$$\text{DTF}_{j\to i}(f) = \frac{|H_{ij}(f)|^2}{\sum_{m=1}^{k} |H_{im}(f)|^2} \tag{4.34}$$

The above equation defines a normalized version of DTF, which takes values from 0 to 1, giving a ratio between the inflow from channel j to channel i to all the inflows to channel i. To make the denominator independent on frequency, i.e., account for inflows to channel i in any frequency a function called full frequency DTF was proposed:

$$\text{ffDTF}_{j\to i}(f) = \frac{|H_{ij}(f)|^2}{\sum_{f} \sum_{m=1}^{k} |H_{im}(f)|^2} \tag{4.35}$$

Sometimes one can abandon the normalization property and use values of elements of transfer matrix which are related to causal connection strength. The non-normalized DTF can be defined as:

$$\text{NDTF}_{j \to i}(f) = |H_{ij}(f)|^2 \qquad (4.36)$$

It was shown by means of modeling that non-normalized DTF is related to the causal connection strength between channels [275]. DTF is an estimator that provides spectral information on the transmitted activity and is robust with respect to noise. Figure 4.1 shows an example where the directedness is determined correctly for signals where the variance of noise was 9 times as high as that of the signal. However, DTF shows direct and indirect transmissions; in case of cascade flow $a \to b \to c$, it also indicates flow: $a \to c$.

dDTF

In order to distinguish direct from indirect flows direct directed transfer function (dDTF) was introduced [300]. The dDTF is defined as a multiplication of the ffDTF by partial coherence ($\kappa_{ij}^2(f)$). The dDTF showing direct propagation from channel j to i is defined as:

$$\text{dDTF}_{j \to i}(f) = \text{ffDTF}_{j \to i}(f)\kappa_{ij}^2(f) \qquad (4.37)$$

$\text{dDTF}_{ij}(f)$ has a non-zero value when both functions $\text{ffDTF}_{ij}(f)$ and $\kappa_{ij}^2(f)$ are non-zero; in that case there exists a causal relation between channels $j \to i$ and that relation is direct. The DTF and dDTF for the same simulation scheme are presented in Figure 4.2. One can see that dDTF shows only the direct flows, whereas in the case of DTF cascade flows are also present, e.g., from channel 1 to 3 and 5.

Time-varying DTF (SDTF)

In biomedical applications, it is often of interest to grasp the dynamical changes of signal propagation, which means we have to apply a very short measurement window. However, it deteriorates strongly statistical properties of the estimate. To fit the MVAR model properly, the number of data points in the window should be bigger than the number of model parameters. For MVAR, the number of parameters is pk^2 (where k – number of channels, p – model order), number of data points is kN, so the condition for proper model fitting is that $\frac{kp}{N}$ should be small, preferably smaller than 0.1. The same rule holds for all estimators based on MVAR. In the case of time-varying processes, when we would like to follow the transmission dynamics, the data window has to be short.

To increase the number of the data points we may use multiple repetitions of the experiment. We may treat the data from each repetition as a realization of the same stochastic process. Then the number of data points is $kN_S N_T$ (where N_T is number of realizations, N_S – is number of data points

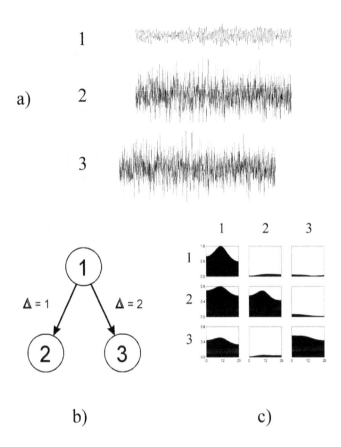

FIGURE 4.1
Performance of DTF for noisy signals. a) 1—stochastic signal, 2 and 3— signals delayed by, respectively, one and two samples with white noise of variance 9 times greater than the signal variance added. b) Simulation scheme, c) results of DTF; in each box DTF as a function of frequency, on the diagonal power spectra; propagation from channels labeled above to the channels labeled at the left of the picture.

in the window), and their ratio to the number of parameters (pk^2) effectively increases. Based on this observation, we can divide a non-stationary recording into shorter time windows, short enough to treat the data within a window as quasi-stationary. We calculate the correlation matrix between channels for each trial separately. The resulting model coefficients are based on the

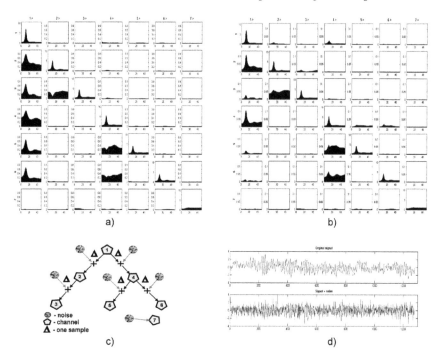

FIGURE 4.2
Comparison of DTF and dDTF (a) In each box DTF as a function of frequency, on the diagonal power spectra; propagation from channels labeled above to the channels labeled at the left of the picture; b) dDTF is shown using the same convention as in a); c) simulation scheme; d) signals used in the simulation.

correlation matrix averaged over trials. The correlation matrix has a form:

$$\hat{R}_{ij}(s) = \frac{1}{N_T} \sum_{r=1}^{N_T} R_{ij}^{(r)}(s) = \frac{1}{N_T} \sum_{r=1}^{N_T} \frac{1}{N_S} \sum_{t=1}^{N_S} X_{i,t}^{(r)} X_{j,t+s}^{(r)} \qquad (4.38)$$

$R_{ij}^{(r)}$ is the correlation matrix for short windows of N_S points, and r is the index of the repetition.

The averaging concerns correlation matrices for short data windows—data are not averaged in the process. The window size choice is always a compromise between the quality of the fit (depending on the ratio between the number of data points and the number of model parameters) and time resolution. The MVAR coefficients are obtained using the above-described procedure for each short data window, and then the estimators characterizing the signals, e.g., power spectra, coherences, short-time DTFs (SDTFs). By applying the sliding window, multivariate estimators may be expressed as functions of time and frequency, and their evolution in time may be followed.

The alternating approach to a sliding window is an adaptive method that relies on calculating time-varying MVAR model coefficients using the Kalman filter algorithm (cf. 3.2.2). This method estimates the changing in time parameters of the multivariate autoregressive model by observing measurements of the system output [13, 643]. A further improvement was recently proposed as a Self-Tuning Optimized Kalman filter (STOK). This novel adaptive filter embeds a self-tuning memory decay and a recursive regularization to guarantee high network tracking accuracy, temporal precision, and robustness to noise [461]. These authors provide codes for computation of STOK in MATLAB and Python, as supplementary material.

4.3.2.4 Partial Directed Coherence

The partial directed coherence (PDC) was introduced by Baccalá and Sameshima [29] in the following form:

$$P_{j \to i}(f) = \frac{A_{ij}(f)}{\sqrt{a_j^H(f)a_j(f)}} \tag{4.39}$$

In the above equation $A_{ij}(f)$ is an element of $\mathbf{A}(f)$ matrix—a Fourier transform of MVAR model coefficients $a_j(f)$ is j-th column of $A(f)$, and the $(.)^H$ denotes the Hermitian transpose. The PDC from j to i represents the relative coupling strength of the interaction of a given source (signal j), with regard to some signal i as compared to all of j's connections to other signals. From the normalization condition it follows that PDC takes values from the interval $P_{j \to i} \in [0, 1]$. PDC shows only direct flows between channels. Although it is a function operating in the frequency domain, the dependence of $\mathbf{A}(f)$ on the frequency has not a direct correspondence to the power spectrum. Figure 4.3 shows PDC for the simulation scheme presented in Figure 4.2 c).

The definition of PDC was re-examined by the authors, who considered the fact that PDC dependence on a signal's dynamic ranges, as modified by gains, obscures the PDC ability to correctly pinpoint the direction of information flow [27]. They proposed the so-called generalized partial directed coherence (GPDC) defined by the formula:

$$GPDC_{j \to i}(f) = \frac{\frac{1}{\sigma_i} A_{ij}(f)}{\sqrt{\sum_{k=1}^{N} \frac{1}{\sigma_k^2} A_{ki}(f) A_{ki}^*(f)}} \tag{4.40}$$

where $\sigma_i = \mathbf{V}(i, i)$, the residual variance of the ith channel (4.15). The modification counteracts the impact of different noise variances in the input signals on GPDC.

More essential drawbacks of PDC were pointed out in [534], namely:

i) PDC is decreased when multiple signals are emitted from a given source,

ii) PDC is not scale-invariant, since it depends on the units of measurement of the source and target processes,

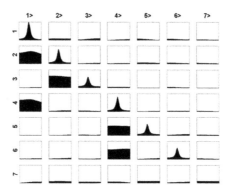

FIGURE 4.3
The PDC results for the simulation shown in Figure 4.2 c). The same convention of presentation as in Figures 4.1 and 4.2.

iii) PDC does not allow conclusions on the absolute strength of the coupling.

Point iii) is illustrated in Figure 4.4. When the activity is emitted in several directions, PDC shows weaker flows than when the same activity is emitted in one direction only. Another feature of PDC is a weak frequency dependence. The PDC spectrum is practically "flat", whereas the DTF spectrum (especially for non-normalized DTF) reflects the signal's spectral characteristics. This feature is also visible in Figure 4.4 (first column, panel D vs. B). An example of the application of DTF and PDC to experimental data (EEG) will be given in Sect. 5.1.6.3.

The authors proposed the renormalization of PDC similar to the one used in the definition of DTF, which helped alleviate the above problems.

$$\text{rPDC}_{j\to i}(f) = \frac{A_{ij}(f)}{\sqrt{\sum_{k=1}^{N}\left|\hat{A}_{ik}(f)\right|^2}} \tag{4.41}$$

4.3.2.5 Directed Coherence

Directed coherence (DC) was introduced in [26] in order to mitigate possible effect of different noise variances in the input channels of connectivity measure. DC is similar to DTF, namely it is based on transfer function of MVAR model and is normalized in respect to the inflows to the destination channel:

$$\text{DC}_{j\to i}(f) = \frac{\sigma_j H_{ij}(f)}{\sqrt{\sum_{k=1}^{N}\sigma_k^2|H_{ik}(f)|^2}} \tag{4.42}$$

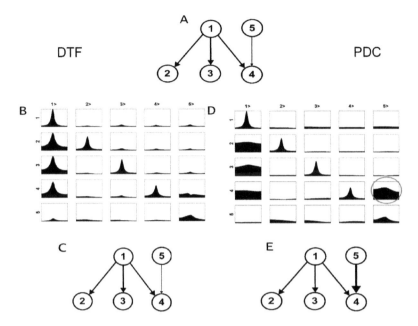

FIGURE 4.4
Comparison of DTF (panel B) and PDC (panel D) for simulation scheme
A. Resulting schemes of propagation for DTF and PDC below corresponding
panels. Thickness of arrows proportional to the flow intensities. The convention of presenting DTF/PDC is as in Figure 4.5. Note that the weak flow from
channel 5 is enhanced by PDC.

where $\sigma_j = \mathbf{V}(j,j)$, the residual variance of the jth channel (4.15). In analogy to dDTF and ffDTF direct Directed Coherence dDC and full frequency
Directed Coherence may be defined:

$$\text{ffDC}_{j\rightarrow i}(f) = \frac{\sigma_j H_{ij}(f)}{\sqrt{\sum_{f=f_1}^{f_{N_f}} \sum_{k=1}^{N} \sigma_k^2 |H_{ik}(f)|^2}} \tag{4.43}$$

$$\text{dDC}_{j\rightarrow i}(f) = |\kappa_{ij}^2(f)||\text{ffDC}_{j\rightarrow i}(f)| \tag{4.44}$$

Measures of connectivity defined in Sect. 4.3.2 may be conveniently estimated in the framework of the MVAR model. Alternatively, they may be
computed by a non-parametric method by performing spectral estimation followed by spectral decomposition [121]. However, autoregressive models provide better spectral estimates—unbiased, free of windowing effect, and they
have better statistical properties than Fourier spectra (Sect. 3.3.2.3).

MATLAB tutorial: MVAR.

The LiveScript `matlab/c4/Ch4_03_MVAR_tutorial.mlx` presents the computation and comparison of the MVAR based functions.

4.4 Non-Linear Estimators of Dependencies between Signals

4.4.1 Kullback-Leibler Entropy, Mutual Information

Kullback-Leibler (KL) entropy introduced in [309] is a non-symmetric measure of the difference between two probability distributions P and Q. KL measures the expected number of extra bits required to code samples from P when using a code based on Q, rather than using a code based on P. For probability distributions $P(i)$ and $Q(i)$ of a discrete random variable i their KL divergence is defined as the average of the logarithmic difference between the probability distributions $P(i)$ and $Q(i)$, where the average is taken using the probabilities $P(i)$:

$$D_{\mathrm{KL}}(P\|Q)) = \sum_i P(i) \log \frac{P(i)}{Q(i)} \tag{4.45}$$

For continuous random variable x, KL-divergence is defined by the integral:

$$D_{\mathrm{KL}}(P\|Q)) = \int_{-\infty}^{\infty} p(x) \log \frac{p(x)}{q(x)} dx \tag{4.46}$$

where p and q denote the densities of P and Q. Typically P represents the distribution of data, observations, or a precise, calculated theoretical distribution. The measure Q usually represents a theory, model, description, or approximation of P. The KL from P to Q is not necessarily the same as the KL from Q to P. KL entropy is also called information divergence, information gain, or Kullback-Leibler divergence.

Mutual information is based on the concept of the entropy of information (see Sect. 3.3.1.2). For a pair of random variables x and y the mutual information (MI) between them is defined as:

$$\mathrm{MI}_{xy} = \sum p_{ij} \log \frac{p_{ij}}{p_i p_j} \tag{4.47}$$

where p_{ij} is the joint probability that $x = x_i$ and $y = y_j$. This measure essentially tells how much extra information one gets on one signal by knowing the outcomes of the other one. If there is no relation between both signals,

MI_{xy} is equal to 0. Otherwise, MI_{xy} will be positive attaining a maximal value of I_x (self information of channel x) for identical signals (Sect. 3.3.1.2).

MI is a symmetric measure, and it doesn't give any information about directionality. It is possible to get a notion of direction by introducing a time lag in the definition of MI_{xy}.

4.4.2 Transfer Entropy

The transfer entropy was introduced by Schreiber [538], who used the idea of finite-order Markov processes to quantify causal information transfer between systems I and J evolving in time. Assuming that the system under study can be approximated by a stationary Markov process of order k, the transition probabilities describing the evolution of the system are: $p(i_n|i_{n-1},\ldots,i_{n-k})$. If two processes I and J are independent, then the generalized Markov property $p(i_n|i_{n-1}^{(k)}) = p(i_n,|i_{n-1}^{(k)}, j_{n-\tau}^{(l)})$ holds, where $i_{n-1}^{(k)} = i_{n-1},\ldots,i_{n-k}$ and $j_{n-\tau}^{(l)} = (j_{n-\tau},\ldots,j_{n-\tau-l})$ is the number of conditioning states from process I and J, respectively. Schreiber [538] proposed to use the Kullback-Leibler entropy (equation 4.45) to quantify the deviation of the transition probabilities from the generalized Markov property. This resulted in the definition of transfer entropy (TE):

$$\mathrm{TE}_{J \to I}(\tau) = \sum p(i_n, i_{n-1}^{(k)}, j_{n-\tau}^{(l)}) \log \frac{p(i_n|i_{n-1}^{(k)}, j_{n-\tau}^{(l)})}{p(i_n|i_{n-1}^{(k)})} \tag{4.48}$$

where τ is the delay of the transfer. The TE can be understood as the excess amount of bits that must be used to encode the information of the state of the process by erroneously assuming that the actual transition probability distribution function is $p(i_n|i_{n-1}^{(k)})$ instead of $p(i_n|i_{n-1}^{(k)}, j_{n-\tau}^{(l)})$. When the sample size is small it is practical to choose $k = 1$ and $l = 1$. With this settings we obtain:

$$\begin{aligned}\mathrm{TE}_{J \to I}(\tau) &= \sum p(i_n, i_{n-1}, j_{n-\tau}) \log \frac{p(i_n|i_{n-1}, j_{n-\tau})}{p(i_n|i_{n-1})} & (4.49)\\ &= \sum p(i_n, i_{n-1}, j_{n-\tau}) \log \frac{p(i_n, i_{n-1}, j_{n-\tau})p(i_{n-1})}{p(i_{n-1}, j_{n-\tau})p(i_n, i_{n-1})} & (4.50)\end{aligned}$$

TE quantifies the statistical dependence between two time-series, with no assumption about their generation processes, which can be linear or non-linear. Despite the apparent simplicity of the estimators, the practical calculation of TE from experimental signals is not an easy task. In practice, estimation of the probability density functions (PDFs) is the crucial and challenging step, and the different estimators of TE approach this problem differently. The most straightforward idea involves obtaining the histograms of the series of outcomes and finding, say, p_i as the ratio between the number of samples in the i-th bin of the histogram and the total number of samples. To get an accurate estimate of this measure by using histogram-derived probabilities, one needs to have a large number of samples and small bins [495]. When taking small bins of the same size for each variable, the values $p_{ij} = 0$ may occur. These difficulties and other problems, including a high probability of systematic errors in the calculation of MI, were discussed in [467]. A more practical approach is the kernel method for PDF approximation. It relies on assuming that each data point is drawn from a distribution modeled by a kernel. The PDF is estimated then as a sum of kernels placed at each data sample. In the three-dimensional space of $i_n, i_{n-1}, j_{n-\tau}$ the joint probability at any given point $\tilde{i}_n, \tilde{i}_{n-1}, \tilde{j}_{n-\tau}$ can be estimated as:

$$p(\tilde{i}_n, \tilde{i}_{n-1}, \tilde{i}_{n-\tau}) \approx \frac{1}{N} \sum_{m=1}^{N} \frac{1}{h_{i_n}, h_{i_{n-1}}, h_{j_{n-\tau}}}$$

$$K\left(\frac{\tilde{i}_n - i_{n,m}}{h_{i_n}}\right) K\left(\frac{\tilde{i}_{n-1} - i_{n-1,m}}{h_{i_{n-1}}}\right) K\left(\frac{\tilde{j}_{n-\tau} - j_{n-\tau,m}}{h_{j_{n-\tau}}}\right) \quad (4.51)$$

where j indexes the N data points and h is a bandwidth for the given dimension. Widely used is the Gaussian kernel $K(u) = \frac{1}{\sqrt{2\pi}}e^{-0.5u^2}$. The remaining probabilities in (4.50) are obtained by marginalizing (4.51).

Another approach was introduced in [261] as an extension of the D-V algorithm to three-dimensional space for transfer entropy estimation. After ordinal sampling[2], the D-V algorithm recursively partitions the three-dimensional space defined by $v_n, v_{n-1}, u_{i-\tau}$ cubes of varying sizes. Initially, the entire space is sliced into 8 equal cubes, where the boundaries are at the mid-points in the three dimensions. Using the 8 cubes, the following χ^2 statistic is computed to test the null hypothesis that data points are evenly distributed across the 8 cubes:

$$s_{\chi^2} = \sum_{c=1}^{8} (M_c - \mu_M)^2, \quad (4.52)$$

where $M_1, M_2, ..., M_8$ are the numbers of data points contained in each of the 8 cubes, and μ_M is the total number of data points divided by the number of cubes. If $s_{\chi^2} > \chi^2_{95\%}(7))$ (at a 5% significance level and 7 degrees of freedom), then the null hypothesis is rejected, and each of the 8 cubes is further partitioned into 8 smaller cubes in the same manner, and the recursion continues. If the null hypothesis is not rejected, the recursion is stopped. Cubes that contain no data point do not contribute to the transfer entropy estimation. The partitioning process results in a finite number of cubes, L. By approximating each probability in (4.50) by fractions, $\mathrm{TE}(\tau)$ can now be estimated using the

---

[2]Ordinal sampling substitutes the values in time series X and Y with their ranks in sorted X and Y. The ranks are integers ranging from 1 (smallest value) to N (largest value).

partitions as follows:

$$\mathrm{TE}_{U \to V}(\tau) = \sum_{c=1}^{L} \frac{\eta_c}{N} \log \frac{\eta_c \eta_c^{v_{n-1}}}{\eta_c^{v_{n-1},u_{n-\tau}} \eta_c^{v_n,v_{n-1}}} \tag{4.53}$$

where N is the total number of data triplets $(v_n, v_{n-1}, u_{i-\tau})$, η_c is the number of data points in the cth cube, and $\eta_c^{v_{n-1}}$, $\eta_c^{v_n,v_{n-1}}$, and $\eta_c^{v_{n-1},u_{n-\tau}}$ are the numbers of data points in the entire data that are greater than or equal to lower bounds and less than upper bounds of the cth partition with respect to the dimensions in the superscript.

MATLAB demo: TE estimated with the KDE and adaptive partitioning.

The LiveScript `matlab/c4/Ch4_04_TE.mlx` presents the computation and basic properties of TE based functions from `https://physionet.org/content/tewp/1.0.0/` accompanying the paper [261].

TE (4.50) was extended to multivariate version by Montalto et al. [410]. They considered M signals. If the information transfer from J to I is of interest, the remaining $M - 2$ signals are denoted as $\mathbf{Z} = \{Z^k\}_{k=1,\ldots,M-2}$. Then, the TE from J to I conditioned on $\mathbf{Z}$, may be defined as:

$$\mathrm{TE}_{J \to I|Z} = \sum p(I_n, I_n^-, J_n^-, \mathbf{Z}_n^-) \log \frac{p(I_n|I_n^-, J_n^-, \mathbf{Z}_n^-)}{p(I_n|J_n^-, \mathbf{Z}_n^-)} \tag{4.54}$$

where $I_n, J_n, \mathbf{Z}_n$ are the samples taken at the present time n, and the superscript $^-$ denotes all the past samples of a given signal. The sum extends over all the phase space points forming the trajectory of the composite system that generates the signals. The TE quantifies the information provided by the past of the process J about the present of the process I that is not already provided by the past of I itself or any other process included in $\mathbf{Z}$. The Authors of [410] provided a convenient MATLAB toolbox implementing this technique at: `http://dx.doi.org/10.6084/m9.figshare.1005245`, and the manuscript contains the tutorial of application to simulated and real data.

4.4.3 Generalized Synchronization and Synchronization Likelihood

Generalized synchronization (GS) is a concept rather than a measure, since following the original idea of GS introduced in [513], different formulas for calculation of GS have been introduced [104, 467]. GS measure is based on the theory of chaotic systems and makes use of the embedding theorem (Sect. 3.5), which is applied to the signals x and y. GS quantifies how well one can predict the trajectory in phase space of one of the systems knowing the trajectory of the other; alternatively, it quantifies how neighborhoods (i.e., recurrences) in one attractor map into the other [494]. In brief, one calculates for each sample x_n the squared mean Euclidean distance to its k neighbors – $R_n^k(x)$. The y conditioned squared mean Euclidean distance $R_n^k(x|y)$ is defined by replacing the nearest neighbors of x_n by the equal number of time samples of the closest neighborhood of y_n. Then the $GS^k(x|y)$ interdependence measure may be defined as [12, 495]:

$$GS^k(x|y) = \frac{1}{N} \sum_{n=1}^{N} \frac{R_n^k(x)}{R_n^k(x|y)}. \tag{4.55}$$

The GS measures based on the normalized difference of $R_n^k(x|y)$ and $R_n^k(y|x)$ are asymmetric and may give the information on the directionality of the interaction. However, one should be careful in drawing the conclusions on the driver and driven system [468].

Synchronization likelihood (SL) [568] similarly to generalized synchronization is based on the theory of chaotic systems and makes use of the embedding theorem (Sect. 3.5) and concept of Euclidean distance. SL is defined as the conditional likelihood that the distance between embedded vectors Y_i and Y_j will be smaller than a cutoff distance r_y, given that the distance between X_i and X_j is smaller than a cutoff distance r_x.

$$SL = \frac{2}{N(N-w)P_{ref}} \sum_{i=1}^{N} \sum_{j=i+w}^{N-w} \Theta(r_x - |X_i - X_j|)\Theta(r_y - |Y_i - Y_j|) \tag{4.56}$$

The vertical bars represent the Euclidean distance between the vectors. N is the number of samples, w is the Theiler correction for autocorrelation [595], and Θ is the Heaviside function. In the case of maximal synchronization SL is 1; in the case of independent systems, it is a small, but non-zero number, namely P_{ref}. This small number is the likelihood that two randomly chosen vectors Y (or X) will be closer than the cutoff distance r. P_{ref} is a parameter that has to be set. The concept of synchronization likelihood described here is closely related to the definition of the mutual information based upon the correlation integral. In the case of two time series the time-dependent mutual information is related to the synchronization likelihood by a simple formula:

$$MI_{XY} = \log_2(SL/P_{ref}) \tag{4.57}$$

The difference between both measures is that MI is normalized and SL is not.

4.4.4 Phase Synchronization (Phase Locking Value)

Two non-linear coupled oscillators may synchronize their phases, even if their amplitudes remain uncorrelated. This phenomenon, called phase synchronization, may be detected by analyzing the phases of the time series. Namely the phases may be found from analytical signals obtained by application of the Hilbert transform to the experimental signals (Sect. 3.4.1). Synchronization means that the phase locking condition applies for time t:

$$\phi_{n,m}(t) = |n\phi_x(t) - m\phi_y(t)| \leq constant, \qquad (4.58)$$

where $\phi_x(t)$, $\phi_y(t)$ are the phases of signals x and y (equation 3.67, Sect. 3.4.1) and n, m are small natural numbers, most often $n = m = 1$.

Several indexes of phase synchronization were introduced. Mean phase coherence, also called phase locking value (PLV) [316, 495], is defined by the formula:

$$\text{PLV}_{n,m} = \frac{1}{N} \left| \sum_{t=0}^{N} \exp\left(i\phi_{n,m}(t)\right) \right|, \qquad (4.59)$$

where N is the number of time instants. From the above formula[3], one can find how the relative phase is distributed over the unit circle. If the two signals are phase synchronized, the relative phase will occupy a small portion of the circle, which means high phase coherence.

Other measures of phase synchronization (PS) are based on calculating the distribution of phases [467]. In the so-called stroboscopic approach, the phase of one of the oscillators is observed at those instants, where that of the other one attains a particular value. The above methods require the bandpass filtering of the signals. When the signals are characterized by modulated natural frequency, the relative phase distributions are broad, making the estimation of PS difficult. The problems connected with the construction of histograms, mentioned in the context of finding generalized synchronization, are also present in the evaluation of phase synchronization based on the consideration of phase distributions. PS does not allow determination of the direction of interaction between the channels of a process.

4.4.5 Testing the Reliability of the Estimators of Directedness

Testing the null hypothesis about the lack or presence of causal relations between time series is not straightforward. The analytical determination of the statistical distributions of the estimators of directedness is difficult. The asymptotic distribution of DTF under the null hypothesis of no information flow was derived by Eichler [141]. He proposed a frequency-dependent pointwise significance level that forms an upper bound for the unknown critical value of DTF's asymptotic distribution.

However, in the absence of a general analytical approach, the directedness estimators' significance is usually tested using surrogate data. Namely, the

---

[3]A convenient MATLAB implementation can be found, e.g., here: https://www.mathworks.com/MATLABcentral/fileexchange/31600-phase-locking-value.

obtained results are compared with those computed for the time series with no dependencies between channels.

The most straightforward approach is to randomly shuffle samples of the data series, but such a procedure does not preserve the signals' autocorrelation structures. The use of such a white-noise version of data as a control condition turns the data into a very unlikely realization of the physiological process. Therefore this procedure is not recommended.

A better approach is to use the surrogates, which preserve the time series's spectral properties and randomize only the phases of each signal. The procedure of generating this type of data was described in (Sect. 2.6). The outcome of this kind of test shows whether there is directedness among the set of signals. A similar approach is used to test for non-linearity of the time series. The problem of the construction of bivariate surrogate data was considered in [8]. In multichannel data, the test indicates the presence or lack of dependencies, but it doesn't tell if the character of dependence is linear or non-linear. The generation of surrogate data for multichannel series destroying only the nonlinear part of interdependence is a problem that is not competently solved.

In the case of estimators operating in time-frequency space such as SDTF, the problem of reliability testing is more complicated, since all realizations of the process are used to compute the estimator. In this case, the bootstrap method [670] is applied. It relies on constructing the new artificial time series with the same number of realizations as the original data. The realizations are created by drawing randomly (with repetitions) trials from the original set. Let us assume that the original ensemble of trials consists of M realizations:

$$[r] = [r^1, r^2, r^3, r^4, r^5, \ldots, r^k, r^{k+1}, \ldots, r^M] \tag{4.60}$$

New ensemble of trials may be:

$$[r_{boot}] = [r^3, r^1, \ldots, r^3, r^7, \ldots, r^5, \ldots, r^5, r^M] \tag{4.61}$$

Repeating such procedure for all k channels L times (in practice 100 or more times) we obtain L ensembles of realizations and we can calculate the average $\overline{\text{SDTF}}$ and its variance as:

$$\text{var(SDTF)} = \sum_{i=1}^{L} \frac{(\overline{\text{SDTF}} - \text{SDTF})^2}{(L-1)} \tag{4.62}$$

The method may be used for evaluation of the errors of any estimator based on ensemble averaging.

4.5 Comparison of the Multichannel Estimators of Coupling between Time Series

Among approaches to quantify the coupling between multivariate time series, we can distinguish between linear and non-linear methods as well as approaches based on bivariate and multivariate estimators.

4.5.1 Bivariate versus Multivariate Connectivity Estimators

For a long time, the standard methods of establishing relations between signals have been cross-correlation and coherence. However, in the case of these bivariate measures, we do not know if the two channels are directly coupled, or another channel drives them. This phenomenon is called common drive effect. This point may be elucidated by simulation, illustrating the situation where the activity is measured in different distances from the source (Figure 4.5). The signals in channels 2–5 were constructed by delaying the signal from channel one (the source) by 1, 2, 3, 4 samples, and adding the noise in each step. The coherences were calculated between all signals pairwise. The phases were found for the peak frequency, and assuming the linear phase relations, the delays were calculated. The resulting scheme of propagation presented in Figure 4.5 b) shows many false flows. Similar results were obtained for bivariate AR model (Figure 4.5 c). This situation is common to all bivariate methods; namely, the propagation is found in each case where the phase difference is present [64]. Sometimes, for bivariate measures, even the reverse propagation may be found [313]. The results obtained using the multivariate method (DTF) show the correct scheme of propagation (Figure 4.5 d). Bivariate methods usually produce many false connections. Let us assume transmissions from site 1 to locations 2, 3, 4, and 5. Bivariate methods will show not only connections from 1 to 2, 3 and 4, but also between 2 and 3, 2 and 4, 3 and 4 etc. (Figure 4.5 b, c). In the general case, when activity is transmitted from channel 1 to N channels, which record its activity, we will obtain N true and $N(N - 1)/2$ false connections [59].

MATLAB demo: bivariate and multivariate approach to the estimation of dependence.

The differences in the results obtained by bivariate and multivariate connectivity estimation can be testified in the LiveScript `matlab/c4/Ch4_05_MVAR_vs_bivariate.mlx`.

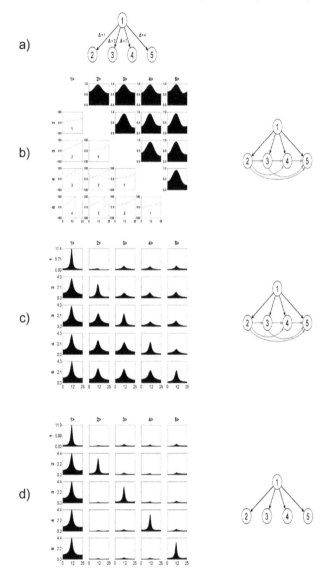

FIGURE 4.5
Illustration of common drive effect. Propagations estimated for simulation
scheme shown in a) by means of: b) bivariate coherences, c) bivariate DTF
model, d) multivariate DTF. In panel b) above the diagonal, the magnitudes of
coherences, below diagonal, phases of the coherences are shown. From phases,
the delays (values printed in each box) were calculated. For c) and d) in
each box DTF as a function of frequency, on the diagonal power spectra;
propagation from channels labeled above to the channels labeled at the left
of the picture. Resulting propagation schemes shown in the right part of each
panel.

4.5.2 Linear versus Non-Linear Estimators of Connectivity

All non-linear methods described, except transfer entropy, are bivariate, so they suffer from the disadvantages pointed out above. It was demonstrated that Granger causality and transfer entropy are equivalent for Gaussian variables [33]. Computation of TE relies on calculation of probability density functions, so it is rather cumbersome (Sect. 4.4.2), besides TE similarly to all non-linear measures is very sensitive to noise. Therefore it seems more reasonable to apply measures based on Granger causality concept robust to noise and based on straightforward calculation of MVAR model.

There were attempts to construct multichannel non-linear extensions of Granger's causality called extended Granger causality by Chen et al. [84]. They applied it to 3 channel simulated non-linear time series, but the advantages over the linear approach were not clearly demonstrated, and the influence of noise was not studied. The method requires the determination of the embedding dimension and neighborhood size, which are difficult to find in an objective and optimal way. The same objections concern other methods that aim to extend non-linear estimators for more than two channels. The problems concerning state-space reconstruction and noise sensitivity become even more serious for a higher number of channels.

In the study devoted to the comparison of linear and non-linear methods of coupling [427], non-linear estimators: mutual information, phase correlation, continuity measure[4] were compared with correlation in case of non-linear signals in the presence of noise. The authors found that any method that relies on an accurate state space reconstruction will be inherently at a disadvantage over measures that do not rely on such assumptions. Another finding was the high sensitivity to the noise of non-linear estimators. The authors conclude: *"We have been as guilty as any of our colleagues in being fascinated by the theory and methods of nonlinear dynamics. Hence we have continually been surprised by robust capabilities of linear CC (correlation) to detect weak coupling in nonlinear systems, especially in the presence of noise"*.

We can conclude that non-linear methods are not recommended in most cases. They might be used only when there is clear evidence that there is a good reason to think that there is a non-linear structure either in data themselves or in the interdependence between them [467]. We have to bear in mind that many systems composed of highly non-linear components exhibit an overall linear type of behavior, such as some electronic devices and brain signals such as electroencephalograms or local field potentials (LFP). This problem will be further discussed in Sect. 5.1. In this context, it is worth mentioning that some linear methods, e.g., DTF, work quite well also for non-linear time series [648].

---

[4]Measure similar to GS testing for continuity of mapping between neighboring points in one data set to their corresponding points in the other data set.

4.5.3   The Measures of Directedness

The estimators that have the potential to show directedness of the information transfer are: (1) correlation, coherence, transfer entropy, phase-locking index, (2) partial phase spectrum (equation 4.11), and (3) the estimators based on the Granger causality principle. The measures from the first group are bivariate (except for the extension of TE), so they suffer from the common drive effect. Additionally, they are not able to detect reciprocal flows (interaction in both directions). Such patterns of propagation may be found using estimators based on the causality principle. The comparison of linear techniques inferring directed interactions in multivariate systems was carried out in [648] and [151]. In the first of these publications, partial phase spectrum and three methods based on Granger causality: GCI, DTF, and PDC, were tested in respect to specificity in the absence of influences, correct estimation of direction, direct versus indirect transmissions, non-linearity of data, and influences varying in time. All methods performed mostly well with some exceptions. The partial phase spectrum failed in the important aspect—estimation of direction. It was due to the large errors in phase estimation for particular frequencies and the inability to detect reciprocal interactions. This might be expected because of large statistical errors of partial phase spectrum (equation 4.27).

GCI failed in the presence of non-linearities. DTF did not distinguish direct from cascade flows, but when such distinction is important, dDTF may be used. PDC for signals with a non-linear component gave correct results only for very high model orders, making the method hardly applicable for strongly non-linear signals because of limitations connected with the number of parameters of MVAR mentioned in Sect. 4.3.2.3.

Granger-based measures: GGC, DTF, dDTF, PDC, GPDC, DC, and DC were tested against signal to noise ratio (SNR) robustness and presence of weak network node (WNN) in [151]. A weak network node occurs when one of the signals set has much lower SNR. GGC was found to be the least robust measure in case of decreasing SNR. DTF and especially dDTF was found as a causality measure which proved to be the most robust against the decreasing SNR. GPDC, DC, and dDC were the most robust in respect of WNN influence. The effect of weak network node depended critically on the situation. Namely, whether the WNN was the source of the activity or a passive sink. WNN influence is important in the first case—of active source. However, in practice, active sources usually have high SNR (unless due to artifact). DTF was the most robust to the presence of weak nodes, as long as they were not active sources of causal flow in the network.

Concerning selectivity in frequency—best frequency resolution was achieved by DTF and dDTF, which was not the case for PDC and GPDC. Their spectral selectivity was improved only for high MVAR orders.

Taking into account the above considerations, we can conclude that the methods estimating direction which perform best are the multivariate methods based on the Granger causality concept. They are robust in respect of noise and the common drive problem, provide the information on causal coupling, and detect reciprocal connections. Depending on the problem, one may choose the most suitable method for his problem, i.e., the method which offers the best frequency resolution or deals well with non-linear interactions. For many applications, DTF, which fulfills the above requirements, might be a method of choice. When the distinction between direct and cascade flows is essential, which is the case for signals detected by subcortical electrodes, dDTF, or GPDC would be suitable.

4.6 Multivariate Signal Decompositions

4.6.1 Principal Component Analysis (PCA)

4.6.1.1 Definition

Principal component analysis (PCA) is a method of decomposition of multi-channel epoched data into linearly independent components; that is, spatially and temporally uncorrelated. Depending on the field of application, it is also named the discrete Karhunen-Loéve transform (KLT), the Hotelling transform, or proper orthogonal decomposition (POD). The geometrical intuition behind this method is the following. We treat samples from all channels at a given time moment as a point in the space of dimension equal to the number of channels. One epoch of multichannel data forms a cloud of points in that space. This cloud can be spread more in some directions than in the others. The measure of the spread is the variance of the points' locations in a given direction. The PCA finds a set of orthogonal axes, a base, such that each consecutive axis spans the directions with consecutively decreasing variance. The projections of the points onto these axes constitute the components. Each component can be visualized as a time series. On the other hand, the original time series can be recovered as a linear combination of these components. Components corresponding to the smallest variance can be neglected, and in this way, a reduction of data dimensionality can be achieved.

4.6.1.2 Computation

PCA can be performed using the singular value decomposition (SVD) algorithm. An epoch of k channels data of length m can be represented as a $m \times k$ matrix $\mathbf{x}$. It is always possible to find three matrices $\mathbf{P}$, $\mathbf{A}$, and $\mathbf{M}$ such that:

$$\mathbf{x} = \mathbf{PAM}^\mathsf{T} \tag{4.63}$$

where $\mathbf{P}$ is the $m \times k$ matrix containing k normalized principal component waveforms ($\mathbf{P}^\mathsf{T}\mathbf{P} = \mathbf{I}$), $\mathbf{A}$ is $k \times k$, diagonal matrix of components amplitudes, $\mathbf{M}$ is a $k \times k$ matrix mapping components to original data such that M_{ij} is the contribution of jth component to i^{th} channel; $\mathbf{M}^T\mathbf{M} = \mathbf{I}$. The SVD algorithm is implemented in MATLAB as a function svd; to find the decomposition of x into $\mathbf{P}$, $\mathbf{A}$, and $\mathbf{M}$ execute: [P,A,M] = svd(x);

When data concerning multiple conditions or subjects are to be analyzed by PCA, the matrix $\mathbf{x}$ is composed by concatenating the matrixes of individual conditions or subjects along the time dimension. Thus for example for N_c conditions we have to decompose matrix $\mathbf{x}$ which has dimension $m \cdot N_c \times k$. The nth row of resulting matrix $\mathbf{P}$ represents N_c concatenated time courses obtained for each condition for the nth component. The corresponding topographical map is contained in the nth column.

4.6.1.3   Possible Applications

PCA can be used as a tool for decomposing multichannel data into a linear combination of components characterized by different spatial, temporal, and amplitude distribution. These components can be used as a preprocessing step in source localization techniques. However, there is no direct correspondence between the principal components and individual sources. A single source may produce a signal that decomposes into several components, while on the other hand, multiple sources can contribute to a single component. In general, the individual components should not be ascribed directly to any physiologically meaningful phenomenon [324].

A similar method, which leads to reducing the dimensionality of the data, is factor analysis (FA). FA is related to PCA but not identical. PCA performs a variance-maximizing rotation (varimax) of the variable space. Thus it takes into account all variability in the data. In contrast, factor analysis estimates how much of the variability is due to common factors. The number of common factors is usually less than the dimension of the original variable space [69]. FA can describe a large set of recorded wave shapes from multiple sensors in a small number of quantitative descriptors. These descriptors or factors can be conceptualized as basic waveforms produced by hypothetical signal generators, which, mixed in correct proportions, would reproduce the original waveforms in the set. FA was used to describe the evoked potentials [258]. FA can be performed in MATLAB using function factoran from the Statistical Toolbox.

4.6.2   Independent Components Analysis (ICA)

4.6.2.1   Definition

Independent component analysis (ICA) is a statistical signal processing method that decomposes a multichannel process into components that are statistically independent. In simple words two components s_1 and s_2 are independent when information of the value of s_1 does not give any information about the value of s_2. The ICA can be represented by a simple generative model:

$$\mathbf{x} = \mathbf{D}\mathbf{s} \tag{4.64}$$

where the $\mathbf{x} = \{x^1, x^2, \ldots, x^n\}$ is the measured n channel signal, $\mathbf{D}$ is the mixing matrix, and $\mathbf{s} = \{s^1, s^2, \ldots, s^n\}$ is the activity of n sources. The main assumption about $\mathbf{s}$ is that the s^i are statistically independent. To be able to estimate the model we must also assume that the independent components have *non-gaussian* distribution [239].

The model implies the following:

- The signal is a linear mixture of the activities of the sources

- The signals due to each source are independent

- The process of mixing sources and the sources themselves are stationary

- The energies (variances) of the independent components cannot be determined unequivocally[5]. The natural choice to solve this ambiguity is to fix the independent components' magnitude so that they have unit variance: $E[s^i] = 1$.

- The order of the components is arbitrary. If we reorder both in the same way: the components in $\mathbf{s}$, and the columns in $\mathbf{D}$, we obtain the same signal $\mathbf{x}$.

The main computational issue in ICA is the estimation of the mixing matrix $\mathbf{D}$. Once it is known, the independent components can be obtained by:

$$\mathbf{s} = \mathbf{D}^{-1}\mathbf{x} \tag{4.65}$$

4.6.2.2 Estimation of ICA

Finding the independent components can be considered in light of the central limit theorem as finding components of least Gaussian distributions. To comprehend this approach, let us follow the heuristic given in [239]. For simplicity, let us assume that the sought independent components have identical distributions. Let us define $y = \mathbf{w}^{\mathsf{T}}\mathbf{x}$. Please note, that if $\mathbf{w}^{\mathsf{T}}$ is one of the columns of the matrix $\mathbf{D}^{-1}$, then y is one of the sought components. With the change of variables $\mathbf{z} = \mathbf{D}^{\mathsf{T}}\mathbf{w}$ we can write $y = \mathbf{w}^{\mathsf{T}}\mathbf{x} = \mathbf{w}^{\mathsf{T}}\mathbf{D}\mathbf{s} = \mathbf{z}^{\mathsf{T}}\mathbf{s}$. This exposes the fact that y is a linear combination of the components s^i with the weights given by z_i. It stems from the central limit theorem that the sum of independent random variables has more gaussian character than each of the variables alone. The linear combination becomes least gaussian when $\mathbf{z}$ has only one non-zero element. In that case y is proportional to s^i. Therefore the problem of estimation of the ICA model can be formulated as a problem of finding $\mathbf{w}$, which maximizes the non-gaussianity of $y = \mathbf{w}^{\mathsf{T}}\mathbf{x}$. Maximizing non-gaussianity of y

---

[5]This is because the multiplication of the amplitude of the ith source can be obtained either by multiplication of s^i or by multiplication of the ith column of the matrix $\mathbf{D}$.

gives one independent component, corresponding to one of the $2n$ maxima[6] in the optimization landscape. To find all independent components, we need to find all the maxima. Since the components are uncorrelated, the search for consecutive maxima can be constrained to the subspace orthogonal to the one already analyzed. In cases when the original data contain strongly correlated channels, it may be useful to perform dimensionality reduction, e.g., by PCA.

4.6.2.3 Computation

The intuitive heuristics of maximum non-gaussianity can be used to derive different functions whose optimization enables the estimation of the ICA model; such a function may be, e.g., kurtosis. Alternatively, one may use more classical notions like maximum likelihood estimation or minimization of mutual information to estimate ICA. All these approaches are approximatively equivalent [239].

There are several tools for performing ICA in MATLAB. The EEGLAB toolbox [365] offers a convenient way to try different algorithms: *runica* [366], *jader* [78] and *fastica* [238]. The *runica* and *jader* algorithms are a part of the default EEGLAB distribution. To use the *fastica* algorithm, one must install the *fastica* toolbox (http://www.cis.hut.fi/projects/ica/fastica/) and include it in the MATLAB path. In general, the physiological significance of any differences in the results of different algorithms (or of different parameter choices in the various algorithms) has not been thoroughly tested for physiological signals. Applied to simulated, relatively low dimensional data sets for which all the assumptions of ICA are exactly fulfilled, all three algorithms return near-equivalent components.

Very important note: As a general rule, finding N stable components (from N-channel data) typically requires more than kN^2 data sample points (at each channel), where N^2 is the number of elements in the unmixing matrix that ICA is trying to estimate and k is a multiplier. As reported in the EEGLAB tutorial[7], the value of k increases as the number of channels increases. In general, it is vital to give ICA as much data as possible for successful estimation. ICA works best when a large amount of basically similar and mostly clean data are available.

4.6.2.4 Possible Applications

The most often reported applications of ICA in the field of biomedical signal analysis are related to EEG and MEG artifact reduction (Sect. 5.1.5) and feature extraction in studies of event-related brain dynamics [446] and as a preprocessing step in search of sources of EEG and MEG activity localization [197]. ICA is applied as well in analysis of heart (Sect. 5.2) and muscle signals (Sect. 5.3).

---

[6]Corresponding to s^i and $-s^i$.

[7]https://eeglab.org/tutorials/ConceptsGuide/ICA_background.html# how-many-data-points-do-i-need-to-run-an-ica.

MATLAB demo: cleaning data with ICA.

The LiveScript `matlab/c4/Ch4_06_ICA.mlx` illustrates the application of ICA for removing typical EEG artifacts.

4.6.3 Common Spatial Patterns

Both PCA and ICA can be seen as special cases of blind source separation (BSS) problem, i.e., a situation where we seek to find both the components and the mixing matrix given only the multichannel recorded signal. They differ in the constraints (lack of linear correlation vs. independence). Another class of BSS arises when we want to classify signals recorded in different conditions, e.g., in an EEG experiment, we aim to differentiate event-related potentials associated with specific stimuli. Below we describe a case relevant for brain-computer interfaces (BCI).

Let's consider a P300-BCI experiment. We have two experimental conditions. Let T mean the trials when the target stimulus was presented and NT the non-target ones when the standard stimulus was shown. We want to find such a linear combination of channels which maximizes the ratio of signals' variance in target relative to non-target trials. To do this we compute projection of the signal $\mathbf{x}(t)$ on a vector $\mathbf{w}$:

$$\mathbf{s}_w(t) = \mathbf{w}^\mathsf{T}\mathbf{x}(t) \tag{4.66}$$

Variance of the obtained signal is:

$$\text{var}(\mathbf{s}_w) = E[\mathbf{s}_w \mathbf{s}_w^\mathsf{T}] \tag{4.67}$$
$$= E[\mathbf{w}^\mathsf{T}\mathbf{x}(\mathbf{w}^\mathsf{T}\mathbf{x})^\mathsf{T}] \tag{4.68}$$
$$= \mathbf{w}^\mathsf{T} E[\mathbf{x}\mathbf{x}^\mathsf{T}]\mathbf{w} \tag{4.69}$$
$$= \mathbf{w}^\mathsf{T} \mathbf{C}_x \mathbf{w} \tag{4.70}$$

where $\mathbf{C}_x$ is the covariance matrix of signal $\mathbf{x}$. Finding the optimal vector $\mathbf{w}$ can be thus stated to maximize the ratio of variance in the T to NT trials. This ratio is known as the Rayleigh coefficient $J(\mathbf{w})$:

$$J(\mathbf{w}) = \frac{\mathbf{w}^\mathsf{T}\mathbf{C}_T\mathbf{w}}{\mathbf{w}^\mathsf{T}\mathbf{C}_{NT}\mathbf{w}} \tag{4.71}$$

The extremum of this coefficient can be found by computing the gradient of $J(\mathbf{w})$:

$$\nabla J(\mathbf{w}) = \frac{2\mathbf{C}_T\mathbf{w}(\mathbf{w}^\mathsf{T}\mathbf{C}_{NT}\mathbf{w}) - 2\mathbf{C}_{NT}\mathbf{w}(\mathbf{w}^\mathsf{T}\mathbf{C}_T\mathbf{w})}{(\mathbf{w}^\mathsf{T}\mathbf{C}_{NT}\mathbf{w})^2} \tag{4.72}$$

setting it to zero we get:

$$\mathbf{C}_T\mathbf{w}(\mathbf{w}^\mathsf{T}\mathbf{C}_{NT}\mathbf{w}) = \mathbf{C}_{NT}\mathbf{w}(\mathbf{w}^\mathsf{T}\mathbf{C}_T\mathbf{w}) \tag{4.73}$$

i.e.:

$$\mathbf{C}_T\mathbf{w} = \frac{\mathbf{w}^\mathsf{T}\mathbf{C}_T\mathbf{w}}{\mathbf{w}^\mathsf{T}\mathbf{C}_{NT}\mathbf{w}}\mathbf{C}_{NT}\mathbf{w} \qquad (4.74)$$

The obtained equation is a generalized eigenvetor problem. The scalar value $\lambda = \frac{\mathbf{w}^\mathsf{T}\mathbf{C}_T\mathbf{w}}{\mathbf{w}^\mathsf{T}\mathbf{C}_{NT}\mathbf{w}}$ is the generalized eigenvalue and $\mathbf{w}$ is generalized eigenvector. To find λ and $\mathbf{w}$ we can use MATLAB function [W, L] = eig(A,B). This function solves the generalized eigenvalue problems of the form $\mathbf{Aw} = \lambda\mathbf{Bw}$ giving a matrix $\mathbf{W}$ containing the eigenvectors $\mathbf{w}$ in the columns and a diagonal matrix $\mathbf{L}$ with the corresponding eigenvalues. The sought optimal vector $\mathbf{w}$ corresponds to the greatest eigenvalue.

MATLAB demo: estimation of source activity with BSS.

The LiveScript matlab/c4/Ch4_07_CSP.mlx demonstrates the application of Common Spatial Patterns method to estimate the activity of simulated sources.

MATLAB demo: application of CSP technique in the SSVEP-BCI.

The LiveScript matlab/c4/Ch4_08_SSVEP.mlx shows an application of the Common Spatial Patterns method to enhance the SSVEP signal hidden in the noisy data from a real BCI calibration session.

4.6.4   Multivariate Matching Pursuit (MMP)

In Sect. 3.4.2.2 a single channel version of matching pursuit algorithm was described. The MP algorithm decomposes the signal into waveforms g_γ (time-frequency atoms). Each atom is characterized by a set of parameters γ. In the case of Gabor functions (equation 3.115) these are $\gamma = \{u, f, \sigma, \phi\}$ where u—time translation, f—frequency, σ—time width, ϕ—phase. The MP algorithm can be extended into multivariate cases by introducing multichannel time-frequency atoms [133, 315]. A multichannel time-frequency atom is a set of functions $\mathbf{g}_\gamma = \{g_\gamma^1, g_\gamma^2, \ldots, g_\gamma^n\}$. Each of the g_γ^i functions is a univariate time-frequency atom. Let $\mathbf{x} = \{x^1, x^2, \ldots, x^n\}$ be the n channel signal. The multivariate matching pursuit acts in the following iterative way:

1. The null residue is the signal $\mathbf{R}^0\mathbf{x} = \{R^0x^1, R^0x^2, \ldots, R^0x^n\} = \mathbf{x}$.

2. In the k-th iteration of the algorithm a multichannel atom $\mathbf{g}_{\gamma_k}$ is selected that satisfies the optimality criterion.

3. The next residuum is computed as the difference of the current residuum and the projection of the selected atom on each channel. For channel i it can be expressed as: $R^{k+1}x^i = R^kx^i - \langle R^kx^i, g_{\gamma_k}^i\rangle g_{\gamma_k}^i$.

An interesting possibility in the MMP framework is that the optimality criterion can be model driven, i.e., it can correspond to the assumed model of signal generation. As an example we can presume that each given component of the measured multichannel signal results from the activity of an individual source, and that the propagation of signal (e.g., electric or magnetic field) from the source to the sensors is instantaneous. In such a case, the most straightforward optimality condition is that in a given iteration k the atom g_{γ_k} is selected which explains the biggest amount of total energy (summed over the channels) with the constraint that all the univariate atoms in individual channels have all the parameters identical, excluding the amplitude which varies topographicaly. Hence atom g_{γ_k} can be expressed as:

$$g_{\gamma_k} = \arg\max_{g_\gamma \in D} \sum_{i=1}^{n} |\langle R^n \mathbf{x}^i, g_\gamma^i \rangle|^2 \quad \text{and} \quad \forall_{i,j} \; g_\gamma^i = g_\gamma^j \tag{4.75}$$

Another example of optimality criterion can be formulated when the multivariate signal is obtained as an ensemble of realization of a process, e.g., multiple epochs containing the realizations of an evoked potential. In such a case the criterion could be analogous to the one given by (equation 4.75), but the constraint of the same phase for all univariate atoms can be relaxed.

5

Application to Biomedical Signals

5.1 Brain Signals

Among electrophysiological signals we can distinguish: signals generated by brain, heart and muscles. These connected with brain activity are: electroencephalogram (EEG), event- related potentials (ERP), electrocorticograms (ECoG), local field potentials (LFP). Since variable electric field generates magnetic field we will consider here also: magnetoencephalograms (MEG) and evoked fields (EF). Brain activity may be probed also by means of functional magnetic resonance imaging (fMRI) and functional Near Infrared Spectroscopy (fNIRS)

Electroencephalogram (EEG) is a record of the electric signal generated by the cooperative action of brain cells, or more precisely, the time course of neuronal extracellular field potentials generated by their synchronous action. EEG recorded in the absence of stimuli is called spontaneous EEG; a brain electric field generated as a response to an external or internal stimulus is called an event-related potential (ERP). EEG can be measured through electrodes placed on the scalp or directly on the cortex. In the latter case it is called an electrocorticogram (ECoG); lately also iEEG (intracranial EEG) abbreviation is used. Electric fields measured intracortically with electrodes implanted in the brain structures were named local fields potentials (LFP). The same electrodes (if small enough) can also record action potentials (spikes). The amplitude of EEG of a normal subject in the awake state, recorded with the scalp electrodes, is 10–100 µV. In the case of epilepsy, the EEG amplitudes may increase by almost an order of magnitude. Scalp potentials are mostly independent of electrode size due to severe space averaging by volume conduction between the brain and scalp. Intracranial potentials amplitude depends on electrode size (smaller electrode—higher potential) and may vary in size over four orders of magnitude.

A variable electric field generates a magnetic field as follows from the Maxwell equations. The recording of the magnetic field of the brain is called a magnetoencephalogram (MEG). The MEG's amplitude is less than 0.5 picotesla (pT), and its frequency range is similar to that of the EEG. Since electric and magnetic fields are orthogonal, radial sources (dipoles oriented perpendicular to the skull) are better visible in EEG and tangential sources in MEG. Due to geometry, radial sources hardly contribute to MEG, but the

DOI: 10.1201/9780429431357-5

advantage of this technique is that the head's structures weakly influence the magnetic field.

Applications of EEG

EEG found application in the investigation of the information processing in the brain and medical diagnosis. In particular, it helps solve the following medical problems and diseases:

- Epilepsy

- Brain tumors, head injury, stroke

- Psychiatric diseases

- Sleep disorders

- CNS disorders, e.g., cerebral anoxia, cerebral inflammatory processes, cerebral palsy

- Creutzfeld-Jacob disease, metabolic, nervous system disorders

- Developmental disorders

- Testing of psychotropic and anticonvulsant drugs

- Monitoring alertness, e.g., controlling anesthesia depth

- Testing sensory pathways

Another EEG application concerns the design of brain-machine interfaces for direct communication between brain and computer and then possibly with specific devices. In the investigation of brain processes, especially perception and cognitive functions usually ERPs are analyzed since they supply information on the brain's reaction to specific external or internal stimuli. ERPs are also used for testing the sensory pathways in psychiatric diseases and developmental disorders, e.g., in dyslexia diagnosis.

Applications of MEG

Magnetoencephalography (MEG), in which magnetic fields generated by brain activity are recorded outside of the head, is now in routine clinical practice throughout the world. MEG has become a recognized and vital part of the presurgical evaluation of epilepsy patients and patients with brain tumors. The MEG technique's significant advantage is that the magnetic field generated inside the brain is affected to a much lesser extent by the conductivities of the skull and scalp than the electric field generated by the same source. In the limiting case of the spherical head model, concentric inhomogeneities do not affect the magnetic field at all. In contrast, they have to be taken into account in the analysis of EEG data [209]. This property of MEG is very advantageous when localizing sources of activity. A current review showing an improvement in the postsurgical outcomes of epilepsy patients by localizing epileptic discharges using MEG can be found in [575]. Magnetic evoked fields (EF) are a counterpart of ERP; they are an effect of event-related brain activity which can be elicited by visual, auditory, sensory, or internal stimuli. Their analysis usually concerns the identification of the brain region responsible for their generation.

The MEG sensors are not directly attached to the subject's head, which implies that during the MEG measurement, the subject's head should not move with respect to the sensors. This requirement limits the possibility of long-session measurements or long-term monitoring.

5.1.1 Generation of Brain Signals

In the brain, there are 10^{11} nerve cells. Each of them is synaptically connected with up to 10^4 other neurons. Brain cells can also communicate through electrical synapses (gap junctions), transmitting current directly. The neurons' electric activity is manifested through action potentials and postsynaptic potentials (PSP). Action potentials occur when the electrical excitation of the membrane exceeds a threshold. As a result, the Na^+ ion channels open. The rapid inflow of Na^+ ions changes the polarization of the neuron's inside from about $-80\,mV$ to about $+40\,mV$. The following cell's repolarization is connected with the outflow of K^+ ions to the extracellular space. In this way, an action potential of a characteristic spike-like shape (duration about $1\,ms$) is created. The action potential's exact time course is shaped by the interplay of many ionic currents specific for the given neuron type.

Postsynaptic potentials are sub-threshold phenomena connected with the processes occurring on the postsynaptic membrane. When the action potential arrives at the synapse, it secretes a chemical substance called a mediator or transmitter, which causes a change in the permeability of the postsynaptic membrane of the target neuron. As a result, ions traverse the membrane, and a difference in potentials across the membrane is created. When the negativity inside the neuron is decreased, an excitatory postsynaptic potential (EPSP) is generated. An inhibitory postsynaptic potential (IPSP) is created when the negativity inside the neuron is increased, and the neuron becomes hyperpolarized. Unlike the action potential, the PSPs are graded potentials; their amplitudes are proportional to the amount of secreted mediator. Postsynaptic potentials typically have amplitudes of 5–$10\,mV$, and a time-span of 10–$50\,msec$. The amplitudes of many postsynaptic potentials have to be superimposed to obtain supra-threshold excitation.

The electrical activity of neurons, producing electric and magnetic fields conforming approximately to that of a dipole, generates currents along the cell membrane in the intra- and extracellular spaces. Macroscopic observation of these fields requires synchronizing the electrical activity of many dipoles oriented in parallel [440]. Indeed, pyramidal cells of the cortex are, to a large degree, parallel, and they are synchronized by common feeding by thalamo-cortical connections [352]. The condition of synchrony is fulfilled by the PSPs, which are relatively long in duration. The contribution from action potentials to the electric field measured extracranially is negligible [442]. EEG comes from the summation of synchronously generated postsynaptic potentials. The contribution to the electric field of synchronously acting neurons is approximately proportional to their number, and of that firing non-synchronously, as the square root of their number. For example: if an electrode records the action of 10^8 neurons (which is typical for a scalp electrode) and 1% of them are acting in synchrony, their contribution will be 100 times bigger than the contribution of asynchronously acting neurons since $\frac{10^6}{\sqrt{10^8}} = 100$.

Scalp potentials are mostly due to sources coherent at the scale of centimeters (roughly 10^8 neurons) with geometries encouraging the superposition of potentials generated by many local sources. Nearly all EEGs are believed to be caused by cortical sources [440] because of the following reasons: cortical proximity to scalp, relatively large sink-source separation in pyramidal cells constituting cortex dipoles, the property of the cortex to produce large cortical layers, high synchrony of pyramidal cells fed by the common thalamic input.

MEG is closely related to EEG. The source of the MEG signal is the same electrical activity of the synchronized neural populations, as is the case for EEG.

The processes at the cellular level (action potential generation) are non-linear. Non-linear properties can also be observed in the dynamics of well-defined neural populations. However, at the EEG level, non-linearities are the exception rather than the rule. The lack of the traces of non-linear dynamics in EEG was demonstrated using surrogate data [2, 565] and by comparison of linear and non-linear forecasting [65]. In the latter work, it was demonstrated that in LFP, recorded from implanted electrodes, the linear forecasting gave the same or better results than non-linear. During epileptic seizures, some EEG epochs of non-linear character were found [480]. Indeed during a seizure, large parts of the brain became synchronized.

Since non-linear methods are very sensitive to noise and prone to systematic errors, as was mentioned in Sect. 3.5.7, one should have a good reason to apply them to the EEG analysis. It is recommended to first check if the signal has a non-linear character and if the linear methods are not sufficient. Most functions can be locally linearized, and thus linear methods may also work quite well for non-linear signals.

5.1.2 EEG/MEG Rhythms

The problem of the origins of EEG rhythmical activity has been approached by electrophysiological studies on brain nerve cells and by the modeling of the electrical activity of the neural populations [168, 352, 653, 104]. It is generally accepted that cooperative properties of networks consisting of excitatory and inhibitory neurons connected by feedback loops play a crucial role in establishing EEG rhythms. Still, there is some evidence that some neurons' intrinsic oscillatory properties may contribute to the shaping of the rhythmic behavior of networks to which they belong. The oscillation frequency depends on the intrinsic membrane properties, the membrane potential of the individual neurons, and the synaptic interactions' strength.

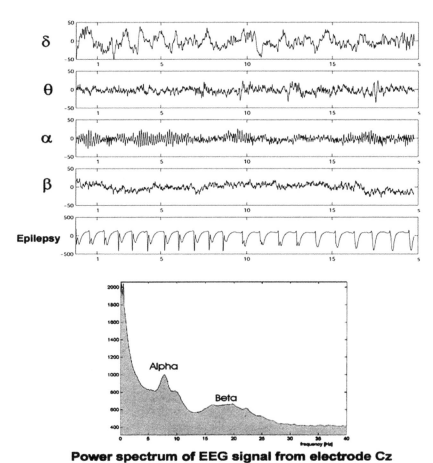

Power spectrum of EEG signal from electrode Cz

FIGURE 5.1
Example of EEG rhythms and EEG power spectrum.

The following rhythms have been distinguished in EEG (the same rhythms can be observed in MEG) [433]: delta (0.1–4 Hz), theta (4–8 Hz), alpha (8–13 Hz), beta (13–30 Hz), and gamma (above 30 Hz) (Figure 5.1). Frequencies of gamma rhythms measured by scalp electrodes usually did not exceed

40–60 Hz because of strong damping of high frequencies by head tissues. High frequency gamma (60–250) Hz, also called high frequency oscillations (HFO), were first observed in ECoG. They may be recorded as well by means of MEG and also in EEG by application of sophisticated modern recording techniques.

The contribution of different rhythms to the EEG/MEG depends on the subject's age and behavioral state, mainly the level of alertness. There are also considerable inter-subject differences in EEG characteristics. EEG pattern is influenced by neuropathological conditions, metabolic disorders, and drug action [433].

- Delta rhythm is a predominant feature in EEGs recorded during deep sleep. In this stage, delta waves usually have large amplitudes (75–200 μV) and show strong coherence all over the scalp.

- Theta rhythms occur in drowsiness and some emotional states, but they are also involved in cognitive and working memory processes. In the latter case, theta waves are associated with gamma activity. The occurrence of theta rhythm can also be connected with the slowing of alpha rhythms due to pathology. Theta waves are predominant in rodents; in this case, the frequency range is broader (4–12 Hz), and waves have a high amplitude and characteristic sawtooth shape.

- Alpha rhythms are predominant during wakefulness and are most pronounced in the posterior regions of the head. They are best observed when the eyes are closed, and the subject is in a relaxed state. They are blocked or attenuated by attention (especially visual) and by mental effort. Mu waves have a frequency band similar to alpha, but their shape resembles the Greek letter μ. They are prevalent in the central part of the head and are related to the motor cortex activity; namely, they are blocked by motor actions.

- Beta activity is characteristic for the states of increased alertness and focused attention, as was shown in several animal and human studies.

- Gamma activity is connected with information processing, e.g., recognizing sensory stimuli and the onset of voluntary movements. High-frequency gamma rhythms are linked to neuronal spikes. They are correlated with the degree of functional activation of the brain and are relevant to cortical computations.

The EEG is observed in all mammals, the characteristics of primate EEG being closest to the human. Cat, dog, and rodent EEG also resemble human EEG but have different spectral content. In lower vertebrates, electric brain activity is also observed, but it lacks the rhythmical behavior found in higher vertebrates.

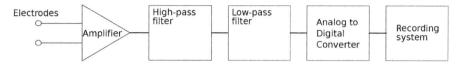

FIGURE 5.2
Block diagram of the recording setup.

5.1.3 EEG Measurement, Electrode Systems

EEG is usually registered through electrodes placed on the scalp. They can be secured by an adhesive or embedded in a special snug cap. A block diagram of the recording setup is shown in Figure 5.2. The first elements of the system are the electrodes and the differential amplifier. The quality of the measurement depends on the ratio between the electrodes' impedance and an input impedance of the amplifier. Modern amplifiers meet this requirement with an input impedance up to 10^{12} Ω. The standard specification that the electrode resistance be less than $5\,\text{k}\Omega$ may be slightly relaxed in such a case.

Prior to sampling, low pass anti-aliasing filters (Sect. 2.2.1.1) are used; high pass filters are applied to eliminate artifacts of lowest frequencies. EEG is usually digitized by an analog-to-digital converter (ADC). The digitization of the signal introduces the quantization error (Sect. 2.2.2), which depends on the covered voltage range and the number of the ADC's bits (for modern apparatus up to 24 bits). The sampling frequency ranges from 100 Hz for spontaneous EEG and several hundred Hz for ERP to several kHz to record short-latency ERP and intracranial activity.

Knowledge of the exact positions of electrodes is essential for interpreting a single recording and comparing results obtained for a group of subjects, hence the need for standardization. The traditional 10-20 electrode system defines 19 EEG electrodes positions (and two electrodes placed on earlobes A1/A2) related to specific anatomic landmarks, such that 10-20% of the distance between them is used as the electrode spatial interval. The first part of the derivation's name indexes the array's row—from the front of the head: Fp, F, C, P, and O (Figure 5.3). In the extended 10-10 system, electrode sites halfway between those defined by the standard 10-20 system were introduced. Nowadays 100 or even 200 electrode systems are used. The number of electrodes providing adequate space sampling is ≥ 128, since inter-electrode distance should be around 20 mm to prevent spatial aliasing [442].

EEG is a measure of potential difference; in the referential (or unipolar) setup, it is measured relative to the same electrode for all derivations. There is no universal consent regarding the best position of the reference electrode. Since currents coming from bioelectric activity of muscles, heart, or brain propagate all over the human body, the reference electrode has to be placed in proximity of the brain: on the earlobe, nose, mastoid, chin, or neck. The

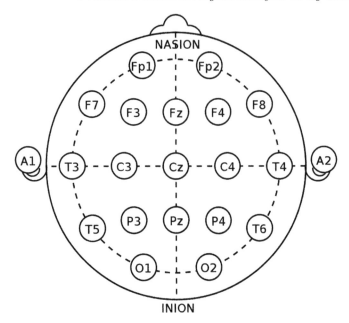

FIGURE 5.3
Electrode locations of the International 10-20 system for EEG recording.

reference electrode placed on the scalp center (which became popular) is a
bad choice when the phase information is considered or propagation of EEG
activity is examined. In the bipolar setup (montage), each channel registers
the potential difference between two particular scalp electrodes. Data recorded
in a referential configuration can be transformed into any bipolar montage for
the sake of display or further processing. The "common average reference"
montage is obtained by subtracting from each channel the average activity
from all the remaining derivations.

The Laplacian operator method is based on the fact that the Laplacian,
calculated as a second spatial derivative of a signal, offers information about
vertical current density and cancels common component due to volume conduction [440, 441]. Hjorth transform [223] is an approximation of the Laplacian
operator method. The Hjorth transform references an electrode E_0 to its four
closest neighbors:

$$J \sim 4E_0 - E_1 - E_2 - E_3 - E_4 \tag{5.1}$$

where J vertical current density at electrode 0, E_i—electric potential at electrode i.

There is no optimal reference recommended for all studies. For many applications, the Laplace operator method may be advised, but it requires a
dense grid of electrodes. For the techniques where the correlations between
channels are essential, as it is for MVAR, the reference electrode should be

"neutral" since the calculations present in the common average approach or Hjorth method introduce additional correlations between signals, which disturbs the original correlation structure.

5.1.4 MEG Measurement, Sensor Systems

The MEG measurement is technically much more demanding than the EEG measurement. This stems mainly from the fact that the magnetic fields generated by the brain are on the order of tenths to hundreds of femtotesla, which is orders of magnitude weaker than the Earth's magnetic field and fields generated by, e.g., moving ferromagnetic objects in the environment. The environmental artifacts are partially removed by a particular spatial setup of measurement coils and partly by magnetically shielded chambers (walls made of a few layers of μ-metal with high magnetic permeability separated by pure highly conducting aluminum layers allowing for the generation of Eddy currents). The practical measurement of the extremely weak fields by MEG is possible thanks to the superconducting quantum interference device (SQUID), which uses quantum effects in superconducting electrical circuits. The SQUIDs are magnetically coupled to the sensor coils, which in principle can be of three types: magnetometer, axial gradiometer, or planar gradiometer (Figure 5.4). In a gradiometer, the sensor coil's inverse winding eliminates the external magnetic fields slowly changing in space.

The sensor configuration of modern whole-head MEG systems consists of either magnetometers, axial or planar gradiometers, or even a combination of them, covering the whole head in a helmet-like fashion.

Magnetometers are superior to gradiometers concerning the possibility of recording signals from more in-depth sources. Apart from their different sensitivity to noise, these sensor configurations differ in the location of the signal's amplitude extremum with respect to its source. Unlike planar gradiometers for which the maximum signal is right above the source, magnetometers (as well as axial gradiometers) show extrema on both sides of the underlying source [209]. In the case of signal analysis on the level of sensors, planar gradiometer representation is easier in interpretation than axial gradiometers or magnetometers since the maximum of the averaged signal directly indicates the location of the activated region.

5.1.5 Elimination of Artifacts

An artifact in EEG can be defined as any potential difference due to an extracerebral source. The careful identification and elimination of artifacts are of utmost importance for EEG analysis. We can distinguish the artifacts of technical and biological origin. The first ones can be connected with: power supply, spurious electrical noise from engines, elevators, tramway traction, etc., bad electrode contact, or detachment. Another kind of artifact has its origin within the subject's body. The most common of them are eye blinks, eye

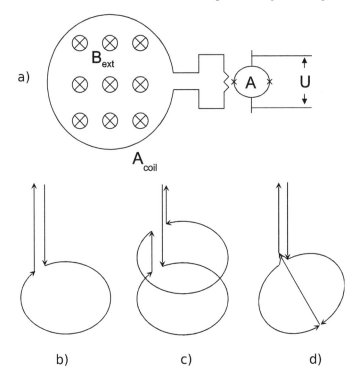

FIGURE 5.4
The schema of the MEG measurement system and types of coils. a) SQUID
coupled to the sensor coil: B_{ext} external magnetic field, A_{coil} area of the sensor
coil, A area of the SQUID coil, b) magnetometer coil, c) axial gradiometer coil,
d) planar gradiometer coil.

movements, muscle activity, and electrocardiogram (ECG). Eye movement
generates an electric signal called an electrooculogram (EOG) because of the
potential difference between the cornea and the back of the eye. Body and
head movements may induce muscles' electrical activity and slow potential
shifts, disturbing low-frequency EEG rhythms. Changes in scalp potentials
may also be caused by respiration, which causes scalp impedance changes and
electrodermographic activity (changes in skin potential connected with the
electric activity of sweat glands). Additionally, the secretion of sweat can affect the resistance of electrode contacts. Figure 5.5 shows examples of different
kinds of artifacts.

Simple filtering of signals to eliminate, e.g., muscle or power line artifacts,
relies on assuming that the EEG signal's interesting part is expressed in a
different frequency band (usually lower) than the artifacts. However, the low-
pass filtering usually disturbs the higher frequency EEG rhythms.

Artifact elimination, when performed on the segmented data, usually involves the rejection of the whole segment (epoch). The simplest methods were based on amplitude thresholds (overflow check), with different scenarios for setting the thresholds. The semi-automatic system for rejecting artifacts is implemented in EEGLAB http://sccn.ucsd.edu/eeglab/ [117]. It includes several routines to identify artifacts based on the detection of extreme values, abnormal trends, improbable data, abnormally distributed data, abnormal spectra. It also contains routines for finding independent components.

If continuous data are considered, the criteria concerning the segment length to be rejected should be set to add ample time before and after the artifact cut-off to avoid the sub-threshold but significant contamination. The elimination of artifacts in the overnight polysomnographic sleep recordings is of particular importance because of the massive volume of data to be inspected—whole night sleep recording as printed on standard EEG paper would be over half a kilometer long. There was an attempt to design the system of artifact elimination in sleep EEG in the framework of the EC SIESTA project [537, 7]. Still, this system was far from being robust and automatic, and it was tested only on selected 90-minutes epochs taken from 15 recordings.

Among the methods, which do not rely on specific models is a class of blind source separation algorithms (BSS), which includes PCA and ICA approaches (Sect. 4.6.1 and 4.6.2). They are based on the assumption that by projecting the data onto orthogonal (PCA) or statistically independent components (ICA), the components corresponding to artifacts can be isolated. Nowadays, the methods based on ICA are favored [366, 118]. The procedure of finding the independent components is described in Sect. 4.6.2. For multichannel EEG or ERP, the input matrix $\mathbf{x}$ (of k rows corresponding to the sensors) is used to estimate the unmixing matrix $\mathbf{D}^{-1}$ that minimizes statistical dependence of the outputs $\mathbf{s}$. We can expect that the activities generated by different sources, e.g., eye or muscle artifacts, will be isolated as separate ICA components. Brain activities of interest can be obtained by projecting selected ICA components back on the scalp. In the back-projection process, the components corresponding to artifacts can be eliminated by setting the corresponding rows of the source matrix $\mathbf{s}$ to zero. In this way, the corrected signals may be obtained.

Several methods addressed the problem of automated identification of artifactual components. Here, we mention two open-source EEGLAB plugins, as they are easy to use and generalize quite well. "Multiple Artifact Rejection Algorithm" (MARA) is based on results described in [645]. MARA uses a supervised machine learning model trained on expert ratings of 1290 components. It uses six features from the spatial, spectral, and temporal domains. Features were optimized to solve the binary classification problem "reject vs. accept". The plugin, together with a tutorial, is available from https://github.com/irenne/MARA. Another plugin we would recommend is

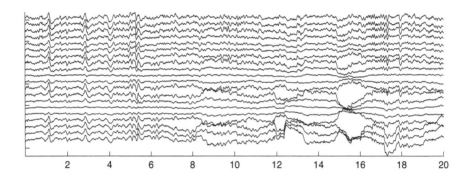

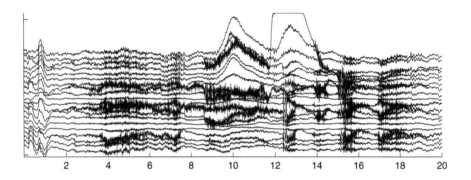

FIGURE 5.5
Examples of different kinds of artifacts. Upper panel, 20 s epoch of EEG contaminated with ECG artifact around seconds 1 and 3, and the eye movement artifact around second 16. Lower panel, muscle and movement artifact causing overflow around second 13.

ICLabel (https://github.com/sccn/ICLabel). It was developed at Swartz Center for Computational Neuroscience and Department of Electrical and Computer Engineering University of California, San Diego, and based on a supervised neural network model. The authors made use of a huge database of crowd-sourced labeled components. The instruction on installation and usage is available at https://sccn.ucsd.edu/wiki/ICLabel. Both plugins support visualization of the components and the probabilities that they belong to various artifact classes. Also, both can be used in GUI or command-driven fashion.

The BSS approach for artifact rejection, called second-order blind inference (SOBI), was proposed in [265]. ICA algorithm assumes that the components are statistically independent at each time point. SOBI considers the relationship between component values at different time lags and insists that these values be decorrelated as much as possible. The remaining correlated components can isolate highly temporally correlated sources, which, according to the authors, is a crucial feature for ocular artifact detection. Another advantage of SOBI put forth in [265] is that it uses only second-order statistics that can be more reliably estimated than higher-order statistics used in the ICA approach.

Please note that if the further analysis aims to find the causal relations (Sect. 4.3.2) between the signals from a multichannel set, the subtraction of the artifacts cannot be used as a preprocessing step since it disturbs the phase relations between the channels of the process. Moreover, one should keep in mind that the unmixing is usually not perfect. There may be some crosstalk between the components; thus, removing the artifact component may also remove some part of the EEG signal. Some residual artifact signals also may persist in the reconstructed "cleaned" signal.

5.1.6 Analysis of Continuous EEG Signals

Since the first recording of EEG from the human scalp by Hans Berger in 1929, clinicians and scientists have investigated EEG patterns by visual inspection of signals recorded on paper charts. Visual analysis is still in use, but the signals are displayed on a computer screen. The complicated character of EEG consisting of rhythmical components and transient structures has promoted early attempts for an automatic EEG analysis. Berger, assisted by Dietch [122], applied Fourier analysis to the EEG series. The first digital analysis of EEG was performed in the time domain [67]. The development of the theoretical background of spectral analysis in [50] promoted the development of signal analysis in the frequency domain, and in 1963 the first paper on digital spectral analysis of EEG appeared [626]. With the development of computer techniques and implementation of fast Fourier transform, spectral analysis became a basic digital tool for EEG analysis since the contribution of characteristic rhythms in brain activity has a significant impact on clinical diagnosis.

Due to its complexity, the EEG time series can be treated as a realization of a stochastic process. Its statistical properties can be evaluated by typical methods based on the theory of stochastic signals. These methods include probability distributions and their moments (means, variances, higher-order moments), correlation functions, and spectra. The estimation of these observables is usually based on the assumption of stationarity. While the EEG signals are ever-changing, they can be subdivided into quasi-stationary epochs when recorded under constant behavioral conditions. Based on several authors' empirical observations and statistical analysis, quasi-stationarity can be assumed for EEG epochs of 10 s length approximately, measured under constant behavioral conditions [433].

EEG signals can be analyzed in the time or frequency domain, and one or several channels can be analyzed at a time. The applied methods involve spectral analysis by Fourier transform (FT), autoregressive (AR) or MVAR parametric models, time-frequency, and time-scale methods (Wigner distributions, wavelets, matching pursuit).

5.1.6.1 Single Channel Analysis

Estimation of power spectra is one of the most frequently used methods of EEG analysis. It provides information about the rhythms present in the signal and can be easily and rapidly calculated using the fast Fourier transform (FFT). The maximum entropy power spectrum may be obtained utilizing the autoregressive model, which can be recommended for the EEG analysis. Notwithstanding the advantages of AR spectral estimate mentioned already (Sect. 3.3.2.2), the transfer function of the AR model represented as a system of parallel filters is compatible with the presence of EEG rhythms. The impulse response function of the AR model is a sum of damped sinusoids (Sect. 3.3.2.2), as was demonstrated by Franaszczuk [165], who introduced the method of parametrization of EEG rhythms called FAD (frequency, amplitude, damping). These parameters determined from the AR model give the frequency and amplitude of the rhythms without the need to plot the spectra. The parameter concerning the damping of rhythms is useful in the modeling of the EEG time series [652]. FAD method was used, e.g., for a study of the stability of electrocortical rhythm generators [407]. It was demonstrated that the AR model's transfer function is compatible with the transfer function of the physiological model of rhythms generation. The FAD parameters were related to the parameters describing neural populations' action connected in a feedback loop [63].

The spectral content of EEG changes during childhood and adolescence attains a stable character about 21 years of age and then may change in old age. There is a preponderance of low-frequency rhythms in the immature brain, and progressively the role of higher rhythms increases. The enhancement of the low-frequency activity in the adult EEG may be a symptom of neuropathology and also learning and performance difficulties. This observation lead [388] to introduce the so-called age quotient based on the contribution to the EEG, rhythms appropriate for each age (determined by regression analysis on a large pool of subjects). The discrepancy above 0.8 in age quotient (a ratio between the age estimated from EEG and the actual one) was considered a sign of possible pathology. In the computer-assisted diagnostic system for neurological and psychiatric diseases—Neurometric [258], the ratio of the slow rhythms delta and theta to the alpha rhythm computed for each derivation through spectral analysis was one of the markers of possible disturbances. Another feature in this system was the degree of asymmetry of spectral power between hemispheres.

The application of the psychoactive or anti-epileptic drugs causes pronounced changes in EEG spectral characteristics. The influence of drugs and the effects of medication are evaluated by calculating spectral power in the characteristic frequency bands and their ratios. In psychopharmacology, the topographical distribution of spectral power is also considered. Usually, the statistically significant differences between spectral power in frequency bands before and after medication are mapped. The average power values in each frequency band and their ratios are also useful markers in depression [93].

In some applications, not only rhythmic activity but also transient structures of EEG are of importance. In this case, the methods of non-stationary signal analysis have to be used. The examples of such transients are sleep spindles or epileptic spikes, which are relevant, e.g., for assessment of the influence of sleep-inducing or anti-epileptic drugs. The optimal method for the detection of transients occurring randomly in signals is matching pursuit. The advantages of the technique in the estimation of the influence of sleep-inducing drugs was demonstrated in [135], for identification of spindles in [673], and detection of epileptic spikes in [134].

5.1.6.2   Mapping

One of the commonly used methods of presenting multichannel brain activity is the topographical display of signals features called mapping. A map may help to make a direct comparison between the topographic distribution of EEG features and an anatomic image obtained, e.g., by the tomographic brain scan. Three types of features are most commonly mapped for clinical applications:

• Direct variable such as amplitude.

• Transformed variables such as total spectral power or relative spectral power in some frequency band.

• The result of a statistical test applied to a given EEG feature.

The amplitude values of the signals or the spectral power in selected bands are frequently displayed. Spatial interpolation between the discrete electrode positions is necessary to obtain a map composed of hundreds of pixels. Different interpolation algorithms are possible, ranging from the simple N-nearest-neighbors electrode algorithm to the more complex spline approximation. The advantage of spline interpolation methods is the easy computation of the second spatial derivative and, thus, the application of the Laplacian operator. In MATLAB, the interpolation of maps can be done, e.g., with the `interp2` function, which supports the nearest neighbor, linear, and cubic spline interpolation. For more sophisticated spline interpolations, the Spline Toolbox may be useful.

One should not assume that the maximum of signal power visible on the map corresponds to the underlying source since it can come from the superposition of coherent sources' activity. The character of the map is strongly dependent on the reference. As was already mentioned, there is no optimal reference system. In the case of mapping, one has to be aware of this limitation. The common average reference system is liable to ambiguities when some of the electrodes pick up similar signals, e.g., occipital alpha rhythm may appear in frontal derivations, and eye movement artifacts may affect posterior electrodes [471].

The recommended representation involves surface Laplacians (Sect. 5.1.3). However, a reliable computation of surface Laplacian requires at least 64 electrodes, and adequate spatial sampling is obtained for more than 128 electrodes. Therefore, quite frequently, an approximation of the Laplacian operator by Hjorth transform is applied. Results obtained by the application of the Laplacian operator may be further improved by deblurring. Deblurring relies on a mathematical model of volume conduction through the skull and scalp, which is used to project scalp-recorded potentials to the vicinity of the superficial cortical surface [328].

Mapping is often performed for the sake of comparison, e.g., to detect changes connected with medication or to find out the possible difference between groups. The maps are usually compared on pixel-by-pixel bases. In case when the distribution of the mapped values is normal, or it can be transformed into an approximately normal one by Box-Cox transformation [66], the t-statistics can be used to differentiate between maps. The second assumption of t-test is the homoscedasticity (the assumption that the variances in both compared groups are equal). In cases where the variances can be different, one should use the Welch's t test [634]. For comparison of values that cannot be transformed into the normal distribution, non-parametric tests should be used. The problem of choice of the appropriate test is discussed in Sect. 2.5.2. To make statistical inference on the maps pixel by pixel, one has to consider the multiple comparison problem. This problem and its possible solutions are discussed in Sect. 2.5.3.

5.1.6.3   Connectivity Analysis of Brain Signals

In brain research, the problem of the determination of connectivity structure has been a subject of intense studies in the last years. We can distinguish anatomical, functional, and effective connectivity. The anatomical connectivity is a pattern of anatomical links. The functional connectivity represents statistical interdependence among brain structures. It does not provide information concerning the inference about the directed coupling between two brain regions. The effective connectivity designates a causal influence in the activity of one brain site over another. It refers explicitly to the impact that one neural system exerts over another, either at a synaptic or population level [170]. Effective connectivity may be assessed by model-driven or data-driven methods. Friston [170] developed the best-known model-driven approach—dynamic causal modeling (DCM). In the DCM framework, nonlinear state-space models are specified, fitted to data, and selected based on the Bayesian model comparison. The models are formulated in terms of differential equations describing the interaction of neural populations generating time series. Parameters in these models quantify the directed influences or effective connectivity among neuronal populations. However, the DCM approach requires a priori assumptions concerning the model's structure. Herein we are dealing with time series analysis; therefore, in the estimation of functional connectivity, we shall concentrate on data-driven approaches, which do not require a priori assumptions concerning systems generating investigated signals.

Functional and effective brain connectivity can be probed by electroencephalography (EEG), magnetoencephalography (MEG), intracranial EEG recordings, fMRI, and functional infrared spectroscopy (fNIRS), each with different spatial and temporal scales. Functional connectivity is commonly assessed by means of cross-estimators described in Sect. 4.1, by means of nonlinear methods (Sect. 4.4), or by multivariate methods (Sect. 4.3.2).

Interdependence between two EEG/MEG signals can be found by a cross-correlation function or its analog in the frequency domain—coherence. In EEG studies, not correlations but rather coherences are usually estimated since they provide the information about the rhythm synchronization between channels. Conventionally, ordinary coherences calculated pair-wise between two signals have been used for EEG as well as for ERP studies. However, for the ensemble of channels taken from different derivations, the relationship between two signals may result from the common driving from another site. This is often a case for EEG since the signals recorded from the scalp are strongly interdependent. The estimation not only of ordinary but also of partial and multiple coherences is recommended [166] to obtain a complete pattern of coherence structure of multichannel EEG.

In Figure 5.6 ordinary (bivariate), partial, and multiple coherences are shown for sleep EEG stage 2 [273]. Multiple coherences have high amplitudes for all derivations and whole frequency range, which means that the system is strongly interconnected. Ordinary coherences decrease monotonically with distance, which is a rather trivial observation. Partial coherences show only direct connections between channels, and the information provided by them is much more selective. The computation of ordinary, partial, and multiple coherences can be easily achieved in the framework of MVAR.

Cross-estimators: correlation, coherence, and bivariate non-linear estimators of dependence between signals are not able to find reciprocal connections, are prone to common drive effect, and yield only functional connectivity. The direction of interactions may be found, e.g., from the phase difference, but not without ambiguities (as explained in Sect. 4.3.1). However, the direction does not imply causality. The effective connectivity, namely, the causal coupling between signals, may be found using measures based on the Granger principle (Sect. 4.3.2), e.g., DTF and PDC. A comparison of the different directionality estimation methods for the same experimental data, namely EEG recorded in an awake state, eyes closed, is illustrated in Figure 5.7. It is known that the primary sources of activity in such a case are located in the posterior parts of the cortex, and some weaker sources occur more frontally. One can see that DTF revealed the most explicit picture indicating the localization of

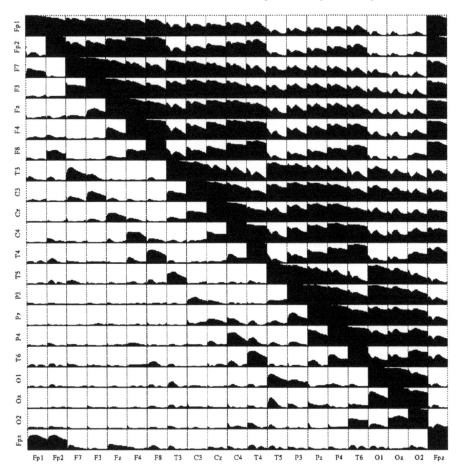

FIGURE 5.6
Coherences: ordinary—above diagonal, partial—below diagonal and multiple—on the diagonal for stage 2 of Sleep EEG. From [273].

the sources of activity (back of the head and some frontal localizations). The anatomical tracts determine the dDTF, as it shows only direct transmissions. In the bivariate case, the picture was quite confusing. Namely, one can even observe the reversal of propagation, specifically from C3 to P3 instead of P3 to C3. The propagation structure obtained using PDC shows rather the sinks, not the sources, so the results are more challenging to interpret. This property results from the normalization of the measure (Sect. 4.3.2.4).

Inspecting Figure 5.7 it is easy to see that, in the case of scalp EEG, it is not so crucial to use an estimator showing only direct (not cascade) flows. It is more important which parts of the cortex communicate than the exact "wiring" scheme. Therefore for estimation of transmissions of EEG registered

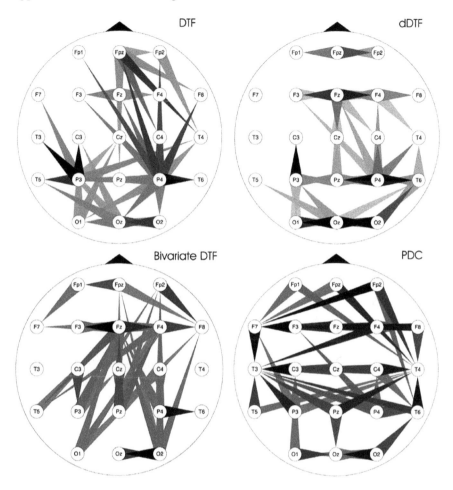

FIGURE 5.7
EEG transmission in awake state, eyes closed obtained by means of different methods.

by scalp electrodes, DTF is recommended. dDTF should be used for the intracranial electrodes, where the direct connections between brain structures are important [300].

DTF found application in different fields of brain research, e.g., sleep studies [273, 47], in epilepsy research [163, 398], finding LFP propagation during locomotion in behaving animals [299], in the study of emotional states [654] and in the diagnosis of psychiatric disorders, e.g., Alzheimer disease [58]. The examples of time-varying DTF (SDTF) applications to EEG and ECoG will be given in Sect. 5.1.7.4.

PDC has been used, e.g., in [149] for the study of thalamic bursting in rats during different awake behavioral states, for investigation of epileptic activity in [28], for a study of disrupted patterns of connectivity in schizophrenia [607], for fMRI based study of connectivity during language processing and EEG based BCI [175]. The properties of PDC were considered in [19] and [583]. The time-varying version of PDC was introduced in [334].

5.1.6.4   Influence of Volume Conduction on Connectivity Measures

Neural activity is transferred through neural tracts; however, oscillating neurons generate electromagnetic fields, which propagate in the brain and may influence the assessment of connectivity. This effect is called volume conduction. The electromagnetic field propagates with the speed of light, so it does not produce phase differences between sensors. The connectivity estimators based on phase differences such as DTF and PDC have no zero phase components, so they mitigate the influence of volume conduction to a large degree. There are some bivariate connectivity measures, such as imaginary coherence [437] or phase lag method, which are immune to zero lag transmissions. However, they suffer from the common drive effect, which is much more disturbing than the effect of volume conduction.

Although in the case of DTF and PDC, the influence of volume conduction is mitigated, mixing cortical sources activity may generate correlated residual noise processes, which violates the assumption of uncorrelated residuals of the MVAR model. Nevertheless, the spatial range of the effect is limited. We may approximate cortical source activity by a dipole layer, and the field of electric dipole decays relatively fast. Nunez and Srnivassan [442] estimated theoretically and experimentally the decay with the distance of coherences due solely to volume conduction. The results showed that at the distance of 7 cm (roughly the distance between electrodes in the 10–20 system), these coherences were close to zero. Therefore, we may assume that mixing cortical source activities due to volume conduction doesn't influence the DTF results significantly. Indeed in practice, results of DTF analysis based on scalp recordings correspond very well with known neuroimaging, intracranial and electrophysiological evidence, including topographic accuracy. Several examples may be found in [360].

Some authors argue that the signals recorded by scalp electrodes should be projected to the source space to remove volume conduction effects before calculating the connectivity pattern. However, the procedure requires solving the inverse problem, which is non-unique and needs assumptions concerning the nature and presumed localization of the sources. Several constraints limit the accuracy of the solution, and the applied methods do not undo mixing thoroughly. Moreover, the computation procedures may change the phases between signals and disturb in this way connectivity structure.

In the preprocessing the brain signals before DTF/PDC applications, care should be taken not to disturb the phases. Therefore no Hjorth transform or Laplacian or common average reference may be used. The application of ICA as a preprocessing step is ruled out. Removing eye artifacts by ICA is not recommended since it may disturb the phase structure between channels.

A comparison of connectivity patterns obtained by ffDTF from signals recorded by scalp electrodes and from presumed sources computed using sLoreta software [460] in three epochs of motor task, i.e., before, during and after movement, [360] showed that for signals from the reconstructed sources, there were no significant differences in connectivity between the three epochs. In contrast, for scalp signals, the desynchronization in the initial epochs and resynchronization after movement were observed as may be expected [475]. The dynamic evolution of propagation during a finger movement task obtained from scalp electrodes by SDTF demonstrated the topographic accuracy of SDTF method, and it is available at `http://fuw.edu.pl/~kjbli` in the form of animation.

We may conclude that the effect of volume conduction in DTF and PDC is largely mitigated, and the projection to the "source space" is not needed.

5.1.6.5 Graph Theoretical Analysis

The application of Graph Theoretical Analysis (GTA) to the brain connectivity measures was inspired by the paper of Watts and Strogatz [630], which introduced the concept of "small world" (SW) organization of networks. GTA-SW formalism assumes the network consisting of nodes (vertices) and edges—connections between them. Every node of the network should be connected with every other node by at least one path. The weights of edges are given equal values. The most commonly used parameters characterizing the network structure were:

cluster coefficient (CC) CC is a measure of the local interconnectedness of a graph. It is a fraction of the neighbors of a node that are also neighbors of each other.

characteristic path lengths (PL) PL is the shortest path between two vertices defined as the number of traversed vertices.

SW networks are characterized by a high CC and a short PL. These parameters are usually normalized to the reference parameters obtained from equivalent random networks. The network is considered a "small world" type if the ratio of the normalized CC to PL is greater than one. The other parameters frequently used to characterize networks are the node degree (the number of edges connected to a given vertex), the average number of edges for vertex, betweeness centrality (the number of the shortest paths that pass through the given vertex), and the global efficiency. Global efficiency was introduced as an inverse of PL, since for nodes that have no connecting path, PL is equal to infinity. We will call the above-described formalism GTA-SW in distinction to other more advanced graph analyses.

GTA-SW is a commonly used approach for post-processing connectivity results obtained by bivariate methods applied to brain signals: EEG, MEG, and BOLD. Nodes are often assigned to sensors. In fMRI, they may be ascribed to voxels, so the recording regime determines network structure.

Although GTA-SW analysis proved to be a convenient technique in several applications, its use for post-processing brain connectivity results raises serious doubts. In the last years, accumulated criticism concerning GTA application may be noticed [60, 456, 222, 224, 271]. Indeed, the results of GTA are critically biased by the following factors:

Nodes choice Large number of nodes makes the network noisy and difficult to manage. In case of a very dense distribution of nodes, they may record highly correlated activity. The problem is especially important in fMRI, where the number of voxels is usually very high; in this case, a brain parcellation—a division of the brain into subunits or regions of interest (ROIs)—is generally applied. The definition of ROIs is challenging since it may influence the network properties heavily.

Setting of connection thresholds The SW parameters are sensitive to thresholds set on connectivity values. Connection thresholds are quite often arbitrarily set to low values, e.g., [613, 333, 567], to meet the assumption that all nodes have to be connected by some path. It is a common practice that for comparison of networks, the threshold values are individually adapted so that the studied networks had the same average number of edges per vertex, e.g., [333]. The GTA parameters are quite sensitive to the choice of thresholds. While for high threshold values, the networks seem to be organized into modules with self-similar properties, the addition of a few more weak connections can make the network a "small-world".

Normalization against random networks The parameters CC and PL are normalized by the ones obtained from random networks to assess the reliability of the SW structure. A random network's construction procedure usually consists of random rewiring of connections, but the number of nodes, links, and degree distribution is typically preserved. Therefore some features of the investigated network are conserved in the reference network. In effect, it shows SW characteristics similar to the ones of the original time series [224].

Method of connectivity estimation The most crucial problem of GTA-SW formalism is connected with input values. Usually, the results of bivariate connectivity measures are used, which are biased by multiple spurious connections due to the common drive effect. Obtained networks are very dense and disorganized. Giving equal weights to all connections leads to a further blurring of a pattern, besides significant information contained in weights is lost [59].

Sometimes the information differing investigated networks are confined in the weights, not in their structure. An example may be the distinction between connectivity patterns in Alzheimer's disease. A multivariate approach showed that the connection strengths' decrease was the most important discrimination feature yielding good classification rates [58]. GTA-SW approach gave divergent results concerning SW parameters, e.g., [567, 566]. Divergent results were also obtained by GTA-SW formalism in the case of schizophrenia [516].

It is difficult to find congruence with imaging and electrophysiological evidence inspecting studies based on bivariate measures and graph theoretical analysis. E.g., in the EEG studies of sleep in which connectivity was estimated using synchronization likelihood and quantified by GTA-SW, no significant changes in the network parameters were found for different sleep stages [333, 156], contrary to the results obtained by the multivariate method (see Sect. 5.1.6.6).

Another example is a finger movement experiment recorded with MEG [35] in which GTA-SW followed wavelet analysis and correlations. No statistically significant changes in GTA parameters depending on frequency were found between rest and movement. Furthermore, no lateralization was observed contrary to the known fact that the movement task is connected with pronounced changes in the topographic and spectral characteristics of brain activity [472, 183, 312].

Considering the pitfalls present in the application of the graph approach to quantification of connectivity and the outcomes of the analysis involving the SW concept, we may question the value of reducing a brain complex network to a small number of summary metrics such as, for example, the small-worldness index [234]. Such an approach is seriously limited since it does not permit localization of an effect on specific circuits. Moreover, conventional GTA-SW analysis does not consider the directedness of activity propagating in the networks—a factor crucial for understanding information processing in the brain.

These limitations may be overcome by utilizing multivariate connectivity measures, which are free of common drive effect. Consequently, the connectivity patterns become relatively sparse and show clear-cut structures. Furthermore, Granger-causality-based multivariate methods provide information on directed connectivity and strengths of causal coupling. There is evidence, confirmed by numerous studies, that the connectivity patterns obtained using multivariate methods based on Granger causality are compatible with the known imaging, electrophysiological, and anatomical evidence, e.g., [273, 183, 312, 52, 55, 338].

Neural networks obtained using multivariate methods are too sparse (disconnected) to apply the GTA-SW formalism, notwithstanding its limitations mentioned above. In particular, CC and PL can hardly be assessed because of the small number of connections yielded by multivariate methods.

The quantification of the sparse directed networks may be achieved by more advanced GTA methods assessing the networks' community structure, i.e., by assortative mixing [428, 429]. In this formalism, first regions of interest (ROIs) are defined, which can be found by clustering procedures or designed on the grounds of anatomical or physiological evidence. The vertices of the network correspond to the ROIs encompassing several sensors. Next, the connectivity matrix $\mathbf{E}$ is defined as a fraction of edges in a network that connects a vertex of type k to one of type l. Indexes k and l do not refer to the sensors but to the modules (ROIs) defined in the framework of assortative mixing. For the modules corresponding to the regions of interest (ROIs), functions F quantifying intra channels connections are calculated. The matrix $\mathbf{E}$ describing the connectivity pattern is expressed by the formula:

$$E_{kl} = \sum_{\substack{n \in \text{channels} \\ \text{in module } k}} \sum_{\substack{m \in \text{channels} \\ \text{in module } l}} F(n \to m) \tag{5.2}$$

Applying the formalism of assortative mixing, the elements of matrix E_{kl} are represented by the connectivity measures F either inside the module (E_{kk}), or between the modules (E_{kl}). In an undirected network matrix $\mathbf{E}$ is symmetric in its indices, i.e., $E_{kl} = E_{lk}$, while in directed networks it may be asymmetric. Mixing is highly assortative when the diagonal elements of the matrix E_{kk} are significantly higher than the off-diagonal ones. It corresponds to the situation of strongly connected modules, with weaker bonds between them. An example of the application of assortative mixing to the analysis of EEG connectivity during working memory task is given in Sect. 5.1.7.4.

5.1.6.6 Sleep EEG Analysis

Sleep EEG reveals a characteristic pattern called sleep architecture. It is classically described by division into stages, originally defined in [502] (R&K) as:

- stage 1 (drowsiness),

- stage 2 (light sleep),

- stage 3 (deep sleep),

- stage 4 (very deep sleep),

- REM (dreaming period accompanied by rapid eye movements).

In 2007, the American Academy of Sleep Medicine (AASM) updated the sleep scoring rules and, since then, keeps making adjustments (for the current version, consult The AASM Manual for the Scoring of Sleep and Associated Events at `https://aasm.org/clinical-resources/scoring-manual/`). The AASM scoring manual recognizes four sleep stages:

- Stage N1, formerly stage 1 sleep,

- Stage N2, formerly stage 2 sleep,

- Stage N3, formerly stages 3 and 4 sleep,

- Stage R sleep, formerly stage REM sleep.

The differentiation of the sleep stages involves measurement of several signals: not only EEG, but also electrooculogram (EOG), muscular activity—electromyogram (EMG), and respiration, sometimes also a measurement of blood flow, electrocardiogram (ECG), and oxygen level in blood. This kind of recording is called a polysomnogram. EOG is measured using electrodes placed at the canthi of the eyes. The EOG and EMG help distinguish the REM state connected with dreaming. During the night, there are usually four or five cycles of sleep consisting of a period of NREM (non-REM) sleep, followed by REM. The sequence of sleep stages is usually illustrated in the form of the hypnogram (hypnogram is shown in Figure 5.11). The recognition of stages is based on the contribution of the different rhythms and the occurrence of characteristic signal transients absent in wake EEG, namely: sleep spindles, vertex sharp waves, and K-complexes.

Sleep spindles are rhythmic waves of frequency 11–16 Hz and duration 0.5–2 s; characteristic waxing and waning of amplitude are not always observed. Vertex wave is a compound potential: a small spike discharge of positive polarity preceding a large spike followed by a negative wave of latency around 100 ms and often another small positive spike.

The K-complex, in the AASM manual, is defined as an EEG event consisting of a well-delineated negative sharp wave immediately followed by a positive component standing out from the background EEG with a total duration above 0.5 s, usually maximal in amplitude over the frontal regions. K-complexes and vertex waves are presumably a kind of evoked responses. Sleep stages may be briefly characterized as follows:

Stage N1 (drowsiness) is associated with a decrease of the alpha rhythm (below 50% of an epoch), rhythms in 2–7 Hz frequency band (more than 50% of an epoch), and low amplitude rhythms of 15–25 Hz band. Vertex sharp waves may occur toward the end of stage N1. Sleep spindles, K complexes, and rapid eye movements are never a part of stage N1 sleep. Slow rolling of eye movement may be observed. Submental EMG tone is relatively high; breathing becomes shallow, blood pressure falls, heart rate becomes regular.

Stage N2 is characterized by the predominant theta activity, appearance of K-complexes, and sleep spindles; the slow-wave activity (SWA) should be less than 20% of the epoch.

Stage N3 is scored when more than 20% of the epoch contains delta waves (SWA) of peak-to-peak amplitude greater than 75 μV.

Stage R is characterized by a decrease of EEG amplitude, occurrence of faster rhythms, rapid eye movements, and a loss of muscular activity. Spectral characteristics in REM sleep is polyrhythmic, and basing on EEG only, it is difficult to distinguish R from stage N1.

The sleep pattern changes very much during childhood and adolescence. For newborn babies, REM takes most of the sleep time, and in young children,

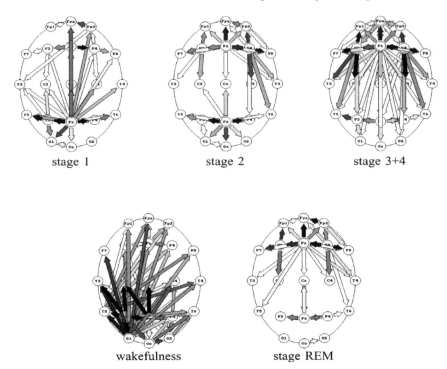

stage 1   stage 2   stage 3+4

wakefulness   stage REM

FIGURE 5.8
The propagation of EEG activity during overnight sleep averaged over nine subjects, estimated by means of DTF integrated in the 0–30 Hz band. Arrows are drawn for DTFs greater than 0.06. Their intensities are reflected by the shade of arrows. From [273].

only REM and non-REM stages can be distinguished. Diminution of deep, slow-wave sleep and increase in wakefulness continues throughout the entire life span after 1 year of age. In old age, the contribution of stage N3 decreases markedly, and the first REM stage appears later in the night.

There is evidence that when sleep becomes deeper, the sources that drive EEG activity move from the posterior regions of the head (prevalent during the awake state with eyes closed) to the centro-frontal regions. In Figure 5.8, the results of the study of activity propagation during overnight sleep obtained by DTF are shown [273]. The patterns of EEG transmissions averaged over nine subjects show a high degree of consistency. The most complicated pattern is visible for wakefulness since subjects differ in exact locations of sources. However, they are always located in the posterior and occipital region. In stage N2, the main sources in frontal and posterior regions correspond to two topographically distinct sources of sleep spindles of different frequencies. In stage N3 (3+4 in Figure 5.8), the primary source of activity is localized over the corpus callosum—the structure wherefrom the nerves are spreading and are conducting low-frequency coherent activity over the whole brain.

A hypnogram describes sleep macrostructure and relies on the time axis division into fixed time epochs (20 or 30 s)—this naturally implies some limitations. In the alternative approaches treating sleep EEG as a continuous process, its microstructure is described as an evolution of spindles and SWA activity or occurrence of transient arousals. They are defined as an abrupt shift in EEG frequency, including theta, alpha, and frequencies greater than 16 Hz, but not spindles. A certain number of spontaneous arousals seems to be an intrinsic component of physiological sleep, but their frequent occurrence may be connected with respiratory sleep disorders, nocturnal myoclonus, and other clinical conditions. The method providing the continuous description of sleep, called the cyclic alternating pattern (CAP), was proposed in [510].

Sleep EEG analysis finds application in the diagnosis of several disorders: insomnias, somnambulism, narcolepsy, epilepsy (some epileptic conditions appear mainly during sleep), depression, dementia, drug withdrawal. In sleep scoring, usually, central EEG derivations C3 or C4 (referenced to electrodes A1, A2 placed at mastoids), and additionally, two EOG channels and an EMG channel are considered. Sometimes occipital electrodes are also taken into account.

The evaluation of sleep pattern connected with the construction of hypnogram, or finding arousals, is a very tedious and time-consuming job involving analysis of a large volume of polysomnographic recordings. Therefore as early as the 1970s, attempts were made to design a system for automatic sleep scoring. The early works dealing with this problem included hybrid systems [558] and pattern recognition techniques [382]. Later, expert systems [500, 307] and artificial neural networks (ANN) found applications in sleep analysis [506]. ANNs with input coming from an AR model were proposed for sleep staging in [457]. The problem was also attacked by model-based approaches [284] and independent component analysis [392]. Wavelet analysis was used for the detection of transient structures in EEG [256] and for the identification of microarousals [79]. The review of the studies concerning digital sleep analysis may be found in [466].

There are several problems in automatic sleep analysis; one of them is insufficient consent between electroencephalographers concerning the classification of stages [308, 438]. Another obstacle is the difficulty in identifying transient structures such as sleep spindles, K-complexes, and vertex waves. The detection of these waveforms is best performed by the methods working in the time-frequency domain. The method providing parametrization in the time-frequency space is wavelet analysis, which was applied, e.g., in [256]. However the parametrization of signal structures in wavelet analysis is bound by the predetermined framework (usually dyadic) in the time-frequency space. Therefore, a given structure may be described by wavelet coefficients belonging to several scale levels in time and frequency (Figure 5.9). The description is not sparse and is not related to the physiological characteristics of the

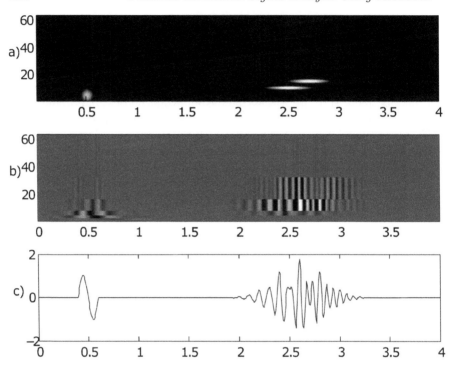

FIGURE 5.9
Comparison of time-frequency representation of sleep structures of a signal (c) consisting of idealized K-complex and two spindles. a) and b) representations obtained using MP and WT (Coiflet3), respectively. Note that MP separates the two spindles overlapping in time and close in frequency.

transient. Matching pursuit is the method, which is maximally adaptive, supplies the rhythmical components and transients' parametrization in the same framework, and characterizes the signal structures using parameters compatible with the clinical definitions. Therefore, we shall discuss this method to a larger extent. The time-frequency distribution of EEG from different sleep stages obtained using MP are shown in Figure 5.10; data structures corresponding to rhythmic and transient components can be easily distinguished. In MP, the signal structures are characterized by the frequency, time-span, time occurrence, amplitude, and phase. It allows constructing filters, based explicitly on the clinical definition, for finding in EEG specific structures of both transient and oscillatory nature. In fact, MP is the only method that allows for the direct application of criteria concerning the percentage of specific activity in the given epoch. One of the first applications of MP to biological signals concerned sleep spindles [62]. High accuracy identification and parametrization of sleep spindles using MP allowed for the distinction of two classes of spindles of different frequency ranges and to-pographic localization [673]. In the following work, the time evolution of SWA and spindle activities was investigated, and an inverse relationship in their magnitudes was found [130]. Macro and microstructure of sleep EEG were considered in [372] and [373]. The identification of K-complexes, detection of deep sleep stages (3 and 4), based directly upon the classical R&K criteria, continuous description of slow-wave sleep, fully compatible with the R&K criteria, and detection of arousals were presented in the framework of the same unifying MP approach (Figure 5.11).

Finally, the results of the works mentioned above were gathered in an open system for sleep staging, based explicitly on the R&K criteria [374]. The system started with detection and parametrization of relevant waveforms and then combining these results (together with information from EMG and EOG signals) into a final decision assigning epochs to sleep stages. Automatic sleep staging was performed in a hierarchical way presented in Figure 5.12. In the first step, each 20-s epoch is tested for muscle artifacts in EEG or EMG derivation, which indicate the movement time (MT). If at least one of the analyzed derivations (C3, C4, or EMG) exceeded a corresponding threshold in more than 50% of the 20 s epoch, the epoch is scored as movement time (MT). In the second step, the algorithm detects slow-wave sleep stages 3 and 4 by applying fixed 20% and 50% thresholds to the amount of epoch's time occupied by slow waves. In the following step, the algorithm detects sleep spindles and K-complexes. If at least one sleep spindle or K-complex occurs in the 20 s epoch, and less than 75% of the epoch is occupied by alpha activity, the epoch is scored as stage 2. If alpha activity occupies above 75%, EOG and EMG signals of this epoch are examined to distinguish stage REM from wake. The performance of the system was evaluated on 20 polysomnographic recordings scored by experienced encephalographers. The system gave 73% concordance with visual staging—close to the inter-expert concordance. Similar results were obtained for other systems reported in the literature [466]. However, these expert systems were tuned explicitly for maximizing the concordance with visual scoring; hence, their power of generalization was doubtful. They were usually based on a black-box approach, so their parameters were difficult to relate to the properties of EEG signals observed in visual analysis. Also, most of these systems are closed-source commercial solutions. On the contrary, a bottom-up approach presented in [374] is flexible and directly related to the visual analysis. The parameters of the system can be easily adapted to the other criteria concerning sleep stage classification.

Analysis of specific structures occurring in sleep EEG may be helpful in medical diagnosis. It was reported recently that: spindle coherence and phase amplitude coupling may be altered in case of disorders such as post-traumatic stress disorders, memory decline and schizophrenia [97].

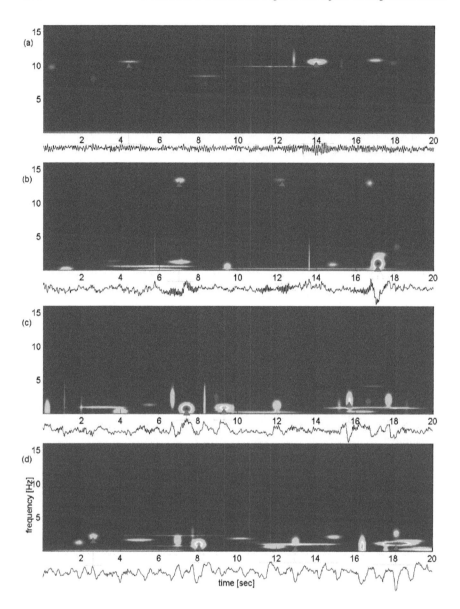

FIGURE 5.10

Time-frequency maps of EEG signal energy for 20-second epochs from different sleep stages shown above the corresponding signals. a) Wakefulness, alpha waves marked by arrows; b) stage 2, two spindles and K-complex marked; c) stage 3, two slow waves, and a K-complex marked; d) stage 4, slow waves marked. From [372].

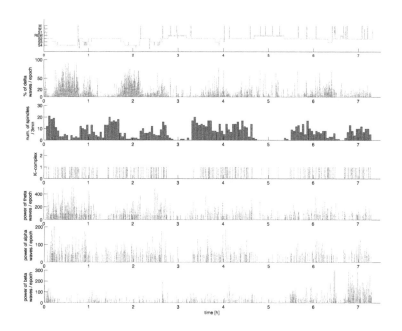

FIGURE 5.11

Time evolution during overnight sleep of the signal structures quantified by means of MP. From top to bottom: hypnogram; percentage of 20 s epoch occupied by waveforms classified as SWA; number of sleep spindles in 3 minutes periods; K-complex occurrence; power of theta waves per 20 s; power of alpha waves per 20 s; power of beta waves per 20 s epoch. By courtesy of U. Malinowska.

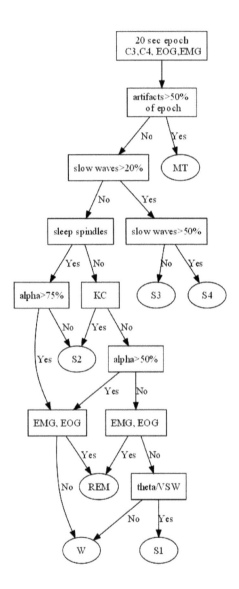

FIGURE 5.12
Block diagram of the sleep staging algorithm. KC—K-complex, MT—movement time, S—sleep stage. From [374].

5.1.6.7 Analysis of EEG in Epilepsy

Epilepsy is one of the most common neurological disorders second only to stroke; it affects about 0.8% of the world's population. The first choice treatment is pharmacological; however, in 25% of patients, seizures are drug-resistant. Epilepsy is manifested by a sudden and recurrent brain malfunction, which originates from neurons' excessive and hypersynchronous activity. The seizures occur at random and impair the normal function of the brain. During the seizure, the electroencephalogram changes dramatically: its amplitude increases by order of magnitude, and characteristic patterns varying in time and frequency appear. The mechanisms of the seizure generation are still not known. It is presumed that seizure occurrence is connected with the decrease of the inhibition in the brain. There are several models of seizure generation, and some of them are compatible with the observed electroencephalographic activity [353].

Electroencephalography is the most useful and cost-effective modality for studying epilepsy. For localization of epileptic foci imaging methods such as single-photon emission computer tomography (SPECT) and high-resolution MRI are useful. However, there are epileptic foci elusive to these imaging methods. In such cases, EEG or MEG is indispensable.

The epileptic seizure characteristics are dependant on the region of the brain involved and the underlying epileptic condition. Two main classes of seizures are distinguished:

1. Partial (focal) seizures. They arise from an electric discharge of one or more localized areas of the brain and possibly secondarily generalized. They may or may not impair consciousness.

2. Generalized seizures. In this case, electrical discharge involves the whole brain. The symptoms could be the absence of consciousness (petit mal) or loss of consciousness connected with muscle contraction and stiffness (grand mal).

Digital analysis of epileptic activity serves different purposes:

• Quantification of the seizure

• Seizure detection and prediction

• Localization of an epileptic focus

The above problems will be addressed in the following paragraphs.

Quantification of Seizures

Investigations of seizure dynamics are conducted to answer some of the most fundamental and yet debatable questions in epileptology, like how the propagating ictal discharges affect the brain's ongoing electrical activity or why seizures terminate. Seizure dynamics have been investigated using many different mathematical methods, both linear and non-linear. A seizure is a non-stationary process. Its spectral properties typically evolve from higher to lower frequencies; therefore, the time-frequency analysis methods are appropriate to estimate the seizure dynamics.

The most advanced method of time-frequency analysis—matching pursuit (Sect. 3.4.2.2) was used in [162] to evaluate seizures originating from the mesial temporal lobe. Periods of seizure initiation, transitional rhythmic bursting activity, organized bursting activity, and intermittent bursting activity were identified (Figure 5.13). The authors showed in a subsequent study [45] that

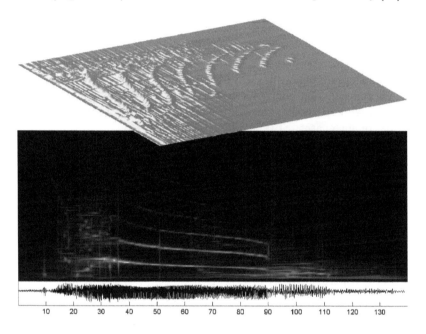

FIGURE 5.13
From the bottom: signal, 2D time-frequency energy distribution obtained by means of MP, the same distribution in 3D. Courtesy of P. Durka, from [127].

the complexity of the EEG signal registered from the location closest to the epileptic focus increases during a seizure.

The non-linear method based on the computation of the correlation dimension D_2 was applied in [479] for analysis of the time course of EEG recorded from the limbic cortex of rats. A comparison of the evolution of a seizure

---

[1]Matching pursuit is a method in which no assumption about linearity or non-linearity of the signals is required.

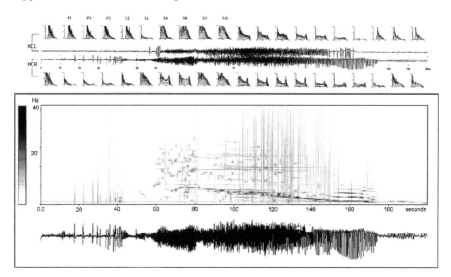

FIGURE 5.14

A comparison of the evolution of a seizure analyzed by means of correlation integral and matching pursuit. Top panel: the time sequence of plots of the correlation integral $\log C(\epsilon, m)$ vs $log(\epsilon)$ (Sect. 3.5.2) obtained from 10-s epochs of the signal from rat hippocampus in the peri-ictal time. Adapted from [480]. Bottom panel: Time frequency representation obtained by means of MP from the HCR signal shown in top panel and repeated at the bottom of the figure. From [61].

analyzed using correlation integral [480] and matching pursuit[1] is shown in Figure 5.14 [61].

At the beginning of the seizure (20–40 s), the occurrence of epileptic spikes resulted in a low value of D_2; in time-frequency energy distribution, it was reflected by very broad-band structures of short duration. During the period of chaotic behavior (60–80 s) characterized by the plot's flatness, we can see a random distribution of time-frequency structures. Interesting is the period (150–170s) when the system tends toward a limit cycle, accompanied by a low value of D_2. We can conclude that the time-frequency distribution obtained by the MP algorithm reveals the dynamics of the signal and explains the behavior of D_2 preventing its misinterpretation.

The problem of spatial synchronization in the pre-ictal and ictal periods has attracted a lot of attention. It is usually approached by assessing the correlation structure of multichannel EEG. Schindler [536] found that the zero-lag correlations of multichannel EEG remain approximately unchanged or, especially in the case of secondary generalization, decrease during the first half of the seizure, then gradually increase before seizure termination. However, zero-lag correlation does not give the full information on synchronization phenomena since there are phase differences between channels connected with the propagation of epileptic activity.

In [535] canonical multivariate discrimination analysis based on singular value decomposition was applied to search for dynamically distinct stages of epileptic seizures. The input values were: total power, the correlation at both zero and arbitrary time lag, average phase-amplitude (calculated by Hilbert transform), phase angle, and amplitude dispersion. The authors distinguished rhythmic partial onset, tonic middle, and clonic terminal activity.

An interesting methodological approach to the assessment of synchronization of interictal and ictal EEG signals was presented in [162]. The authors have used the multichannel autoregressive method of analysis that can be interpreted in stochastic and deterministic framework. As a measure of synchronization they have used a value connected with the goodness of fit of MVAR model:

$$SY = \log\left(\det(\hat{\mathbf{V}})\right) \tag{5.3}$$

where $\det(\hat{\mathbf{V}})$) is a determinant of the residual matrix $\hat{\mathbf{V}}$ of MVAR (Sect. 4.2). For a purely uncorrelated Gaussian normalized white noise, $\hat{\mathbf{V}}$ is a diagonal identity matrix[2], and $SY = 1$ setting the upper bound value for SY. For a purely deterministic linear system or a dynamical system of a periodic or quasi-periodic trajectory, the matrix $\hat{\mathbf{V}}$ represents a covariance matrix of measurement errors, setting the lower bound value of SY. For chaotic or stochastic colored-noise systems, the value of SY will be between these bounds. The quantity SY can be interpreted as a measure of order in the system. There is a close relationship between Shannon entropy and residual variance of an autoregressive process [543]. In the case of a multichannel process, a high correlation between channels increases predictability. If channels are highly correlated, one channel can be predicted using other channels; thus, the number of variables necessary to describe the system dynamics is lower, and the MVAR fits better, resulting in smaller values of SY. The changes to lower values of SY reflect higher spatiotemporal synchronization. The method was tested on a limited number of patients yielding very coherent results. The relatively high and stationary level of SY in the interictal EEG remote from seizure onset reflected much less synchronization. The minimum of SY always occurred shortly after the onset of a seizure, reflecting high regional synchrony. The synchronization level remained high after a seizure for prolonged periods,

---

[2]Signal must be normalized by subtracting the mean value and division by standard deviation.

which may explain in part the phenomena of seizure temporal clustering often observed in patients. It seems that the method has a high potential for explaining the evolution of seizures and can be helpful in seizure detection [264].

Seizure Detection and Prediction

The study of epileptic EEG dynamics can potentially provide insights into seizure onset and help to construct seizure detection and prediction systems. A method capable of predicting the occurrence of seizures from EEG would open new therapeutic possibilities. Namely, long-term medication with antiepileptic drugs could move toward EEG-triggered on-demand therapy (e.g., excretion of fast-acting anticonvulsant substances or electrical stimulation preventing or stopping the seizure). The problem of forecasting seizures is not yet satisfactorily resolved, which raises the question if the prediction is feasible at all.

There are two different scenarios of seizure generation [351]. One implies a sudden and abrupt transition, in which case a seizure would not be preceded by detectable changes in EEG. The model describing this scenario assumes the existence of a bi- or multistable system where the jump between coexisting attractors takes place, caused by stochastic fluctuations [576]. This model is believed to be responsible for primary generalized epilepsy. Alternatively, the transition to the epileptic state may be a gradual process by which an increasing number of critical interactions between neurons in a focal region unfolds over time. This scenario, likely responsible for focal epilepsy, was modeled in [637].

Since the 1970s, many methods have been devised to forecast an epileptic seizure. They are critically reviewed in [415]. The early idea was to use the rate of interictal epileptic spikes as a measure indicating the time of seizure onset. There were some positive reports in this respect [325]. Still, in the systematic study based on large experimental material, it was demonstrated that spiking rates do not change markedly before seizures but may increase after them [191, 281]. Other attempts involved applying autoregressive modeling, which allowed for detecting pre-ictal changes within up to 6 s before seizure onset [509, 521].

In the 1980s, the non-linear methods based on chaos theory were introduced to epilepsy research. They involved calculation of attractor dimension, e.g., [142], Lyapunov exponent [240], correlation density [383]. However, more recently, the optimistic results obtained by non-linear measures have been questioned. Starting from early 2000, several papers based on rigorous methodology and large experimental material showed that the results reported earlier were difficult to reproduce and too optimistic [16, 323, 214, 397]. In a study concerning coupling of EEG signals in epileptic brain, non-linear regression, phase synchronization, generalized synchronization were compared with linear regression [636]. The results showed that linear regression methods outperformed the other tested measures. The re-evaluation of the earlier works led to the conclusion that the presence of non-linearity in the signal does not justify the use of complicated non-linear measures, which might be outperformed by simpler linear ones.

The detection of seizures is especially essential in patients with frequent seizure episodes and neonates. The methods applied for this purpose usually consist of two stages. In the first stage, signal features such as spike occurrence, spike shape parameters, increase of amplitude, or time-frequency characteristics found utilizing wavelet transform were extracted. In the second step, artificial neural networks (ANN) were used to identify seizure onset. The review of the works based on this approach may be found in [526].

Saab and Gotman [518] proposed to use for patient non-specific seizure detection a wavelet transform (Daubechies 4 wavelet). From WT, the characteristic measures were derived, namely: relative average amplitude (the ratio of the mean peak-to-peak amplitudes in the given epoch to the mean of the amplitude of the background signal), relative scale energy (the ratio of the energy in the coefficients in the given scale to the energy of the coefficients in all scales), coefficients of the variation of the amplitude (square of the ratio of the standard deviation to the mean of the peak-to-peak amplitude). The classification was performed based on Bayes conditional formula, which allows the description of the system's behavior based on how it behaved in the past. The a priori probabilities were obtained using the training data, and the a posteriori probabilities served as the indicator of the seizure activity. The reported sensitivity was 76%, and the median value of delay in detecting seizure onset was 10 s.

The patient-specific method for detecting epileptic seizure onset was presented by Shoeb et al.   [547]. They applied wavelet decomposition to construct a feature vector capturing the morphology and topography of the EEG epoch. Utilizing a support vector machine classifier, the vector representa-

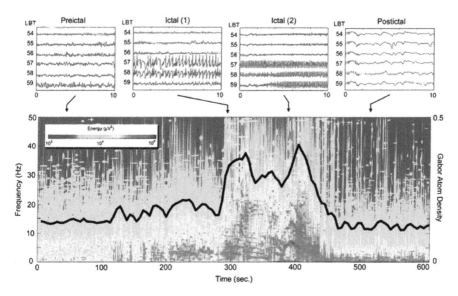

FIGURE 5.15
GAD analysis of a complex partial seizure. Upper panels: EEG traces for 4 different epochs. Lower panel: color-coded time-frequency energy density of the signal from contact LBT 58. Superimposed over this plot is the corresponding GAD (black trace). From [263].

tive of seizure and non-seizure epochs were constructed individually for each patient. The method was validated on 36 subjects. The authors reported 94% sensitivity and average latency in detecting seizure onset of 8 s.

The interesting method of seizure detection based on the decomposition of the signal using the MP algorithm was proposed in [263]. The method relies on the observation that the number of atoms needed to describe the signal rapidly increases at the seizure onset. The authors introduced a measure called GAD (Gabor atom density). GAD was defined as the number of atoms obtained during the MP decomposition divided by the reconstructed time-frequency space size. The criterion for termination of iterative decomposition was based on the energy threshold found empirically for all studied subjects. GAD calculated for moving window together with the time-frequency plot is shown in Figure 5.15. GAD measures the complexity of a signal as the number of elementary components needed to represent it. The method appeared to be very sensitive and specific for detecting intracranial ictal activity. The earliest changes during a seizure have been observed in the channels closest to the seizure onset region.

New approaches in anticipating seizures involve the so-called active algorithms in contrast to passive algorithms described above. The active probe may be intermittent photic stimulation. It was found [269, 459] that in

EEG and MEG signals, before the seizure onset, the phenomenon of phase clustering of harmonically related frequency components occurs. It doesn't mean that the stimulus provokes the seizure in any way. Based on this observation, the phase clustering index (PCI), which reflects the ability of neuronal systems to generate epileptic discharges, was introduced [269]. PCI is zero if the phases are random and tends toward high values if the phases are strongly clustered. The development of seizures was correlated with the PCI's value in the gamma band; therefore, rPCI, i.e., PCI in the gamma band relative to PCIs in the other frequency band, was introduced. Based on rPCI, the time horizon of the occurrence of seizures may be estimated [270], which opens possibilities of counteracting the seizures. It has been shown, using the computational model, that active paradigms yield more information regarding the transition to a seizure than a passive analysis [577]. The method seems to be promising. Still, it requires further experimental and model studies.

The evolution of the approach to seizure prediction leads to the conclusion that before addressing the question of whether a developed method might be sufficient for clinical applications, it has to be tested if the performance of a new approach is better than chance. To this aim, methods for statistical validation are needed. They may rely on comparison with analytical results (derived from random or periodic prediction schemes) [647] or on simulations by Monte Carlo method, e.g., [9, 305, 254]. The statistical significance of results may be tested, e.g., using the concept of seizure time surrogates [8]. To this aim, the seizure-onset times of the original EEG recordings are replaced by the seizure onset times generated by random shuffling of the actual values. If a measure's predictive performance for the original data is higher than the one for a particular quantile of the shuffled ones, then the performance of this measure can be considered as significantly better than the random prediction. In the study of Mormann et al. [414] comprising linear and non-linear methods, the concept of surrogates introduced in [8] was used. The authors found significant predictive performance for the estimators of spatial synchronization, contrary to the univariate methods such as correlation dimension, Lyapunov exponent, and the signal energy, which were unable to discriminate between pre-ictal and ictal periods. It was also found that non-linear measures didn't have higher predictive performance than linear ones.

The following guidelines were formulated [415] to assure the methodological quality of future studies on seizure prediction:

- Prediction should be tested on unselected continuous long-term recordings

- Studies should assess both sensitivity and specificity

- Results should be tested using statistical validation methods based on Monte Carlo simulations. If prediction algorithms are optimized using training data, they should be tested on independent testing data.

These rules should be mandatory for any method, based on the analysis of EEG/MEG signals, aiming to predict or classify seizures.

Localization of an Epileptic Focus

In the case of drug-resistant epilepsy, surgery is a treatment option. Its goal is to remove a minimum volume of brain tissue responsible for seizure generation; therefore, a seizure focus should be determined with maximal precision. To this aim, presurgical evaluations are performed involving EEG, video-EEG monitoring, high-resolution MRI, SPECT. The advantage of the MEG technique over EEG is that the head's tissues very weakly influence the magnetic field. On the other hand, EEG, combined with video monitoring, provides long-term recordings including pre-, post-, and seizure periods.

The application of the MEG technique for presurgical evaluation of epilepsy activity was reported in [157]. In this study, the authors used a single equivalent dipole model for source reconstruction. The localizations obtained for epileptic structures were subjected to hierarchical cluster analysis, and ICA was used to construct an ellipsoid representing the epileptic focus. The ellipsoid volume was compared voxelwise with the resection volume generated from pre- and post-operative MRI scans. The surgical treatment's positive outcome correlated with a short distance between the mass centers of both volumes and high coverage of the MEG ellipsoid by the resection volume.

MEG was compared with long-term EEG in [464]. The authors reported that in a studied group of 105 patients, MEG was inferior to long-term EEG in recording interictal spikes but yielded more circumscribed localizations if spikes were present. In another study [638], the site of surgery was correctly predicted based on MEG in 52%, on interictal scalp video-EEG in 45%, on ictal scalp video-EEG in 33%, and on the invasive EEG in 70%.

At present for localization of epileptic zone combined EEG and fMRI techniques are used, namely EEG informed fMRI which is described in Sect. 5.1.8.1.

The surgical treatment is usually preceded by the application of subdural and depth electrodes in the brain area delineated as responsible for seizures by the EEG, MEG, and imaging techniques. Intracranial EEG allows for more precise localization of epileptic focus. However, some seizures are difficult to localize even with intracranial probes. It was reported in [389] that in patients with partial seizures originating from lateral temporal or extratemporal regions, only 15% of seizures could be localized by visual analysis because of the rapid regional spread of epileptic activity.

The method that helps in this respect, providing insights into EEG signal sources, is directed transfer function. It has been successfully applied in [163] to the localization of mesial temporal seizures foci. The authors demonstrated that the DTF method could determine whether during seizure development, the initial focus continues to be the source of a given activity or whether the other more remote areas become secondary generators.

In another study intracranial EEG recordings were analyzed with a short-time direct directed transfer function, SdDTF, to estimate the directionality and intensity of propagation of high-frequency activity (70–175 Hz) during ictal and interictal recordings. The study revealed prominent divergence and convergence of high-frequency activity propagation at sites identified by epileptologists as part of the ictal onset zone. In contrast, relatively little propagation of this activity was observed among the other analyzed sites. This observation has a potential to be helpful in localization of epileptic foci [297]. Recently the information concerning localization of epileptic zone derived from technical and methodological developments and their combinations was reviewed in [669]. In particular, the importance of looking beyond the EEG seizure onset zone and considering focal epilepsy as a brain network phenomenon was underlined.

5.1.6.8   EEG in Monitoring and Anesthesia

Monitoring Brain Injury by Quantitative EEG

The monitoring technique of brain injury (which may be, e.g., a result of cardiac arrest) is based on the hypothesis that the injury causes unpredictable changes in the statistical distribution of EEG, which reduces the signal's information content. The impairment of brain function decreases its ability to generate complex electrophysiological activity, leading to reduced EEG entropy. Therefore the measure which is commonly used in monitoring brain injury is entropy (equation 3.29, Sect. 3.3.1.2).

Since the EEG of an injured brain is usually non-stationary, the time-dependent entropy based on the sliding window technique was introduced, e.g., [48]. In this technique, the entropy is calculated within each window, and its time course is followed.

The alternative measure called information quantity (IQ) is based on a discrete wavelet transform DWT. In this approach, the DWT is performed within each moving window, and IQ is calculated from the probability distribution of wavelet coefficients. In this way, subband information quantity (SIQ) may be found [602].

The probability $p_n^k(m)$ is obtained from the probability distribution of the wavelet coefficients in the kth subband; m is the bin number in the probability histogram. Subband information quantity in k^{th} subband and in the epoch n may be expressed as [593]:

$$SIQ^k(n) = -\sum_{m=1}^{M} p_n^k(m) \log_2 p_n^k(m) \tag{5.4}$$

In this way the time evolution in the subbands may be followed. It was found that SIQ depends on frequency band [593]. Entropy or SIQ are early markers of brain injury and measure of recovery after an ischemic incident.

Monitoring of EEG During Anesthesia

EEG monitoring is a routine aid to patient care in the operating room. It is used to reliably assess the anesthetic response of patients undergoing surgery and predict whether they respond to verbal commands and whether they are forming memories, which may cause unintentional unpleasant recall of intraoperative events. The algorithms working in both domains—time and frequency—were developed to evaluate EEG during anesthesia. In this state, patients may develop a pattern of EEG activity characterized by alternating periods of normal to high or very low voltage activity. The phenomenological parameter in the time domain called the burst suppression ratio (BSR) was introduced to quantify these changes [497]. BSR is calculated as the fraction of the epoch length where EEG meets the criterion: occurrence of a period longer than 0.5 s with the amplitude not exceeding ±5 μV.

Spectral analysis is used in anesthesia monitoring as well. Specific parameters based on the calculation of spectral density function were introduced, namely: median power frequency (MPF)—frequency which bisects the spectrum, with half the power above and the other half below—and spectral edge frequency (SEF)—the highest frequency in the power spectrum. Spectral power as a function of time is used for tracking the changes of EEG during anesthesia. The methods described in Sect. 3.4.2 may be applied for this purpose. Another measure used in anesthesia monitoring is bicoherence (Sect. 3.3.2.2).

A measure based on the concept of entropy, so called spectral entropy, is applied during anesthesia and critical care. Spectral entropy SE for a given frequency band is given by the formula [497]:

$$SE(f_1, f_2) = -\sum_{f_i=f_1}^{f_2} P_n(f_i) \log P_n(f_i) \qquad (5.5)$$

where $P_n(f_i)$ is a normalized power value at frequency f_i. This measure may be normalized in the range $(0, 1)$ taking into account the number of data points $n(f_1, f_2)$ in the band (f_1, f_2). Normalized spectral entropy SE_N is given by the formula:

$$SE_N(f_1, f_2) = \frac{SE(f_1, f_2)}{\log n(f_1, f_2)} \qquad (5.6)$$

5.1.7 Analysis of Epoched EEG Signals

In the previous section, we discussed signal analysis methods suitable for getting insights into ongoing, spontaneous brain activity.

Let's consider the brain as a system with input and output. We can learn a lot about it, observing and analyzing the system's reactions to the input's specific perturbation. The perturbations are usually caused by stimuli. Studies of the brain's reactions to the stimuli require focusing the analysis on the time surrounding the moment of stimulation. The methods developed for this type of analysis try to quantify the EEG/MEG signal changes provoked by external or internal stimuli. These changes can be globally referred to as event-related potentials (ERP). A subset of them, related to sensory (visual, auditory, somatosensory) stimuli, is commonly called evoked potentials (EP). In the case of MEG signals, the respective names are event-related fields (ERF) and evoked fields (EF). For the sake of simplicity, we shall refer further in this section to the EEG signals: ERP and EP, keeping in mind that the described methods can be applied equally well to the analysis of MEG signals.

The voltage deflections observed in the ERP reflect the reception and processing of sensory information, as well as higher-level processing that involves selective attention, memory updating, semantic comprehension, and other types of cognitive activity. In clinical practice, ERPs are viewed as a sequence of components defined by their positive or negative polarity (in respect to the reference electrode potential), their latencies, their scalp distribution, and the relation to experimental variables. The clinical and research interest in ERPs relies on the fact that they are linked in time with physical or mental events [126].

In clinical neurophysiology, the ERP or EF is described as the response averaged across tenths or hundreds of repetitions. The diagnostic strength of ERP analysis comes from the observations, accumulated over the years, of correlations between the latency and amplitude of positive (P) and negative (N) deflections of the average ERP and the behavioral or clinical state of the subject. According to the published guidelines ([477], p. 141), "the simplest approach is to consider the ERP waveform as a set of waves, to pick the peaks (and troughs) of these waves, and to measure the amplitude and latency at these deflections".

The deflections appearing in the first tenths of a second are the easiest to interpret. They reflect the early stages of sensory processing. For instance, the brainstem auditory evoked potentials (BAEP) consists of 7 waves with latencies between 1 and 12 ms generated in well-defined locations in the auditory pathways. BAEPs can be used as markers of the integrity of the auditory pathway.

The deflections appearing later than about hundred ms are related to the activity of specific brain systems, which, in some cases, have been delineated (e.g., [181, 483, 435]). At these later stages, we should rather consider components, not deflections as such. The activity of a given ensemble of neural populations manifests as a topographically specific distribution of potentials on the scalp. It has a particular time course that may depend on the task being processed. Such a spatio-temporal pattern of activity of a specific ensemble we would call a component. Usually, we do not observe an isolated component but rather a linear mixture of different components related to the hypothetical ensembles activated at various stages of processing the experimental task. The problem of disentangling such a mixture is illustrated in Figure 5.16. The observed ERP trace could be obtained as a mixture of different latent components. Further, change of amplitude of only one of the latent components can result not only in the changes of ERP amplitude but also of the latencies of its peaks (Figure 5.17). Moreover, components are embedded in the so-called ongoing spontaneous activity, which is not related to the task at hand. The most straightforward approach to study the components is an intelligent design of the experiment, such that the component of interest can be isolated or at least well exposed, e.g., by appropriate contrasts between the conditions [356].

The two interesting component's features, the modification of which by experimental conditions can be studied, are magnitude and latency. The most straightforward approach is to measure the amplitude in the ERP's maximum and the time after the stimulus when it occurs. However, this approach is prone to error due to noise that is never completely removed and to jitter effects of latency (c.f. Figure 5.19). The recommended method of measuring the size of a component is to compute the area under the average potential waveform in the interval in which the tested component is expected to occur. Similarly, the suggested method of measuring latency is to evaluate the time location where the area under the ERP curve is divided in half [356].

One of the most commonly considered ERP component is P300. The "300" is a naming convention rather than the latency value since this positive deflection may appear between 300 and 600 ms after the stimuli of different modalities (e.g., visual, auditory, sensory). P300 is thought to reflect processes involved in stimulus evaluation or categorization. It is usually elicited using the oddball paradigm in which low-probability target items are inter-mixed with high-probability non-target (or "standard") items. The topography of the P300 recorded by scalp electrodes reveals a maximum over the midline centro-parietal regions. The most direct way to localize functional activation may be obtained using iEEG signal analysis. In particular, the P300 components were extensively studied to identify distributed networks responsible for their generation [552]. Several studies addressed in the above reference explored the neural substrates of various ERPs elicited by functional activation of motor and sensory cortices. The literature on experimental paradigms for evoking and the interpretation of amplitudes, latencies, and topographies of late components such as P300 (and its subdivision into P3a and P3b), N200, N2pc, N170, P200, N400, P600 is very rich and reviewing it is far beyond the scope of this book. However, from the signal analysis point of view, the problems and techniques for dealing with the ERPs are common to all of them and are described in the following paragraphs.

The main difficulty in the analysis of ERPs is that, compared to spontaneous activity of the brain, the changes provoked by the stimuli are often very subtle; they are by order of magnitude smaller than the ongoing EEG. The analysis aims to separate the activity of the brain not related to the stimuli (called the ongoing or background activity), which is treated as noise, and the activity related to the stimulus, which is the signal of interest. Thus in technical terms, the problem in the quantification of ERPs is the low signal to

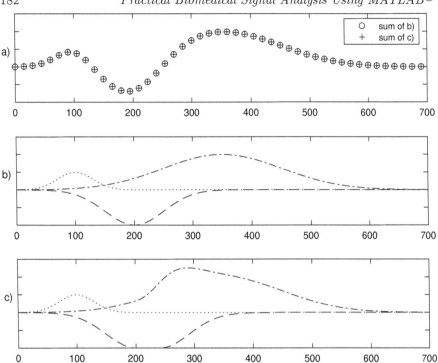

FIGURE 5.16
Illustration of relation between an ERP and latent components. The traces in b) and c) show two sets of time courses of hypothetical latent components; a) shows that exactly the same ERP trace can be obtained by summation of either b) or c). The MATLAB code for this simulation is given in `MATLAB/c5/ERP_summing_latent_components.m`.

noise ratio. The choice of methods to improve the signal to noise ratio depends on the assumptions that one makes about the relation between the signal, the stimulus, and noise. In the following sections, we discuss methods appropriate for analysis of responses that are phase-locked to the stimulus (Sect. 5.1.7.1) and these that are non-phase-locked (Sect. 5.1.7.4).

5.1.7.1   Analysis of Phase-Locked Responses

Preprocessing

Before analysis of ERPs, usually one has to filter the raw data with a high pass filter, e.g., to remove slow drift, and apply a low pass filter, e.g., to remove residual contamination with muscle activity. These operations should be done preferably before extracting epochs containing responses to stimuli to minimize filtering edge effects. The processing of ERP takes place in the

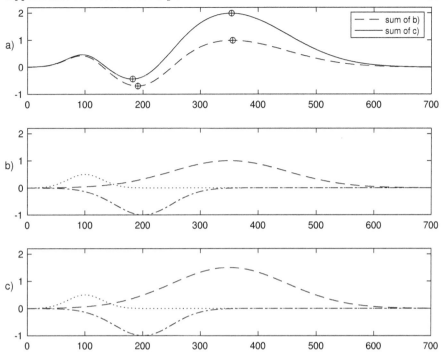

FIGURE 5.17
Illustration of relation between positions of ERP peaks and latent component amplitude. In a), dashed line is the sum of latent components plotted in b) and the solid line is the sum of components shown in c). Note, that the latency of the first minimum in the traces in a) appears earlier when the amplitude of the component marked by dashed line in b) is increased in c). The MATLAB code for this simulation is given in MATLAB/c5/ ERP_latent_components_amplitude.m.

time domain; thus, one should consider the filters' temporal properties, i.e., the step response and the impulse response. Illustration of potential pitfalls related to high pass filtering is presented in Figure 5.18. Note the spurious oscillations and "leakage" of the first component to the baseline period. The recommended high-pass filtering is 0.01 Hz for cooperative subjects and may be shifted to 0.05 or 0.1 Hz for subjects with problems sitting still. The low pass filters should be about one-third of the sampling frequency, or in case of contamination with residual muscle artifacts, can be lowered to about 30 Hz. A very slow drift of baseline can be resolved with so-called baseline correction. It requires defining some period preceding each stimulus's presentation and subtracting its mean value from the whole epoch. This period is often selected as 100 or 200 ms which could encompass one or two cycles of alpha oscillations that may appear in the resting periods between stimuli.

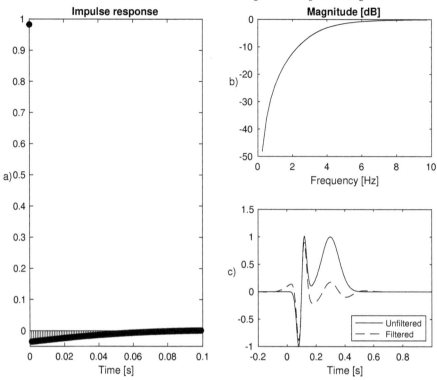

FIGURE 5.18
Illustration of possible disruptive effects of low pass filtering on ERP. We simulated two components centered at 0.1 and 0.3 s. The simulated trace is plotted with a solid black line in c). The filtered signal is plotted with a dashed line. In this case, we applied zero-phase filtering (filtfilt) with Butterworth high pass filter order 2 with cutoff frequency 4 Hz. Note, that due to the impulse response of this filter, spurious oscillations appear in the filtered signal. Moreover, the oscillation extends before the presumed onset of stimulus at 0 s. The MATLAB code for this simulation is given in MATLAB/c5/ERP_filters.m.

Time Averaging

The basic technique used in the analysis of EP is the time averaging of multiple responses, time-locked to the stimulus. The first approach of this type, which resulted in the visualization of the EPs, was reported by Dawson [107]. In his solution, many time-locked responses were photographically superimposed. Today, the averaging of the digitized epochs of EEG, aligned to the stimulus, is easily done using a computer.

The time-averaging method is based on three assumptions:

- The brain response is time-locked (invariably delayed) with respect to the stimulus

- The brain response to the consecutive stimuli is of the same morphology

- The ongoing activity is a stationary, zero-mean noise

In commonly used analysis a simple additive model is assumed where the measured stochastic signal $x(t)$ is a sum of two terms: the deterministic EP signal $s(t)$ and the uncorrelated zero mean noise $n(t)$. According to this model the ith realization of the signal is:

$$x_i(t) = s(t) + n_i(t) \qquad (5.7)$$

Averaging across N repetitions gives:

$$\bar{x}(t) = \frac{1}{N} \sum_{i=1}^{N} x_i(t) = \frac{1}{N} \left(Ns(t) + \sum_{i=1}^{N} n_i(t) \right) \qquad (5.8)$$

The expected value of the averaged signal is:

$$E[\bar{x}(t)] = s(t) \qquad (5.9)$$

since for zero mean noise we have $E[\frac{1}{N} \sum_{i=1}^{N} n_i(t)] = 0$. The variance of $\bar{x}(t)$ is:

$$\sigma_{\bar{x}(t)}^2 = E\left[\left(\frac{1}{N} \sum_{i=1}^{N} n_i(t) \right)^2 \right] = \frac{1}{N} \sigma_{n(t)}^2 \qquad (5.10)$$

since $s(t)$ is deterministic. Therefore the signal to noise ratio in terms of amplitude improves proportionaly to $\sqrt{N}$.

For the potentials appearing in the first tenths of a second after the sensory stimulation, such as BAEP, the simple additive model (equation 5.7) can be regarded as valid [257]. The validity of the model for more prolonged latency potentials is doubtful. Nevertheless, due to its simplicity, its applications in the practice of EP analysis is overwhelming. A model closer to reality can be formulated assuming that the response is a stochastic process $s_i(t)$. Then the model (equation 5.7) becomes

$$x_i(t) = s_i(t) + n_i(t) \qquad (5.11)$$

This model implies that the response may not be completely described by the mean value $\bar{x}(t)$ but also by higher statistical moments (Sect. 2.1) as, e.g., variance.

Indeed, it should be stressed that the averaging procedure is not as simple as it may appear at first glance. The simple additive model in which an EP is a sum of deterministic signal and uncorrelated background noise holds only in a few ideal cases. One can expect that there are interferences between the neural activity evoked by the stimulus and the ongoing activity resulting in the reorganization of the latter [529]. Despite this fact, one may still accept the simple additive EP model, but he must be aware of the violation of the assumptions. It should be realized that the signal, in reality, is not deterministic but stochastic. Thus it is recommended to evaluate the variance of the EP along with its mean value. The assumption of "noise" as an uncorrelated signal is also a significant simplification. It is necessary to consider the statistical characteristics of the ongoing activity, as shown in the next paragraph. In any case, it is always suggested to include some pre-stimulus time in the EEG epochs to be averaged to obtain some estimate of the mean and variability of the ongoing activity.

Influence of Noise Correlation

In quantification of EP, the main issue is to increase the signal-to-noise ratio to minimize the contamination of EP by the background EEG activity. Equation 5.10 shows that the ratio of signal variance to the noise variance for N repetitions of the stimulus is $\frac{1}{N}$. This result holds for the uncorrelated noise samples. However, this often is not true, e.g., in the presence of strong rhythmic background activity like alpha rhythm. In such a case, subsequent samples of $n(t)$ are not independent. The degree of dependence of noise samples is given by its autocorrelation function, which in this case has the following form [433, chap. 50]:

$$R_{xx}(\tau) = \sigma^2 \exp(-\beta|\tau|) \cos(2\pi f_0 \tau) \tag{5.12}$$

where σ^2 is the variance of noise, f_0 is the mean frequency of rhythmic activity, β/π is the corresponding bandwidth, and τ is the delay parameter. In this case the EP variance takes another form [564]:

$$\sigma^2_{\bar{x}(t)} = \frac{\sigma^2}{N} \left[\frac{1 - \exp(-2\beta T)}{1 - 2\exp(-\beta T)\cos(2\pi f_0 T) + \exp(-2\beta T)} \right] \tag{5.13}$$

where T is the inter-stimulus period. It is important to note that rhythmicity in the background activity will influence the signal-to-noise ratio. It can be seen from the above equation that the variance ratio tends to $\frac{\sigma^2}{N}$ as βT becomes large. A related problem is whether the stimulus should be periodic or aperiodic. Aperiodic stimulation can lead to reduced EP variance [592]. Periodic stimulation can result in generation of a rhythmic activity with frequency corresponding to the repetition rate of the stimulus.

Variations in Latency

One of the forms of randomness in a single trial response is the variation in latency. The latency jitter may lead to a distortion of the averaged EP. The distortion may result in an underestimation of the averaged EP peak

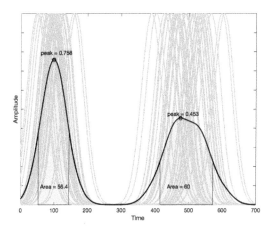

FIGURE 5.19

Illustration of the effect of latency variation on the amplitude and ERP deformations. We simulated 30 identical gaussian curves centered around latency 100±30, and another 30 centered around latency 500±50. The individual realizations are plotted in light gray. The average is plotted in black. The area under the peaks measured in the range of width-at-half-height is shaded, and the value of that area is printed. The peaks are marked with circles, and their height is displayed. Note that increased variance of the location of individual realizations of the components influences the height, spread, and shape of the average. The MATLAB code for this simulation is given in MATLAB/c5/ERP_latenceies_amplitude.m.

amplitude and its greater spread over time (Figure 5.19). A more robust estimation of the latent component magnitude can be obtained by calculating the area under the peak than the height of the peak. In the example from Figure 5.19 we get the ratio of the peak areas equal 0.94, while the ratio of the peak amplitudes is 1.67. In more severe cases, latency jitter may render the component undetectable. A possible consequence of this is that a significant difference in amplitude of an average EP recorded in two experimental conditions can result from a difference in latency jitter between conditions. Moreover, suppose the distribution of latency jitter is skewed. In that case, the difference in latency jitter between conditions will also lead to a difference in the latency of the average ERP's deflections. The extent to which latency jitter affects ERPs averaged in the time domain varies as a function of the duration of the ERP amplitude deflection. Indeed, for the same latency variance, a long-lasting deflection will be relatively more repeatable across trials than a short-lasting deflection and, consequently, less distorted by time-domain averaging. For longer latency of brain response, latency jitter is greater, so short-lasting components with long latency can hardly by observed [416].

We can assume that in equation 5.11, the signal's variability is due only to the variable latency, and for each trial i, the shape of the response $s(t)$ is invariant. It implies that there exists a time translation Δt_i such that:

$$s_i(t) = s(t + \Delta t_i). \tag{5.14}$$

The first approach to correction for the latency variation was proposed by Woody [650]. In his method, the Δt_i was estimated using the cross-correlation function of the response i with a template (i.e., the initial estimate for the average EP). The required translation was evaluated as the lag of the maximum of cross-correlation. The individual trials are then readjusted by Δt_i and averaged, giving the new estimate of average EP. This estimate was used as a template in the next step of the process. This iterative procedure leads to a better estimate of the average EP. Woody also showed that the magnitude of the cross-correlation function could be used to sort the single-trial EP to different sub-classes. The refinement of Woody's method relied on taking into account the statistical properties of the ongoing EEG activity [395].

A latency correction method, involving additionally a classification step, was proposed in [474]. According to this technique, single-trial EPs are cross-correlated with different templates. The maxima of cross-correlation functions are used to estimate the amplitude and latency of particular components. Next, these trials that fall within a certain amplitude or latency limits are averaged together. In this way, different classes of selected averages are obtained that may correspond to different subsystems or physiological conditions. Averaging of all trials as one class may lead to a considerable information loss.

Habituation

The observed trial-to-trial variability in amplitude and latency of the components of an evoked potential may arise from two major sources. One is systematic and depends on the changes in reaction to the repetitive stimulation, and the other is stochastic. Systematic changes in response to repeated stimulation are a general property of the nervous system. New stimuli first elicit an arousing response, named the orientation response [561]; it consists of changes in many autonomic variables, such as the skin conductance reaction, phasic heart rate changes, and it also involves a reaction of the cerebral cortex. Habituation is, in a broad sense, defined as a decrease of response as a function of stimulus repetition, and it is a constant finding in almost any behavioral response. The simplest mathematical model of habituation is an exponential decay in case of regularly repeated stimulus [561]. The decay rate depends on the physical properties of the stimulus, its relevance, the interstimulus interval, and the properties of the neural systems involved in the reaction. Habituation is not the only possible reaction to a sequence of stimuli. The increases in the responses during the first stimuli related to a sensitization process have been described in the literature [596, 200]. Moreover, these authors proposed that response to repeated stimulation is the interaction of sensitization and habituation.

There are, in general, two possible approaches to study these systematic changes over time. One is the adjustment of the experimental paradigm, and this is described further in this paragraph. Second is the development of methods able to extract single-trial responses; these are described in the following paragraph.

In the classical works on habituation/sensitization, we can find two possible paradigms: sub-ensemble averages and block averages. In the case of sub-ensemble averages, recording sessions were subdivided into consecutive subsets consisting of a few successive trials. In each of the subsets, the ensemble average is computed. The time evolution of these sub-ensemble averages is used to assess the habituation/sensitization process. This approach is successful only if changes between trials are much slower than the time-span of the number of trials included in each sub-average. Another method to study the trial-to-trial changes is to repeat a sequence of trials as a block, each sequence starting after an assumed recovery time. Then, corresponding responses from each block are averaged. This second method assumes that consecutive blocks are equivalent, and the habitation/sensitization processes are the same in each block. Moreover, both sub-ensemble and block-averaging require long recording sessions, and thus general arousal changes are likely to occur during the experiment [493].

5.1.7.2 In Pursuit of Single Trial Evoked Responses

The averaging techniques do not allow for direct access to variations of ERP latencies and amplitudes. In particular, short lasting changes, which may provide relevant insights into cognitive operations, are blurred or even masked by averaging processes, whereas single trial analysis would allow a reliable extraction of signal characteristics from a single ERP sequence. From this point of view, single-trial analysis of ERP is the ultimate goal of almost all related research, crucial for understanding subtle changes in individual brain responses to a repeated stimulus.

Several techniques have been proposed in order to improve access to individual ERPs. In each case specific assumptions must be made about the nature of the response.

Wiener Filters

Many approaches involve filtering of the single-trial traces, in particular using techniques based on the Wiener formalism (Sect. 3.1.4), which provides an optimal filtering in the mean square error sense [650, 625, 125]. However, these approaches have the common drawback of considering the signal as a stationary process and, since the ERPs are compositions of transient responses with different time and frequency signatures, they are not likely to give optimal results.

Model Based Approach

A model based approach was proposed in [81]. The background EEG activity was modeled as an AR process and event-related activity as autoregressive moving average process ARMA. In this way it was possible to find a filter extracting single ERP. However, the method relied on average ERP in the identification of ARMA model, and as such was not free from the assumptions pointed out in Sect. 5.1.7.1. A similar approach was presented in [620]. The method relies on the assumption that the measured activity can be separated into its evoked and spontaneous parts. A compound state-space model trying to incorporate the observable properties of both parts is formulated on the basis of additivity of the two components. Within this model, spontaneous activity is described as an autoregressive process, while the EP is modeled by an impulse response of a parametrically described system. Based on the state-space representation of the model, a Kalman filter for the observation of the system's state can be designed which yields estimates for both activities.

Time-Frequency Parametric Methods

The discrete wavelet decomposition (Sect. 3.4.2.2) is particularly suitable to parameterize the transients which are time locked to the stimuli. The procedures aiming at extraction of single-trial ERP based on discrete wavelet decomposition were proposed in [34]. In the applied approach the individual trials were first split into two parts: one which was assumed to contain components related to the stimulus (ERP epoch) and the second, which was assumed to contain only the spontaneous EEG activity (reference epoch). Both datasets were decomposed by dyadic wavelet transform. The coefficients of the basis functions that comprise the detail functions were then analyzed by regression and discriminate analysis to identify the coefficient best suited for distinguishing ERP from spontaneous EEG. Finally, the coefficients that most significantly differentiated the ERP epochs from the reference epochs were selected and used for reconstruction of the underlying single-trial ERPs. These reconstructed ERPs had significantly improved signal to noise ratio. The authors verified the method on a set of 40 auditory potentials. The correct classification to the group of ERP epochs was obtained for 34 cases. It was demonstrated that by means of DWT and discriminant analysis it is possible to construct a method capable of extracting components strongly related to the ERP with minimal assumptions. Nothing has to be assumed about the particular form of the components. A similar method based on wavelet denoising, although less advanced in the selection of relevant wavelet coefficients, was successfully applied to the study of the habituation effect of rat auditory evoked potentials [493].

A further development of the idea of parametric decomposition application for extracting single-trial ERPs, was proposed in the framework of multivariate matching pursuit (Sect. 3.4.2.2 and 4.6.4). The flexible parametric approach offered by MMP allows for explicit implementation of assumptions made about the ERPs. For instance, one may assume that the ERP morphology is constant and that only the amplitude varies across trials [548]:

$$x_i(t) = a_i s(t) + n_i(t) \tag{5.15}$$

where a_i is the amplitude of ERP in trial i. This assumption is equivalent to imposing constraints on the parameters $\gamma = \{u, f, \sigma, \phi\}$ (equation 3.115) of the Gabor functions such that all parameters, except the amplitude weight $a_i = \langle R^i x, g_{\gamma_i} \rangle$, are kept identical across all trials, yet not set a priori to any particular values. This approach guarantees that only these features, which consistently occurred across all trials, emerge from the multi-trial decomposition. Relaxing the constraint of a constant phase allows for accounting for the jitter and some limited shape variability of s_i:

$$x_i(t) = a_i s_i(t) + n_i(t) \tag{5.16}$$

Both approaches were successfully validated in the of analysis of habituation influence on amplitude and in the latter case also on the latency of auditory M100 evoked magnetic field. At first glance, the relaxation of the phase constraint may be seen as a favorable reduction of the bias of the estimation; however, the algorithm becomes more prone to individual noise patterns in each trial [549]. Therefore, neither of the two approaches is superior to the other one in an absolute sense. They should be used according to the particular conditions of the paradigm, taking into account the inherent assumptions and limitations of the two methods.

So-called consensus matching pursuit algorithm, tried to explain the trial-to-trial variability in the parameter space by means of a certain restricted variation of the Gabor functions approximating each trial independently within a certain neighborhood of the so-called consensus atom [41]. This consensus atom is selected by means of a specific voting procedure, and is the most representative atom for all trials. This approach constitutes a further step in releasing constraints on the parameters of the Gabor functions, since it allows adjustment of amplitude, scale, frequency, and phase at least to a certain degree, which is controlled by a Gaussian kernel.

ERP Topography

So far we discussed the evaluation of the time course of the ERP. A lot of valuable information is obtained by considering ERP topography since the distribution of the current sources in the brain is reflected in the spatial pattern of evoked potential, which can be visualized in the form of an ERP map.

The first problem one has to face when analyzing multichannel EEG data, concerns reference. EEG can only be measured as a difference of potentials, and as such the measured waveform differs when the reference electrode changes.This may influence the interpretation of the results [419].

One popular solution to this problem is the use of global field power (GFP) introduced by Lehmann and Skrandies [332]. GFP is computed at any moment in time as the standard deviation of voltage across the set of channels. The GFP tells how steep on average are gradients of potentials across the electrode montage. In this sense the GFP is a measure of the strength of activation at a given time point. It can be evaluated with the standard ERP parameters like: latencies, area, peak amplitude, etc.[3] The GFP extrema can be used in order to identify brain activity components in a reference invariant way.

We should note that the spatial distribution of potentials on the head surface is reference invariant[4], since if the reference potential shifts up or down the whole "potential landscape" shifts by the same amount. It is important to keep in mind that the absolute locations of the potential maxima or minima on the scalp do not necessarily reflect the location of the underlying generators. This is due to the propagation of the intracranial signals by volume conduction. One has to realize this fact in order to avoid confusion in the interpretation of EEG data.

One possibility to obtain reference-free waveforms, which exploits the reference invariance of topography, is the computation of the so-called current source density (CSD) waveforms. The CSD or Laplacian derivation is obtained by estimation of the 2nd spatial derivative across the electrode montage (Sect. 5.1.3). In this way, CSD derivations are intrinsically based on spatial gradients in the electric field at the scalp. This procedure eliminates to some extent the problem of reference-dependance inherent to voltage waveforms and it also minimizes contributions of volume conduction within the plane of the scalp, thus effectively working as spatial "sharpening" (high-pass) filters. However this kind of reference is not proper for methods relying on estimation of correlation matrix between the channels of the process, i.e., for the MVAR model. The Laplacian operator disturbs the correlation structure of the multivariate set of signals, and hence the phase dependencies.

It is important to note that the changes in the topography of the electric field at the scalp can only be caused by changes in the configuration of the underlying intracranial sources (excluding artifacts such as eye movements, muscle activity, etc.).[5] Mapping of electrical brain activity in itself does not constitute an analysis of the recorded data but it is a prerequisite to extracting quantitative features of the scalp recorded electrical activity. In the

---

[3] It is important to note that because GFP is a non-linear transformation, the GFP of the group-average ERP is not equivalent to the mean GFP of the single-subject ERPs.

[4] Please note that this does not hold for maps of power.

[5] The opposite need not be the case, since there is an infinite number of current sources configurations inside the volume conductor that give the same distribution of potentials on the surface.

second step, in order to find the differences between experimental conditions or between groups of subjects, the derived topographical measures must be subjected to statistical tests.

Due to volume conduction, potentials measured on the scalp are spatially blurred. The linear superposition of electric fields allows to assume that each electrode records a signal which is a linear mixture of all the underlying sources of electrical activity (not only from the brain, but also from muscle, cardiac, eye movement etc.). Successful separation of the signals from different sources gives possibilities of more correct functional interpretation of the different components of recorded brain responses. In general there are three approaches to the problem of separating the sources. Two of them operate in the sensor space and will be discussed in this paragraph, the third approach is the solution of the inverse problem. Methods concerning solution of the inverse problem are beyond the scope of signal processing, therefore they will not be considered here.

The methods operating in the sensor space do not require an explicit head model. They include: principal component analysis (PCA, see Sect. 4.6.1) and blind source separation (BSS), most often implemented as independent component analysis (ICA, see Sect. 4.6.2). Linear data decompositions, such as PCA or ICA, separate multichannel data into a sum of activities of components, each comprised of a time course of its activity in every trial and a single scalp map, giving the strength of the volume conducted component activity at each scalp electrode. Note that polarity seen in the ICA and PCA maps of topographical component distribution has no real meaning. The actual potential distribution is the product of the map and the value of the component in a given moment in time.

The goals of PCA and ICA are different. Let us consider for a k channel signal that the k values of potential measured at a certain moment in time are coordinates of a point in the k-dimensional space. The PCA extracts temporally orthogonal directions (axes) in that space.[6] The components are the projections of the original data onto these axes, with the first principal component representing the maximum variation in the data; the second principal component is orthogonal to the first and represents the maximum variation of the residual data. This process of component extraction is repeated several times, and since the original variables are correlated only a small number of principal components accounts for a large proportion of the variance in the original data [556, 557].

The goal of ICA, on the other hand, is to find components in the data whose activities are statistically as distinct from one another as possible, meaning that their signals have the least possible mutual information. Minimizing mutual information implies not only decorrelating the component signals, but also eliminating or reducing their higher-order joint statistics. This stronger statistical independence allows ICA to relax the orthogonality of the component

---

[6]Corresponding to necessarily orthogonal scalp maps.

scalp maps, which is physiologically more plausible, and to separate phase-uncoupled activities generated in spatially fixed cortical domains (or non-brain artifact sources).

If the scalp maps associated with different activities are not orthogonal, PCA will mix portions of them into one or more principal components, rather than separating them into different components, as is the case in ICA. Thus, if the recorded data are in fact the sum of (nearly) independent signals from spatially confined and distinguishable sources, PCA will lump, and ICA will split the source activities across resultant signal components.

Data reduction by PCA may be efficient for compression, but the higher flexibility of ICA decomposition, with spatially overlapping scalp maps allowed, may result in components which are easier to interpret in the physiological context. ICA can be applied to EEG epochs averaged in the time domain [368] or to non-averaged concatenated EEG epochs [367]. In both cited studies ICA allowed separating late cognitive ERPs into distinct, independent constituents.

Applying ICA one has to be aware that an obtained independent component (IC) does not necessarily represent an anatomically defined source of activity. Indeed, an IC is defined by its temporal independence relative to the other sources of activity. If the activities within two distinct regions of the brain strongly covary, they will be represented within a single component. In case of unconstrained ICA, the total number of estimated ICs equals the total number of recording electrodes. If that number is considerably greater than the true number of independent sources contributing to the recorded signals, ICs containing spurious activity will appear because of overfitting. Contrarily, if the total number of ICs is considerably smaller than the true number of underlying sources, valuable information will be lost because of underfitting. This important limitation (that the number of independent components is arbitrarily defined) could explain why, until now, ICA has not been able to clearly and unequivocally disentangle event-related electrical brain responses into physiologically meaningful independent constituents [416]. Another problem that has to be considered when applying ICA is the amount of available data needed to correctly estimate the unmixing matrix (see Sect. 4.6.2).

5.1.7.3 Applications of Cross-Frequency Coupling

Coupling between two neural oscillations can be in the form of the phase, amplitude, or frequency of signals. Phase-amplitude coupling (PAC), in which the amplitude of a fast oscillation covaries with the phase of a slow oscillation, is recently attracting researchers. PAC potentially provides a mechanism for synchronization and interaction between local and global processes across wide cortical networks [160]. The methodology for evaluation of PAC was discussed in Sect. 3.4.3. Here we review some of the practical applications of this technique.

The review [539], based on results on many spatial and temporal scales ranging from multiunit activity through LFP to EEG, proposes low-frequency oscillation and coupling of higher frequency activity to them as a plausible mechanism of attentional selection of task-relevant networks of neural ensembles.

Tort et al. [606] demonstrated that PAC, between theta (4–12 Hz) and gamma (30–60 Hz), observed in the hippocampus CA3 region of rats, was augmented as they learned to associate items with their spatial context. Furthermore, this PAC's strength was positively correlated with the increase in performance accuracy during learning sessions.

A robust coupling, with task-dependent topography, between the high- and low-frequency bands of ongoing electrical activity was observed in the human iEEG recordings [77].

The idea that working memory in humans depends on a neural code using phase information was supported by [20]. Further studies on local PAC in the human hippocampus recording during the formation of new episodic memories, measured as the heterogeneity in the distribution of the mean high-frequency amplitude over the low-frequency phase intervals, confirmed this hypothesis [331]. In particular, slow theta activity (2.5–5 Hz) modulated gamma-band activity (34–130 Hz). Moreover, the PAC magnitude correlated with successful encoding and gamma activity was coupled with the trough of slow theta oscillation. The authors of [220] provided evidence that similar effects could also be detected in the MEG signal. Specifically, they showed that items in different sequence positions exhibit greater gamma power at distinct phases of a theta oscillation. Furthermore, this segregation was related to successful temporal order memory.

Coupling between delta or theta (2–5 Hz) phase and high gamma (70–250 Hz) amplitude was suggested to play a role in the allocation of visuospatial attention by coordinating information flow within and between frontal and parietal areas [579]. In the ECoG recordings from subdural electrode arrays during a spatial-cueing task, subjects allocated visuospatial attention to either the right or left visual field and detected the appearance of a target. Changes in the coupling strength between the low-frequency phase and high gamma amplitude predicted the reaction times to detected targets on a trial-by-trial basis. In this study, PAC was analyzed using the phase-locking value [316, 621] for electrodes with a significant increase of high gamma power after trial onset as compared to baseline. Also, in a visual searching task, the intensity and preferred phase of coupling between broadband power and band-limited rhythms ($\delta, \theta, \alpha, \beta$, and γ), was shown to follow numerous coupling motifs [405].

There is a growing interest in understanding CFC patterns since they may be relevant for diagnosing and eventually treating various disorders or in designing preventive strategies.

Cross channel PAC, measured by MI (see Sect. 3.4.3.2), was successfully used to determine the seizure onset zones (SOZs) [201]. In this case, for all pairs of channels of a subdural electrode matrix, MI was computed. The low-frequency phase was obtained for one channel and the high-frequency amplitude for another one. Based on the eigenvalue decomposition of the matrixes of cross-channel MIs, the authors identified electrodes located in SOZs. The correlation with the post-operative outcomes validated the correctness of selection.

The study of PAC in the subjects with focal epilepsy during sleep revealed that the coupling is stronger in seizure onset zones (SOZs) than in normal regions even in the interictal period after excluding sharp activities [6]. The coupling intensity was the highest in a deep sleep (stage N3) and the lowest in rapid eye movement sleep.

De Hemptinne et al. [110] reported one of the earliest indications that increased PAC in primary motor cortex local field potential recordings could be used as a biomarker of Parkinson's disease [110]. They demonstrated that coupling between beta rhythm (13–30 Hz) phase and gamma-amplitude (50–200 Hz) in the primary motor cortex is amplified compared with the coupling in patients with craniocervical dystonia and humans without a movement disorder. Further, these authors [111] demonstrated that acute therapeutic deep brain stimulation transiently reduced phase-amplitude interactions and the time course of the changes induced in PAC correlated with reducing parkinsonian motor symptoms.

Further support for the importance of correlations between high-frequency oscillations to beta oscillations coupling in Parkinson disease was presented in [616]. Specifically, within the subthalamic nucleus, PAC correlated positively with the severity of motor impairment.

Decreased PAC may be a marker of deficits in amyloid precursor proteins (APP). APP-knockout mice show cognitive deficits. Comparative studies of freely moving healthy and APP-knockout mice have shown that the lack of APP leads to reduced theta-gamma coupling in the parietal cortex and hippocampus but not in the prefrontal cortex, although in all these regions, the spectral power in these bands in both types of mice was comparable [665]. These observations are relevant because APP is significantly involved in the destructive process in Alzheimer's disease, and early measurement of APP deficits may be valuable in diagnosis.

5.1.7.4   Analysis of Non-Phase-Locked Responses

A given sensory or internally generated stimulus or a motor response can provoke not only the time-locked event-related response, but can also induce changes in the ongoing EEG/MEG activities. These effects can be quantified by observing the relative changes: (i) in the power of specific frequency bands—known as event-related synchronization ERS or desynchronization ERD [475], (ii) the changes in the phase distribution—known as phase resetting [529], (iii) phase synchrony [143], and (iv) phase clustering [269].

Event-related Synchronization and Desynchronization

To introduce the ERD and ERS measures we have to realize that in the spontaneous ongoing brain activity (recorded as EEG) rhythmic components are present (see Sect. 5.1.2). The measure of ERD is defined as the relative decrease and ERS as the relative increase of power in a given frequency band relative to some baseline level. Thus mathematically both effects can be described by the same formula:

$$ERD/ERS = \frac{S_f - R_f}{R_f} \qquad (5.17)$$

with ERD relating to negative values and ERS to the positive values; S_f is the band power in the analysis time; and R_f is the band power in the baseline time. The baseline time should be selected in such a way that the signal can be considered as not affected by the processes under investigation. The baseline epoch should contain only the spontaneous activity.

The physiological interpretation of ERD/ERS is explicitly expressed in their names. The EEG signal can be observed mainly due to synchronous activity of neurons under the electrode which corresponds to the area of cortex of the order of a few squared centimeters. The amplitude of the rhythmic oscillation is proportional to the number of synchronously active neurons (Sect. 5.1.1). In this sense the ERD and ERS observed in EEG measure the level of synchronization of huge ensembles of cortical neurons. However, one should be aware of the spatial scale of the physiological processes reflected in the ERD/ERS. Especially in the case of ERD one can imagine two scenarios leading to decrease of power measured on the EEG level. First, the simplest, is that in the whole patch of cortex the oscillations in a given frequency band disappeared; perhaps the frequency of oscillation had shifted to some other frequency band. The second scenario is that, during the baseline time, an ensemble of neural populations oscillates synchronously in a given rhythm. Then during the information processing or mental activity the ensemble splits into many smaller ensembles, each oscillating with its own frequency and phase. In the superposition of their activity we would observe a decrease of power of the rhythm in respect to that which was present during baseline time.

In the following paragraphs we shall review methods developed for quantification of ERD/ERS. In order to determine the statistically significant ERD/ERS values one needs an appropriate number of repetitions of the experiment.

Classical Frequency Band Methods

The standard method for evaluation of ERD/ERS was proposed by Pfurtscheller and Aranibar [473]. It can be described by the following algorithm:

- Let $x_f(t; i)$ be the band-pass filtered signal in trial $i \in \{1, \ldots, N\}$, and the baseline time $t_b \in \{t_1, \ldots, t_k\}$

- Square the filtered signal samples to obtain instantaneous band-power

$$S_f(t;i) = x_f^2(t;i) \tag{5.18}$$

sometimes the mean of the data across trials is subtracted before squaring to remove the ERP component, giving so called inter-trial variance:

$$S_f(t;i) = (x_f(t;i) - \bar{x}(t,f))^2 \tag{5.19}$$

- Compute averaged band-power by averaging instantaneous band-power across trials:

$$\bar{S}_f(t) = \frac{1}{N} \sum_{i=1}^{N} S_f(t;i) \tag{5.20}$$

- Average the power in the baseline time

$$R_f = \frac{1}{N} \sum_{t \in t_b} \bar{S}_f(t) \tag{5.21}$$

- Compute the relative power change

$$ERD/ERS(t) = \frac{\bar{S}_f(t) - R_f}{R_f} \cdot 100\% \tag{5.22}$$

- Smooth the relative power change by moving average or low-pass filtering or complex demodulation using the Hilbert transform.

  Illustration of the above algorithm is shown in Figure 5.20.

Time-frequency Methods

The frequency bands at which the ERD/ERS effects are observed and ERD/ERS unique temporal and topographic patterns related to the functional brain activation vary considerably between subjects. In the classical approach one has to try out a number of possible frequency bands in the search for the ones that display the most pronounced effects—so called reactive frequency bands. As was pointed out in [472] it is important to examine event-related changes in signal energy from the broader perspective of the entire time-frequency plane.

In order to accomplish this, one needs to estimate the energy density of the signal in the time-frequency space. In the literature many methods were proposed for this purpose, e.g.:

- Spectrogram [364],

- Bandpass filtering in overlapping bands [193],

- Scalogram [587],

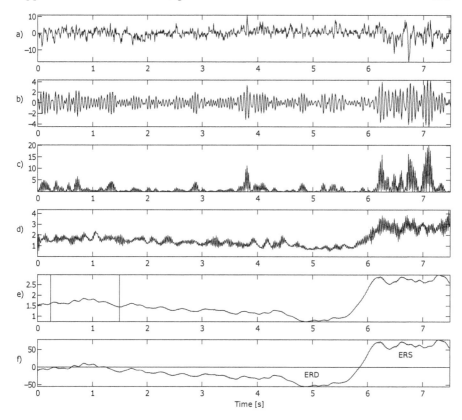

FIGURE 5.20
Illustration of the classical ERD/ERS estimation for the beta band. a) One
of the N EEG epochs; b) the above signal filtered in the 15–25 Hz band; c)
instantaneous power of the filtered signal; d) instantaneous power averaged
over N epochs; e) smoothed version of the signal in (d), the reference time
marked by two vertical lines; f) ERD/ERS time course—average instantaneous
power relative to the mean value in the reference time.

- Smoothed pseudo Wigner-Ville transform [317]),

- Estimate of energy density derived from the matching pursuit (MP) parameterization [129].

The mathematical basis, properties, and MATLAB routines to compute the
above mentioned estimators of the energy are described in Sect. 3.4.2. Different estimates of the time-frequency distribution of signal energy offer different trade-offs between temporal and spectral resolution. For example, the
scalogram has high temporal resolution and low frequency resolution at high
frequencies, low temporal resolution and high frequency resolution at low

frequencies. The presence of cross-terms in some of the time-frequency estimators must be taken into account when interpreting the ERD/ERS in the time-frequency space. In contrast, the matching pursuit (MP) parameterization provides optimal time-frequency resolution throughout the time-frequency plane. Different estimators of energy density were compared in [674]. The results yielded by different estimators gave compatible results, however those obtained by MP procedure provided more detailed information of the time-frequency structure of the ERD/ERS.

The time frequency estimator of energy density $E(t, f)$ can be used to evaluate the ERD/ERS in the time-frequency plane in accordance with the general definition (5.17):

$$ERD/ERS(t, f) = \frac{\langle E(t, f)\rangle_{tr} - B(f)}{B(f)} \tag{5.23}$$

where $\langle E(t, f)\rangle_{tr}$ is the energy density at (t, f) averaged across trials, and $B(f)$ is the mean energy of baseline at frequency f averaged across trials.

The important question to be answered is: which of the ERD/ERS effects are significantly above the fluctuation level? The statistical problem related to that question is the assessment of significance in the statistical maps (see Sect. 2.5.3.2). In [132, 674] a massive univariate approach with the multiple comparisons problem controlled with FDR (Sect. 2.5.3.3) was proposed as an efficient solution to that question, and will be described below.

For the null hypothesis of no changes at the given time-frequency coordinates, we have to reduce the resolution to time-frequency boxes ($\Delta t \times \Delta f$). There are two reasons for decreasing the time-frequency resolution:

- The time and frequency resolution are bounded by the uncertainty principle, which for the frequencies defined as the inverse of the period (Hz) gives: $\Delta t \times \Delta f \geq \frac{1}{4\pi}$. The lower bound is only reached by the Gabor functions

- In real data there are big variations of energy estimates due to different kinds of noise. Increasing the product $\Delta t \times \Delta f$ reduces the variance of energy estimate

Results from [132], suggest that $\Delta t \times \Delta f = 1/2$ gives robust results. This introduces a discretization of the time-frequency plane into resels (resolution elements) $r(i, j)$. In order to obtain $E_n(i, j)$ — energy in resels $r_n(i, j)$ (subscript n denotes the trial) — we integrate[7] energy density $E_n(t, f)$:

$$E_n(i, j) = \int_{i \cdot \Delta t}^{(i+1) \cdot \Delta t} \int_{j \cdot \Delta f}^{(j+1) \cdot \Delta f} E_n(t, f) dt df \tag{5.24}$$

---

[7] In the spectrogram equation (3.85) and scalogram equation (3.96), the energy is primarily computed on the finest possible grid and then the integration is approximated by a discrete summation. In case of MP the integration of equation (3.118) can be strict. The procedure was described in detail in [132].

At this point, we may proceed to testing the null hypothesis of no significant changes in $E_n(i, j)$. The distribution of energy density for a given frequency is not normal. However, in many practical cases it can be transformed to an approximately normal one using an appropriate Box-Cox transformation [66]. The Box-Cox transformations are the family of power transformations:

$$BC(x, \lambda) = \begin{cases} \frac{x^\lambda - 1}{\lambda} & if \quad \lambda \neq 0 \\ \log(x) & if \quad \lambda = 0 \end{cases} \tag{5.25}$$

For each frequency j the λ parameter is optimized by maximization of the log-likelihood function (LLF)[237] in the reference period:

$$\lambda_{\text{opt}}^j = \max_\lambda \left\{ LLF(\lambda) \right\} = \max_\lambda \left\{ -\frac{m}{2} \log \sigma_{BC(x,\lambda)}^2 + (\lambda - 1) \sum_{k=1}^m \log x \right\} \tag{5.26}$$

where m is the length of data x, $x \in \{E_n(i, j) : i \in t_b, n = 1, \ldots, N\}$. The optimal λ_{opt}^j is then used to transform all the resels in frequency j. Standard parametric tests can be applied to the normalized data. However, we cannot a priori assume equal variances in the two tested groups. The known solution to the problem of variance heterogeneity in the t-test is Welch's [633] correction of the number of degrees of freedom. The test can be formulated as follows for each of the resels (i, j). The null hypotheses $H_0^{i,j}$ and alternative hypotheses $H_1^{i,j}$:

$$H_0^{i,j} : \langle X(j) \rangle_{\text{b,tr}} = \langle X(i,j) \rangle_{\text{tr}} \tag{5.27}$$

$$H_1^{i,j} : \langle X(j) \rangle_{\text{b,tr}} \neq \langle X(i,j) \rangle_{\text{tr}} \tag{5.28}$$

where: $\langle X(j) \rangle_{\text{b,tr}}$ is the normalized energy in the baseline time averaged across baseline time and trials, and $\langle X(i,j) \rangle_{\text{tr}}$ is the normalized energy in resel (i, j) averaged across trials. The statistics is:

$$t(i, j) = \frac{\langle X(j) \rangle_{\text{b,tr}} - \langle X(i,j) \rangle_{\text{tr}}}{s_\Delta} \tag{5.29}$$

where s_Δ is the pooled variance of the reference epoch and the investigated resel. The corrected number of degrees of freedom ν is:

$$\nu = \frac{\left(\frac{s_1^2}{n_1} + \frac{s_2^2}{n_2} \right)^2}{\frac{\left(\frac{s_1^2}{n_1} \right)^2}{n_1 - 1} + \frac{\left(\frac{s_2^2}{n_2} \right)^2}{n_2 - 1}} \tag{5.30}$$

where s_1 is the standard deviation in the group of resels from the reference period, $n_1 = N \cdot N_b$ is the size of that group, s_2 is the standard deviation in the group of resels from the event-related period, and $n_2 = N$ is the size of that group.

If we cannot assume $X_n(i, j)$ to be distributed normally, we estimate the distribution of the statistic t from the data using (5.29), separately for each frequency j:

1. From the $X(i, j)$, $i \in t_b$ draw with replacement two samples: A of size N and B of size $N \cdot N_b$

2. Compute t as in (5.29): $t_r(j) = \dfrac{\langle X_A \rangle - \langle X_B \rangle}{s_\Delta}$, where s_Δ is pooled variance of samples A and B.

3. Repeat steps 1 and 2 N_{rep} times.

The set of values $t_r(j)$ approximates the distribution $T_r(j)$ at frequency j. Then for each resel the actual value of (5.29) is compared to this distribution:

$$p(i, j) = 2 \min \{ P\left(T_r(j) \geq t(i, j)\right), 1 - P\left(T_r(j) \geq t(i, j)\right) \} \qquad (5.31)$$

yielding two-sided $p(i, j)$ for the null hypothesis H_0^{ij}. The relative error of p is (c.f. [138])

$$err = \frac{\sigma_p}{p} = \sqrt{\frac{(1 - p)}{pN_{rep}}} \qquad (5.32)$$

Although the above presented method is computationally intensive, at present it causes no problems in most of the standard applications. However, corrections for multiple comparisons imply very low effective critical values of probabilities needed to reject the null hypothesis. For the analysis presented in [132] critical values of the order of 10^{-4} were routinely obtained. If we set $p = 10^{-4}$ in (5.32), we obtain a minimum $N_{rep} = 10^6$ resampling repetitions to achieve 10% relative error for the values $p(i, j)$.

In [132] either parametric or resampled statistical tests were applied to energies in each resel separately. However, the very notion of the test's confidence level reflects the possibility of falsely rejecting the null hypothesis. For example, a confidence level of 5% means that it may happen in approximately one in 20 cases. If we evaluate many such tests we are very likely to obtain many such false rejections. This issue is known in statistics as the issue of multiple comparisons, and there are several ways to deal with it properly.

To get a valid overall map of statistically significant changes we suggest the approach chosen in [132] that is a procedure assessing the false discovery rate (FDR, proposed in [43]). The FDR is the ratio of the number of falsely rejected null hypotheses (m_0) to the number of all rejected null hypotheses (m). In our case, if we control the FDR at a level $q = 0.05$, we know that among resels declared as revealing a significant change of energy, at most 5% of them are declared so falsely. [44] proves that the following procedure controls the FDR at the level q under positive regression dependency, which can be assumed for the time-frequency energy density maps:

1. Order the achieved significance levels p_i, approximated in the previous section for each of the resels separately, in an ascending series: $p_1 \leq p_2 \leq \cdots \leq p_m$

2. Find

$$k = \max_i \left\{ p_i \leq \frac{i}{m} q \right\} \qquad (5.33)$$

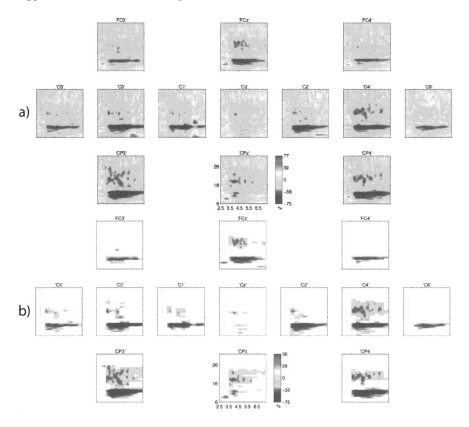

FIGURE 5.21
An example of time-frequency analysis of ERD/ERS (left hand movement imagination). The t-f maps of selected electrodes are located topographically. Horizontal axis—time [s]; vertical—frequency [Hz]. Onset of imagination marked by a dotted line. a) ERD/ERS map, b) significant effects (Sect. 5.1.7.4).

3. p_k is the effective significance level, so reject all hypotheses for which $p \leq p_k$.

Resels $r(i,j)$ are marked as significant if the null hypothesis $H_0^{i,j}$ can be rejected using the significance level p_k for the probabilities $p(i,j)$ of the null hypothesis (5.27). An example of ERD/ERS time-frequency maps together with the assessment of their significance obtained with the above described procedure is illustrated in Figure 5.21.

ERD/ERS in the Study of iEEG

The ERD/ERS methodology was also applied to iEEG signals. The most advanced method providing the highest and most adaptive time-frequency resolution—matching pursuit was used, e.g., in [674] for evaluation of ERD/ERS responses from cortex during hand movement and for iEEG analysis during speech perception [499].

The ERD/ERS methodology applied to the iEEG made it possible to study high gamma responses (HGR), which are difficult to record in standard EEG. They were first described in detail in human ECoG recorded in sensorimotor cortex [98]. HGRs have a broadband spectral profile, which range from 60 Hz to 200 Hz with the majority of event-related energy in the band 80–150 Hz. The ubiquitous occurrence of HGRs which have been found in motor, somatosensory, auditory, and visual cortex [552] during functional activation suggests that they are a general electrophysiological index of cortical processing, since they are correlated with the degree of functional activation. Detailed time-frequency analysis of HGRs from LFP of somatosensory cortex revealed that they are temporally tightly linked to neuronal spikes [498].

Recently high frequency gamma components have been measured also in scalp EEG [31, 102] and in MEG [100, 202], but iEEG and LFP give the best opportunity to study HGR and especially to correlate their characteristics with the spike occurrence.

Event-related Time-varying Functional Connectivity

The methods described in previous paragraphs concentrated on the analysis of ERD/ERS that is on the variations of power related to a certain event. Another phenomenon, which is equally interesting, is related to the event-related variation of couplings between the different distant populations. In this respect the methods derived from multivariate auto-regressive model (Sect. 4.2) can be very useful.

Phenomenon involving the phase information, namely a time varying directional connectivity between neural population is recently the focus of interest. In the past the bivariate methods usually based on coherences were used, however the demonstration that they give misleading results (Sect. 4.5) turned attention to the methods based on extension of Granger causality defined in the framework of MVAR.

When applying MVAR one has to keep in mind that the number of model parameters should be preferably smaller by an order of magnitude than the number of samples in the data window. Number of MVAR parameters is pk^2 (where p—model order, k—number of channels), number of data points is kN (where N—number of points in the window). Effectively we get a condition: $pk/N < 0.1$.

In order to get a time-varying estimate we need to use a short time window, which is in contradiction with the above relation. The number of data points may be effectively increased by means of ensemble averaging over realizations when they are available from repeated trials of the experiment. To solve the problem of the time-varying estimate two approaches may be distinguished: sliding window (SDTF) or time-continuous fit of MVAR model, which may be performed adaptively, e.g., by means of a Kalman filter. Usually in both approaches ensemble averaging is applied.

The Kalman filter approach was extended for multivariate non-stationary processes by Arnold et al. [13]. Adaptive DTF (ADTF) based on Kalman filter was used, e.g., in [643] to study time-variant propagation of simulated epileptic seizures. The comparison of SDTF and ADTF showed for limited number of channels similar performance of both methods. However, in the case of the Kalman filter the computation time is higher and rises fast with the number of channels and number of realizations [276]. Taking into account that the estimation of the errors is usually based on bootstrap methods, which involves repetition of computations hundreds of times, the computation time in the Kalman filter approach (much longer than in case of SDTF) can be a serious drawback, which hampers its wide application. In [276] it was also reported that the Kalman method has difficulties in adaptation for a large number of channels.

Another adaptive method of MVAR fitting is the least mean square (LMS) algorithm. The adaptation capability was found to be better for its modification—recursive least-square (RLS) algorithm with forgetting factor [409]. This algorithm takes into consideration a set of EEG epochs and the RLS estimation is controlled (similarly to Kalman filter) by the value of adaptation factor. The algorithm was initially used for estimation of multivariate coherences [409]. It was also applied for estimation of bivariate Granger causality in the Stroop test[8] [219] and in [18].

An example of application of time-varying formulation of PDC was a study of a foot movement task [114]. Time-dependent MVAR parameter matrices were estimated by means of the RLS. The MVAR model was fitted to the signals representing the current density on cortex found by solving the linear inverse problem [198]. The dense connections obtained between 16 regions of interest formed a complicated pattern and to find the meaningful transmissions between the regions of interest the theoretical graph indexes [430] have to be applied.

The procedure of finding current density in the cortex by solving the inverse problem requires assumptions concerning source localization, is non-unique and may disrupt the phases between the channels disturbing in this way connectivity patterns. Taking into account the fact that DTF and PDC estimators are practically not influenced by volume conduction (which is a zero phase propagation), and fast decay of electromagnetic field of dipole layer with the distance the operation of finding the current density on cortex seems to be unnecessary.

---

[8]In the Stroop test congruent and non-congruent stimuli are compared, e.g., word blue written in blue with the word blue written in red.

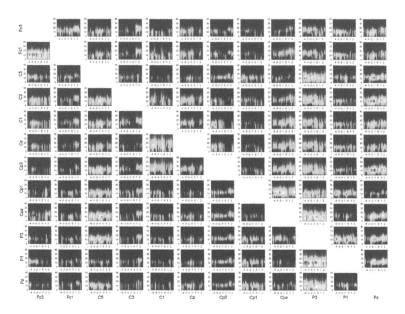

FIGURE 5.22
Propagation of EEG activity in the left hemisphere during right finger movement. In each panel SDTF as a function of time (horizontal axis) and frequency (vertical axis). The flow of activity is from electrode marked under the column to the electrode marked at the relevant row. Red—the highest intensity, blue—the lowest. From [183].

The determination of the dynamic propagation during performance of finger movement and its imagination [183, 184, 312] may serve as an example of the application of short-time DTF to the scalp EEG. The established evidence [475] is that during the movement the decrease of activity in alpha and beta bands (ERD) in the primary motor area corresponding to the given part of the body is observed (in case of a hand movement in the contralateral hemisphere), and increase of beta activity called beta rebound follows. In the gamma band, brief increase during movement was reported [475]. These findings corresponded very well with the results obtained by means of SDTF. Figure 5.22 shows the SDTFs as functions of time and frequency for a right finger movement experiment. The gaps in the propagation of alpha and beta activity for electrodes overlying left motor cortex (most pronounced for electrode C3) and subsequent fast increase of propagation in beta band more frontally are compatible with spectrally, temporally, and topographically specific ERD/ERS phenomena.

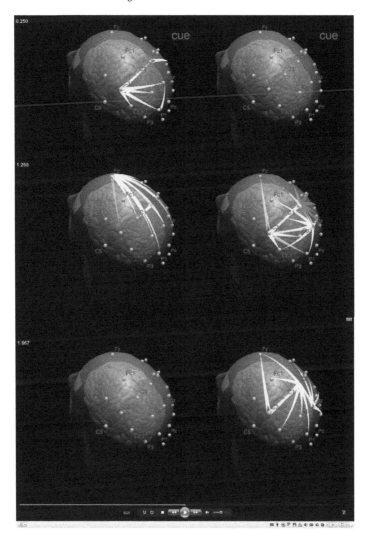

FIGURE 5.23
Snapshots from the animation representing transmission in the gamma band during right finger movement (left column) and its imagination (right column) obtained by SDTF. The intensities of flows represented as colors of arrows (red the strongest). Numbers indicate the time in seconds after the cue presentation.

By virtue of selectivity of SDTF to phase difference (neglecting zero phase or random phase dependencies) the evolution in the gamma band may be studied (after filtering of prevailing alpha activity). In Figure 5.23 the EEG propagation in the gamma band during right finger movement and its imagination, obtained by SDTF, is shown. In case of movement the short burst of gamma propagation from C3 followed by propagation from frontal electrodes was observed. In case of movement imagination this propagation started later and a cross-talk between different sites overlying motor area and other sensorimotor areas may be noticed. (The dynamics of propagation may be observed in animations available at: `https://www.fuw.edu.pl/~kjbli/DTF_MOV.html`). This kind of transmission is compatible with the notion that a more difficult task requires involvement of the several sensorimotor areas and it is in agreement with neurophysiological hypotheses concerning the interactions of brain structures during simple and complicated motor tasks.

Another application of SDTF concerned evaluation of transmission during cognitive experiments. In the continuous attention test (CAT) geometrical patterns were displayed and the subject had to react to two subsequent identical pictures by pressing the switch (situation target) and do nothing when the patterns were not identical (situation non-target). The results confirmed the engagement of pre-frontal and frontal structures in the task and elucidated the mechanisms of active inhibition [52]. Figure 5.24 shows the snapshots from the movie illustrating transmissions. Only significant flows showing increase or decrease of propagation in respect to reference period are shown. They were calculated according to the procedure devised in [298]. In Figure 5.24 one can observe for the non-target situation the flow from electrode F8 (overlying right inferior cortex) to electrode C3. This kind of transmission was present in six out of nine studied subjects. In three subjects the propagation for non-target was from the Fz electrode (overlying pre-supplementary motor area). Both above mentioned structures are connected with movement inhibition. One can conjecture that the observed propagation was the expression of withdrawal from the action of pressing the switch. In case of target the propagation from the C3 electrode initiating the motor action was observed. The obtained results supported the hypothesis of an active inhibition role in motor tasks played by right inferior frontal cortex and pre-supplementary motor area. The animations of propagation during a CAT test are available at URL `https://www.fuw.edu.pl/~kjbli/CAT_MOV.html`.

The results obtained using SDTF in experiments involving working memory were compatible with fMRI studies on the localization of the active sites and supplied the information concerning the temporal interaction between them [70, 56].

Transmission of EEG activity during working memory (WM) task concerning memorizing letters and relations between them was analyzed using DTF and SDTF and was followed by application of assortative mixing formalism. In agreement with fMRI study [70], SDTF revealed the existence of main centers of EEG activity in frontal and posterior regions, which intermittently exchanged information using bursts of EEG propagation mainly in theta and gamma bands. This kind of dynamical transmission was observed for visual [60] and auditory WM task [359, 274]. Analysis of dynamical

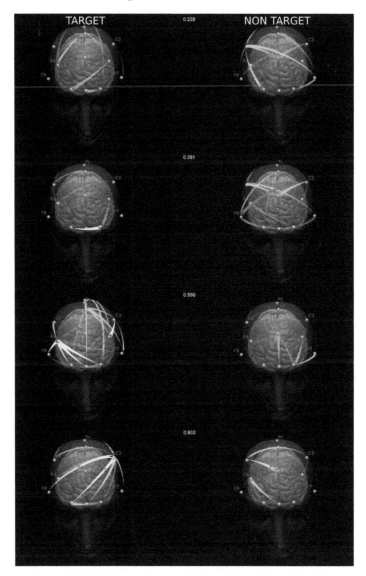

FIGURE 5.24

The snapshots from the animation representing the significant changes of propagation during the CAT test obtained by SDTF. Red and yellow arrows—increase of propagation, blue—decrease of propagation. At the right—target; at the left—non-target. In the middle—time after the stimulus in seconds. Please note in the non-target case at 0.381 s strong increase of propagation F8 → C3 connected with active inhibition of movement; and in the case of target, at 0.803 s, propagation from C3 connected with command to move the finger.

information processing in the WM task showed frequency-specific information transfer occurring transiently between distant neural populations.

The quantitative assessment of connectivity structure in WM tasks was based on the assortative mixing approach (see Sect. 5.1.6.5). For the modules corresponding to the regions of interest (ROIs), DTF functions concerning intra-module and inter modules propagations were calculated. According to the formalism of assortative mixing, the elements of the matrix E_{kl} were represented by the propagations (DTFs) either inside the module (E_{kk}) or between the modules (E_{kl}). The **E** matrix representing the connectivity pattern was calculated using the formula :

$$E_{kl} = \sum_{\substack{n \in \text{channels} \\ \text{in module } k}} \sum_{\substack{m \in \text{channels} \\ \text{in module } l}} \text{DTF}(n \to m) \tag{5.34}$$

where $\text{DTF}(n \to m)$ denotes the value of $\text{DTF}_{n \to m}(f)$ integrated in the given frequency range. Then the matrix **E** was averaged over subjects and normalized in such a way that the sum of all its elements was 1. Inspecting the elements of matrix, E_{kl}, we may infer, if our case corresponds to the situation of highly assortative mixing $E_{kk} > E_{kl}$, which means stronger coupling inside the modules than between the modules. Using the above method, the existence of a modular structure of brain networks with a higher connection density within modules than between them was demonstrated [60]. The quantitative analysis based on statistical distinction of intra-module and extra-module causal couplings revealed high similarity of propagation patterns for both modalities.

The above described experiments point out that by means of SDTF estimated from scalp recorded EEG a very clear-cut and consistent evidence concerning the activity propagation may be found. The findings are compatible with the known facts and at the same time they have supplied new evidence concerning the information processing in the brain.

Functional Connectivity Estimation from Intracranial Electrical Activity

The study of intracranial EEG (iEEG) gives the opportunity to record the electrical activity, practically free of artifacts, directly from the brain structures. In the past, this research was mostly concentrated on the analysis of spike trains; the role of slow cortical activity was mostly neglected. Nowadays, the role of oscillatory activity of neural networks has become more and more recognized. There is a consensus that the higher cortical functions depend on dynamic interplay between spatially distributed multiple cortical regions coupled by the transmission of oscillatory activity, e.g., [553, 75].

The most popular method to study the interaction between cortical areas is coherence (in spite of the limitations of bivariate coherence). To investigate the dynamics of interactions between brain structures, temporal fluctuations in coherence were studied, e.g., in [73], and oscillatory synchrony was considered in [586]. Recently attention has focused on the directionality of interactions between brain regions. Among the methods used for the study of directionality, those based on Granger causality seem to be the most appropriate.

For the analysis of signals recorded from specific brain structures it is important to determine the direct interaction. Direct directed transfer function (Sect. 4.3.2.3) was introduced in [300] for the purpose of investigating the propagation of LFP recorded from electrodes chronically implanted in the brain of a behaving animal. The patterns of propagation were recorded from four brain structures involved in processing of emotions. The patterns of transmission were studied for an animal walking on a runway and for walking on a runway accompanied by a stressful stimulus (bell ringing). More flows between the brain structures appeared in the case of the stressful stimulus and most of them were reciprocal.

The investigation of the intracranial human EEG is limited to cases of patients who are being prepared for surgical intervention. The limitations in the iEEG research are caused by the fact that the electrode placement is dictated solely by clinical concerns. However, the area covered by electrodes is usually not confined to the location of the diseased tissue because this location is not precisely determined before implantation and also the neighboring areas have to be checked to be sure that surgery will not seriously disturb sensory, motor, speech, or cognitive functions. iEEG provides unprecedented opportunity for studying indices of cortical activation, since it is characterized by high temporal resolution and mesoscopic spatial resolution that is intermediate between the macroscopic scale of EEG/MEG and multiunit recording of neuronal activity.

iEEG allows for investigation of high-gamma activity, which is hard to observe using scalp electrodes. Causal interaction between signals in high-gamma range (above 60 Hz) for the word repeating task were studied in [298]. The authors introduced a new measure—short-time direct directed transfer function (SdDTF), which combined the benefits of directionality, directedness, and short-time windowing. This function appeared to be an effective tool for analyzing non-stationary signals such as EEG accompanying cognitive processes. The performance of the function was tested by means of simulations, which demonstrated that the SdDTF properly estimates directions, spectral content, intensity, direct causal interactions between signals, and their time evolution. To evaluate event-related changes by SdDTF, that is, event-related causality (ERC), a statistical methodology, described in Sect. 5.1.7.4, was developed to compare prestimulus and poststimulus SdDTF values.

In order to quantitatively describe the transmissions, ERC for high gamma activity was integrated in the 82–100 Hz range, which was empirically derived based on the mean ERC over all time points and all pairs of analyzed channels. Figure 5.25 shows the magnitudes of the interaction in the form of arrows of different widths. The most prominent identified connections involved: in the first phase (listening) flows from the auditory associative cortex to mouth/tongue motor cortex, and in the second phase (repeating of the word)

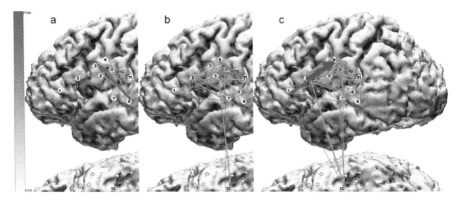

FIGURE 5.25
Integrals of ERC for frequency range 82–100 Hz calculated for three stages of an auditory word repetition task. (a) Auditory perception stage, (b) response preparation, (c) verbal response. Arrows indicate directionality of ERC. Width and color of each arrow represent the value of the ERC integral. Color scale at the left. For clarity, only integrals for event-related flow increases are shown. From [298].

propagation from the Broca's area (responsible for speech) to mouth/tongue motor cortex.

One of the problems important for neuroscience is the relation between the spike trains and local field potentials. For the spike train evaluation the methods developed in the field of point processes analysis are customarily applied [68]. Nevertheless there is a possibility of using the broad repertoire of stochastic continuous signal analysis methods, described in this book, to spike trains.

In the approach proposed in [295] the spike trains were processed in the following way: the spikes were low-pass filtered by an order 1 Butterworth filter with cutoff frequency at 10% of Nyquist frequency (the filtering procedure was applied as zero phase filter; Sect. 3.1); then 10% of stochastic noise uncorrelated with the signal was added in order to make the spike train better match the stochastic character of the AR model. The procedure is illustrated in Figure 5.26.

The described approach was used in the experiment where LFP was recorded from hippocampus and spike trains from the supramammilliary nucleus (SUM) of a rat with the aim of finding the dynamic coupling between the structures in a situation when the sensory stimulus was applied. The MVAR model was fitted to the spike signal from SUM (transformed in the above described way) and the hippocampal LFP; then the SDTF functions were estimated. The temporal dynamics of the direction of influence revealed sharp reverses in the direction of the theta drive in association with sensory-elicited

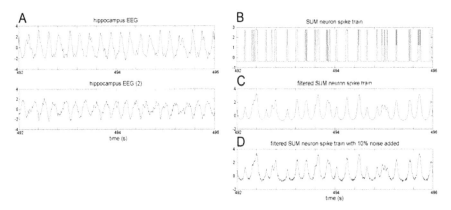

FIGURE 5.26
Transforming spike train into continuous signal. A, LFP recorded from hippocampus, B, standardized spike train, C, low-pass filtered spike train, D, low-pass filtered spike train with 10% noise added. From [295].

theta rhythm. It was found that in this situation the subpopulation of SUM neurons contains information predicting future variations in the LFP rhythm in hippocampus. In contrast, during slow spontaneous theta rhythm it was the SUM spike train that can be predicted from the hippocampal activity.

The described procedure of turning the spike trains to the continuous signals opens new perspectives of application of the methods of signal processing to point processes. In particular this approach may be useful for spike train analysis and for the investigation of the relations between point processes and continuous signals.

Statistical Assessment of Time-varying Connectivity

Functional connectivity is expected to undergo rapid changes in the living brain. This fact should be taken into account when constructing the statistical tool to test the significance of the changes in the connectivity related to an event. Especially, it is difficult to keep the assumption that during the longer baseline epoch the connectivity is stationary. ERC involves statistical methods for comparing estimates of causal interactions during pre-stimulus baseline epochs and during post-stimulus activated epochs that do not require the assumption of stationarity of the signal in either of the epochs. Formally the method relies on the bivariate smoothing in time and frequency defined as:

$$Y_{f,t} = g(f,t) + \varepsilon_{f,t} \qquad (5.35)$$

where $g(f,t)$ is modeled as penalized thin-plate spline [515] representing the actual SdDTF function and $\varepsilon_{f,t}$ are independent normal random ($N(0,\sigma_\varepsilon^2)$)

variables. The thin-plate spline function can be viewed as spatial plates joined at a number of knots. The number of knots minus the number of spline parameters gives the number of the degrees of freedom.

To introduce the testing framework proposed in [298], we first introduce some notations. Denote by $f_1, \ldots, f_m$ the indexes of frequencies where f_m is the number of analyzed frequencies. Also, denote by $t = t_1, \ldots, t_n$ the time index corresponding to one window in the baseline, where t_n is the total number of baseline windows. Similarly, denote by $T = T_1, \ldots, T_n$ the time index corresponding to one window in the post-stimulus period, where T_n is the total number of post-stimulus windows. The goal is to test for every frequency f, and for every baseline/stimulus pair of time windows (t, T), whether $g(f, t) = g(f, T)$. More precisely, the implicit null hypothesis for a given post-stimulus time window T at frequency f is that:

$$H_{0,f,T} : g(f, t_1) = g(f, T) \text{ or } g(f, t_2) = g(f, T) \text{ or }, \ldots, g(f, t_n) = g(f, T)$$
(5.36)

with the corresponding alternative

$$H_{1,f,T} : g(f, t_1) \neq g(f, T) \text{ and } g(f, t_2) \neq g(f, T) \text{ and }, \ldots, g(f, t_n) \neq g(f, T)$$
(5.37)

To test these hypotheses a joint 95% confidence interval for the differences $g(f, t) - g(f, T)$ for $t = t_1, \ldots, t_n$ is constructed. Let $\hat{g}(f, t), \hat{\sigma}_g^2(f, t)$ be the penalized spline estimator of $g(f, t)$ and its associated estimated standard error in each baseline time window. Similarly, let $\hat{g}(f, T), \hat{\sigma}_g^2(f, T)$ be the penalized spline estimator of $g(f, T)$ and its associated estimated standard error in each post-stimulus time window. Since residuals are independent at points well separated in time, the central limit theorem applies and we can assume that for every baseline/stimulus pair of time windows (t, T)

$$\frac{(\hat{g}(f, t) - \hat{g}(f, T)) - (g(f, t) - g(f, T))}{\sqrt{\hat{\sigma}_g^2(f, t) + \hat{\sigma}_g^2(f, T)}} \sim N(0, 1)$$
(5.38)

approximates a standard normal distribution. A joint confidence interval with at least 95% coverage probability for $g(f, t) - g(f, T)$ is:

$$\hat{g}(f, t) - \hat{g}(f, T) \pm m_{95} \sqrt{\hat{\sigma}_g^2(f, t) + \hat{\sigma}_g^2(f, T)}$$
(5.39)

where m_{95} is the 97.5% quantile of the distribution

$$MAX(t_n, T_n) = max_{t \in \{t_1, \ldots, t_n\}, T \in \{T_1, \ldots, T_n\}, f \in \{f_1, \ldots, f_m\}} |N_{t,T,f}|$$
(5.40)

where $N_{t,T,f}$ are independent $N(0, 1)$ random variables. This is equivalent to applying the Bonferonni correction for $t_n T_n f_m$ tests to control the family-wise error rate (FWER).

The utility of the ERC approach was demonstrated through its application to human electrocorticographic recordings (ECoG) of a simple language task [298]. ERC analyses of these ECoG recordings revealed frequency-dependent interactions, particularly in high gamma ($>60\,\text{Hz}$) frequencies, between brain regions known to participate in the recorded language task, and the temporal evolution of these interactions was consistent with the putative processing stages of this task.

5.1.7.5 Analysis of EEG for Applications in Brain-Computer Interfaces

Brain-computer interface (BCI) is a system that enables a user to control it using only the activity of the brain, without the need to activate muscles. It is envisioned as a technology that can enable severely motor-impaired users, like in locked-in syndrome, to use assistive appliances [363]. It is being developed towards improving diagnosis and monitoring of patients with disorder of consciousness (DOC) [124]. Finally, there are also attempts to apply it as a new input device for gaming [286].

A BCI is a closed-loop system. It consists of a user whose mental activity is measured, a measurement system, a signal processing algorithm, and a classifier; and lastly, an appliance that executes actions meant by the user and giving him feedback. In this chapter, we will focus only on the signal processing part.

Among the techniques of measuring brain activity like fMRI, NIRS, MEG, EEG, and iEEG, only the last two are of practical interest as they:

- measure electrical activity, thus operate in the time scale relevant for real-time interaction,

- are potentially mobile,

- can be affordable for individual users.

Further, we will focus on non-invasive measurements, i.e., EEG based BCI. Current literature is dominated by three BCI classes, i.e., based on P300, steady-state visual evoked potential (SSVEP), and mental imagery (MI). Let us shortly recall what the core effect utilized in each of them is.

P300 is an event-related wave observed in the EEG recorded from the centro-parietal locations with latency in the range 300–600 ms after stimulus. It is enhanced when an attended (target) stimulus is perceived. Usually, the input-feedback part of the interface consists of a display that highlights items in random order, and the user counts the number of hits of the target item. If a strong enough P300 wave related to one of the items is detected, action connected with this item is executed.

SSVEP is evoked by observing light flickering with a constant frequency. The frequency observed by the user can be detected in the EEG signal from parieto-occipital locations. Thus the input part of the SSVEP-BCI consists of a display with several fields simultaneously flickering with different frequencies. If the user observes one of them, let's call it f_0, with attention, the corresponding set of harmonic frequencies, i.e., $f_0, 2f_0, 3f_0, \ldots$, gets augmented in the EEG signal, which allows for detection of the intended field.

Finally, mental imagery, often motor imagery, causes a modulation in the amplitude of specific brain rhythms in particular brain regions. For example, in motor imagery, mu-rhythm over the motor cortex is modulated differently, depending on the limb and type of movement being imagined.

Common Elements

In all the three types of EEG-BCI, the main problem, from the signal processing point of view, is the low signal-to-noise ratio of the command related alternations in the EEG signal vs. the ongoing brain activity and noise. The requirement of operating as close as possible to real-time enforces specific ways to deal with the SNR improvement. It practically excludes, e.g., averaging over realizations, which, as we discussed in previous sections, is typically used in the case of ERPs. The most useful are here time domain techniques and spatial filtering.

The second issue typical for BCI is the huge inter and intrasubject variability of the signal features. Even for the same user, electrodes' application differs slightly from session to session due to small differences in electrode placement and the contacts' impedance. Also, in performing the tasks used for evoking the effects utilized in the interface, the user brain activity changes because of learning. This effect is especially pronounced in the mental imagery BCI (MI-BCI). Therefore, currently, most BCI systems have to begin the operation by performing a calibration procedure. Trials recorded in the calibration session are used to set-up spatial and time-domain filters and train the classifier. Only after calibration, the system is switched to the online, operational phase.

Time-domain Filtering

The BCI system's specificity is that it should operate on the EEG signal stream or short-time buffers in the online phase. Therefore we would prefer causal filtering. In MATLAB, this can be done with the `filter.m` function. The final conditions of the filter delays should be set as the initial conditions for the filter delays for the subsequent sample or buffer. See, for example, demo `MATLAB/c5/BCI_filter.mlx`

Spatial Filtering

Spatial filters, obtained in a data-driven manner, have become a key component of BCI classification pipelines. We have described the common spatial patterns (CSP) dedicated to ERP (P300-BCI), band power features (MI-BCI), and frequency-following activity, like in SSVEP-BCI in Sect. 4.6.3. To recall these ideas consider the live scripts `MATLAB/c4/Ch4_07_CSP.mlx` and `MATLAB/c4/Ch4_08_SSVEP.mlx`.

A special spacial filter developed to enhance the ERPs in P300-BCI settings is known as xDAWN [505]. Conceptually, it consists of two steps. First, a time-locked part of EEG response to the target stimuli is estimated using a least-square method. It has the advantage over a simple trial-based averaging, as it takes into account the fact that the responses can overlap if the inter-stimulus-interval is shorter than the evoked response. Second, a spacial filter, which maximizes the signal to signal-plus-noise ratio, is computed for the ERP estimated in the first step. Although xDAWN takes into account the overlapping responses, designing a P300-BCI interface, we should consider that the possible improvement of the SNR by filters such as xDAWN may be diminished by the psychophysiological decrease of the P300 amplitude due to shortening of the inter-target-intervals [190].

Riemannian Geometry

The important obstacle in making BCI more usable is the need for frequent calibration. A CSP or xDAWN filter has to be estimated after each application of electrodes. An exciting direction of development is the search for features that are invariant in this respect. The assumption that a given mental state correlates with a specific coactivation of an ensemble of EEG sources suggests using the covariance matrix as a relevant feature. Covariance matrix is a symmetric positive defined (SPD) matrix. Riemannian geometry delivers tools to compute distance (5.41) between SPD matrixes:

$$D^2(\mathbf{C}_1, \mathbf{C}_2) = \sum_n \log^2 \lambda_n(\mathbf{C}_1^{-1}\mathbf{C}_2) \tag{5.41}$$

where $\lambda_n(\mathbf{M})$ denotes the nth eigenvalue of matrix $\mathbf{M}$. Distanse D can be used to compute the centre of mass $\hat{\mathbf{G}}$ of a set $\{\mathbf{C}_1, \ldots, \mathbf{C}_K\}$ of K SPD matrices:

$$\hat{\mathbf{G}} = \arg\min_{\mathbf{G}} \sum_k D^2(\mathbf{C}_k, \mathbf{G}) \tag{5.42}$$

The authors of [32] proposed two classifiers using the Riemannian geometry approach. Riemannian minimum distance to mean (RMDM) classifier uses calibration data to estimate the geometric means of covariance matrixes corresponding to different classes. In the online phase, it computes the distance of the current covariance matrix to all of the geometric means and returns the label of the closest one. The second method maps the covariance matrices onto the Riemannian tangent space, where matrices can be vectorized and treated as Euclidean objects. Then, after the decrease of dimensionality, a classification by linear discriminant analysis (LDA) is performed.

Since the Riemannian distance is invariant both to matrix inversion and to any linear invertible transformation of the data, any mixing applied to the EEG sources does not change the distances between the observed covariance matrices, which enabled researchers to set up calibration-free adaptive ERP-BCIs using simple subject-to-subject and session-to-session transfer learning strategies [631]. For a comprehensive review of the Riemannian approaches in BCI, please refer to [92, 656].

A practical issue to consider when applying the methods based on covariance matrixes is the compromise between the need to robustly estimate: the covariance matrix, the number of channels, and the time window length used for the estimation. The constraints arise from the fact that the number of coefficients grows as the square of the number of channels, and the appropriate window length is restricted by the time-scale of the processes characteristic of the effects utilized in the interface.

Selection of Important Features

The training of BCI classifiers faces one common problem. It is a small number of available examples for model fitting, as most of the approaches rely on data obtained by the system calibration prior to online use. For this reason, in the BCI, classifiers with good generalization, even for limited size training sets, are utilized, such as naive bayes, SVM, Lasso, and LDA with shrinkage.

It is also necessary to reduce the feature space dimension to a minimum to train the classifier on a small number of examples. The simple approach measures the relation between each feature and the classes, e.g., by the mutual information between each feature and the target variable [314] (cf. Sect. 4.4.1).

Research Directions

The main direction of current research on BCI interfaces is the search for a solution that, on the one hand, ensures high efficiency of recognizing user intentions and at the same time does not require calibration [354]. It is expected that a combination of transfer learning and adaptive classifiers may have such characteristics. However, it should be remembered that in the BCI system, a user plays a vital role and that he also learns, and the activity of his brain while solving tasks adapts over time. Therefore, BCI should be viewed as a system composed of two learning elements. Thus the possible adaptation on the side of the classifier must be matched to the speed and scope of the user's adaptation [314].

5.1.8   fMRI Derived Time Series

Functional magnetic resonance imaging (fMRI) evaluates brain activity by detecting changes associated with hemodynamic response related to energy consumption by brain [517].

At present, from fMRI, it is possible not only to acquire the localizations of the active regions in the brain but also to get the brain activity time series from specific locations.

fMRI, which was primarily applied for localization of the structures involved in cognitive and perceptual processes in the brain, is currently increasingly used to study the temporal dynamics of brain activity and determine the interaction between its structures. However, fMRI's temporal resolution is not matched with the speed of information processing by the brain, so quite often, fMRI experiments are paralleled with EEG recording.

The MRI technique is based on the properties of specific atomic nuclei able to absorb and emit radiofrequency energy when placed in an external magnetic field. In an MRI scanner, a strong magnetic field forces the nucleis' spins to align with that field and precess (spin on their axes) in an orderly direction. When a radiofrequency electromagnetic impulse is applied, the nuclei are stimulated, and spin out of equilibrium, straining against the pull of the magnetic field. When the radiofrequency field is turned off, the MRI sensors detect the energy released as the nuclei realign with the magnetic field. The signal decay process is characterized by T1 (spin echo) or T2 (gradient echo) relaxation. For spatial localization, gradient magnets are applied. They increase the magnetic field in a given direction (x, y, z), changing the frequency of the precession, which serves the spatial encoding in three dimensions.

In the human body, the hydrogen atoms are most abundant, and protons are mainly used for structural imaging. In fMRI, hemoglobin blood molecules are used to generate a detectable radiofrequency signal. Hemoglobin without bound oxygen molecules is called deoxyhemoglobin or reduced hemoglobin (HbR). It is paramagnetic because of the high spin state ($S = 2$) of the heme iron. In contrast, oxygen-bound hemoglobin HbO has low spin ($S = 0$) and is diamagnetic. HbR in red blood cells makes their magnetic susceptibility different from the diamagnetic plasma in the blood, hence induces a difference in magnetic susceptibility between the blood and the surrounding tissue. When the concentration of deoxyhemoglobin changes in the blood, the relaxation process is modified, and one can see these changes in fMRI. The signal intensity that varies with deoxyhemoglobin content has been named Blood Oxygenation Level Dependent (BOLD) and is used in functional study of the brain. The BOLD signal depends on blood flow, which is contingent on the metabolic load change caused by the neuronal activation.

The relation of the BOLD signal to neural activity, i.e., neurovascular coupling, involves a complex sequence of cellular, metabolic, and vascular processes. Various neural processes require energy, provided by a cerebral metabolism depending on a constant supply of glucose and oxygen. These two energy substrates are supplied by cerebral blood flow, which delivers glucose and oxygen to neural tissue through the web of blood vessels. These processes determine the BOLD signal, which detects alterations in levels of deoxygenated hemoglobin and cerebral blood volume [74, 106]. A local increase in neuronal activity induces a substantial increase in cerebral blood flow, mediated by regional autoregulation mechanisms, resulting in a net decrease in HbR concentration; hence a positive BOLD response [255]. Another interesting problem is whether fMRI can differentiate between small activity changes in large cellular populations and large changes in small populations. In the case of an MRI apparatus of the typical resolution, several neural populations of different activity patterns may be scanned, making this problem difficult to resolve.

BOLD signal changes reflect the synaptic activity driving neuronal assemblies, not the neurons firing [349]. Coupling of BOLD signal to synaptic activity allows usage of the fMRI signal to probe functional responses in the brain, even though the response time of the BOLD signal connected with metabolic changes is of the order of several seconds. Hence, it is much slower than the underlying neuronal processes. The change in the BOLD signal caused by neuronal activity (the hemodynamic response) lags the neuronal events by a couple of seconds since it takes some time for the vascular system to respond to the brain's demand for glucose. The hemodynamic response typically peaks at about 5 seconds after the stimulus. If the neurons stay active, the peak spreads to a flat plateau. When activity comes to an end, the BOLD signal falls below the baseline and then recovers.

The sampling time—basic time resolution parameter determines how often a particular brain slice is excited and recovers. Time resolution could vary from the very short (500 ms) to the very long (3 s), depending on the experimental conditions, particularly on the magnetic field strength.

When analyzing the interrelation between fMRI signals the large number of voxels requires a reduction of the number of fMRI time series to be analyzed, which can be achieved by considering signals from particular voxels representing particular brain areas or regions of interest (ROIs). They can be chosen as specific anatomical or functional structures or can be found using parcellation procedures, which split the brain network into clusters. ROIs (nodes of brain network) should represent brain regions with coherent patterns of extrinsic anatomical or functional connections. Namely, they should be built by grouping voxels of similar anatomical characteristics and having the same functional connections pattern toward all the other brain regions. Additionally, a correct definition of the ROIs should be derived in a way that completely covers the cortex's surface or of the entire brain. Moreover, individual ROIs should not spatially overlap. The most common approach to define the ROIs is to choose the active clusters from the general linear model (GLM) [411] and devise as ROIs the spatial average of fMRI time series over the voxels belonging to the same cluster. However, these types of approaches can neglect eventual anatomical and functional inhomogeneity existing within a single ROI. The alternative method of parcellation, which can be recommended, is based on considering the community structure of networks [430, 429].

The problem connected with high dimensionality of fMRI signal sets, namely relatively short time-series measured over thousands of voxels is especially challenging in case of application to BOLD signals parametric models such as MVAR used for connectivity estimation. The solution to this problem was proposed by Valdes-Sosa et al. [612] who introduced sparse MVAR(sMAR) model. The model was estimated in a two-stage process involving penalized regression which was followed by pruning of unlikely connections by means of the local false discovery rate. The performance of the approach was tested by means of simulations and application of the sMAR model for determination of connectivity patterns for distinguishing emotions.

We can distinguish two approaches in fMRI signal analysis: model-driven and data-driven methods. To the first category belong: structural equation modeling (SEM) [501] and dynamic causal modeling (DCM) [172, 173]. Both approaches involve the construction of models based on several assumptions about the data which, must be met for inferences to be valid.

Here we shall consider data-driven approaches. Pre-processing steps of fMRI data include image reconstruction from sampled data, motion and slice-timing correction, spatial smoothing, normalization, slow-drift removal, and normalization to standard space (e.g., MNI [147]), if a group analysis is required [559, 574]. Among the data-driven approaches, the methods of fMRI signal analysis frequently used are: 1) Independent Component Analysis (ICA) (e.g.: [396, 290, 38] 2) covariance analysis, i.e., a spatial map of correlation scores e.g.: [49, 116] 3) seed method, which relies on examination of how the functional connections of a particular brain region (seed) are correlated against the time series of other areas: [94, 532]. The two most widely applied in fMR-based brain network study methodologies: seed-based correlation analysis (SCA) and ICA, were revised in [91] along with examples of their use, advantages, and disadvantages. Seed-based, ICA-based, and cluster-based methods were considered in the review of resting-state fMRI functional connectivity [614]. The Authors promoted graph theoretical analysis for network study. However, more recently, this method's flaws and pitfalls in the context of brain signals studies were demonstrated (see Sect. 5.1.6.5).

A key challenge in determining the interaction between brain structures based on fMRI is determining effective connectivity within brain networks. To this aim, methods based on the Granger causality principle may be applied. Granger Causality Analysis (GCA), contrary to DCM, does not use any biophysical model. The measures based on multivariate Granger causality: dTF and PDC were successfully applied to BOLD data for studying brain connectivity, e.g., [120, 528]. An open-source software toolbox GMAC (Granger multivariate autoregressive connectivity) implements multivariate spectral Granger causality analysis in a MATLAB framework `https://www.nitrc.org/projects/gmac_2012` [589].

5.1.8.1 Relation between EEG and fMRI

The BOLD signal depends on the oxygen uptake, which is connected with neuronal activity, however in contrast to EEG, the BOLD response is an indirect and delayed metabolic correlate of neuronal process [347]. Moreover, the neurovascular coupling has been found to vary across brain regions [418, 187], and modifications have also been identified in certain pathological conditions. The relationships between the evoked changes in brain electrical activity and measured BOLD responses are described by HRF—hemodynamic response function (impulse response function of the neurovascular signal), characterized by its magnitude, peak latency and width of the response. The recovering of the neuronal source signal may be performed by deconvolving the fMRI signal with HRF.

There are several approaches to modeling HRF; they include applying a single canonical HRF, the use of a basis set of smooth functions, the usage of flexible basis sets such as finite impulse response models, and non-linear estimation of smooth reference functions with multiple parameters. The choice of HRF is not an easy task since the models vary in respect of efficiency, bias, and parameter confusability. The review and comparison of the above approaches may be found in [341].

The relationship between EEG and fMRI may be disturbed by the local decoupling phenomena. The neuronal population whose electrical activity generates the EEG signal is not necessarily co-localized with the vascular tree that provides the blood supply to these neurons and gives rise to BOLD. The BOLD signal may be due to the neural population's electrical activity generated by the so-called closed field, which does not show in EEG. In addition to pre- and post-synaptic electrochemical dynamics, several physiological processes require energetic support, for example, neurotransmitter synthesis, glial cell metabolism, etc. Additionally, we have to remember that while BOLD reflects the number of active neurons, EEG amplitude depends primarily on the number of neurons acting synchronously. The neural population's synchronous action gives rise to the EEG rhythms whose contribution to the spectrum depends on a specific task or behavior.

The simultaneous recording creates problems connected with the technical artifacts. The most severe EEG signal artifacts are associated with noise due to the varying magnetic fields; also, ballistocardiogram effects are present in EEG. These problems may be approached by sharing the same protocol by interleaved acquisitions. However, the distinct environments in which EEG and fMRI are typically acquired can present very different and potentially confounding spurious stimuli, also training or habituation effects may occur. Especially some paradigms such as resting-state experiments and study of trial-by-trial fluctuations require simultaneous sessions [262].

In EEG data acquired simultaneously with fMRI, reduction of gradient and pulse artifacts has to be performed. At present, effective algorithms for the elimination of these MRI disturbances in EEG exist, usually, through average template subtraction [5], possibly complemented with principal component analysis (PCA)-based techniques [432]. Different algorithms provide varying, often frequency-specific, balances between noise reduction and signal preservation [169]. Technical solutions associated with acquisition, artifact reduction, and safety are approached in: [335, 199, 326, 417, 1]. During simultaneous EEG/fMRI recording, the presence of EEG hardware degrades fMRI image quality [1], although to a substantially lesser degree compared with the effects of MRI influence on the EEG signal [417].

Establishing a relationship between the rhythmical electroencephalographic activity and the BOLD signal is important for understanding information processing in the brain. A heuristic model relating hemodynamic changes to the spectral characteristic of ongoing EEG activity was elaborated in [289]. The BOLD signal was assumed proportional to the rate of energy dissipation, where dissipation is proportional to the effective connectivity and temporal covariance of the trans-membrane potentials. It followed from the model that as neuronal activation increases, there is a concomitant increase in BOLD signal and shift in the spectral power toward higher frequencies. This result found some experimental confirmation [189, 385].

Negative correlations of fMRI activity in visual cortex with occipital alpha EEG (8–12 Hz) were interpreted in terms of an idling character of this rhythm indicating cortical inactivation [189, 413]. The studies concerning resting state confirmed these findings [658, 283]. Increased BOLD was linked to decreased alpha and beta power and increased gamma activity. In the motor task, negative covariation between alpha, beta EEG rhythms, and BOLD signals were found as well [658]. However, the problem of BOLD relation with brain rhythms is highly complex since different LFP frequency bands showed different, distinct, context-dependent correlations with BOLD [434, 530, 533, 531, 639].

The link between BOLD and neural activity may be established by the multimodal analyses of EEG/BOLD time series [53, 511, 236, 262]. The multimodal analyses of EEG/BOLD signals based on simultaneous measurement of both modalities are undertaken to utilize complementary properties of both signals: good time resolution of EEG and good spatial resolution of fMRI to enhance EEG analysis by fMRI or vice versa. Namely, the topographical information gained from fMRI may help solve the EEG inverse problem, or time information from EEG may indicate the epochs to be analyzed topographically by fMRI.

Four approaches to the analysis of simultaneously acquired EEG/ERP-fMRI can be distinguished, namely: 1) fMRI informed EEG for constrained source localization, 2) EEG or ERP-informed fMRI analysis, 3) parallel independent component analysis (ICA) and 4) joint ICA application for matching temporal sources from the EEG with spatial sources from fMRI [53].

The idea behind fMRI informed EEG is that the localization of potential EEG generators can be determined from the fMRI maps of activation. These potential sources corresponding to local fMRI maxima are then seeded to locations in the brain, subsequently allowing to derive the neural activity for each of these locations. Consequently, the well-defined spacially fMRI data make it possible to estimate the corresponding time courses of brain activity at high time resolution [640, 641].

In the EEG source reconstruction problem, two approaches may be used: equivalent dipole assumption and optimization of distributed source models, which compute a reconstruction of neuroelectric activity at each point in a 3D grid of possible current sources. Bledowski et al. [51] acquired EEG and fMRI data from subjects undergoing an oddball visual paradigm. Dipole "seeds" were positioned at the foci of clusters exhibiting paradigm significant, oddball-specific BOLD responses, and inverse solutions were then obtained for the averaged ERPs elicited by each stimulus condition. EEG inverse solutions can be further enhanced by specifying not only dipole positions but also their orientations [400, 617].

Babiloni et al. [24] in the study involving: EEG, MEG, and fMRI) used a subject's multicompartment head model (scalp, skull, dura mater, cortex) to design a multi-dipole source model. Determination of the priors in the resolution of the linear inverse problem of cortical current density was performed using information from the hemodynamic responses of the relevant cortical areas. The method was applied to estimate the time-varying cortical current density for movement-related activity in the selected regions of interest.

In the case of distributed source models, the statistical maps of the fMRI results can be used to confine the putative source space by providing the probability of a particular region being the origin of the electrophysiological signal [208, 101, 450]. In [610] and [23] statistically significant percentage increase of the BOLD signal during the task compared to the rest-state provided the norm in the source space for the solution of the inverse problem.

Studies concerning EEG informed fMRI aim to identify BOLD correlates of EEG events. In the case of epilepsy, the early applications involved spike-triggered acquisition mode whereby each fMRI scan was acquired following detection of spike e.g.: [578]. In a more advanced approach [345] the semi-automatic system was proposed based on spatio-temporal clustering of inter-ictal EEG events. The clusters of interictal epileptic discharges were correlated to BOLD activations. In a second step, signal space was used to project scalp EEG onto dipoles corresponding to each cluster. This allowed for the identification of previously unrecognized epileptic events, the inclusion of which increased the experimental efficiency.

The general idea behind ERP-informed fMRI analysis is to correlate active brain regions identified by fMRI with amplitude modulation of individual ERP. In the experiment concerning the oddball auditory paradigm, in the first step, the data were de-noised using wavelets and ICA [140]. Then, single-trial N1, P2, and P3 amplitudes of ERP were found. These amplitude vectors were convolved with the hemodynamic response function and used as regressors to find the BOLD time course. Different approaches to EEG-informed fMRI, including univariate and multivariate methods, are reviewed in [1].

In the simultaneous ERP and fMRI analysis, the recorded signals are spatially and temporally mixed across the brain since they are volume-conducted and temporally extended by the hemodynamic response. The data in both modalities are generated by multiple, simultaneously active overlapping neural populations. In solving problems where specific hypotheses regarding spatial and temporal relationships are lacking or are ill-specified, blind source separation methods are useful. To address the issue of mutual ERP-fMRI relation ICA approach was proposed in [139]. The ICA was used in parallel: to recover spatial maps from fMRI in terms of spatial Independent Components—(sIC) and time-courses from ERP in terms of time-related Independent Components—(tIC). Then the components were matched across modalities by correlating their trial-to-trial modulation. Inspecting the results, only one tIC component was found, which predicted one sIC component's time-course selectively. There were no other covariances between components corresponding to both modalities. A similar approach was used in a visual detection task by [404].

An alternative approach is to fuse the ERP and fMRI signals in a common data space. In [413] joint independent component analysis was proposed to analyze simultaneous single-trial ERP-BOLD measurements from multiple subjects. The authors presented the results based on simulated data, which indicated the feasibility of the approach. It seems that there is a long way to obtain a satisfactory integration of EEG and fMRI modalities. There are already various mathematical approaches for multimodal data fusion, the selection of an appropriate fusion model requires tailoring to the specific problem.

5.1.9 Near-Infrared Spectroscopy Signals

Near-Infrared Spectroscopy (NIRS) is a technique that uses the light of infrared (IR) wavelengths (700–2500 nm) to obtain a picture of the composition of the analyzed substance. Molar absorptivity in the IR range is small compared to visible light, which enables the penetration of the IR light in the tissue. Functional near-infrared spectroscopy (fNIRS) is a non-invasive imaging method that uses near-infrared (700–900 nm) light for quantification of chromophores (light-absorbing molecules) concentration.

Most fNIRS applications concern brain functioning, but fNIRS is also used to monitor peripheral vascular disease and can be useful in oncology since tumor presence is connected with increased blood volume resulting from increased vessel density and enhanced levels of deoxyhemoglobin. Here we will focus on fNIRS application to the study of brain activity.

The beam of photons from light-emitting diode penetrating a biological tissue is partly scattered and partly absorbed by the chromophores. A photon's trajectory could vary when crossing tissue. A few photons will reach the photodetector without undergoing any dispersion or absorption; some will be absorbed by chromophores, others scattered out will not reach the photodetector, and the remainder will make it, but by traveling a path longer than the geometrical distance between light source and detector. In the so-called reflectance geometry technique, a photodetector captures the light wave resulting from the interaction with the chromophores, following a banana-shaped path back to the surface of the skin (see Figure 5.27). In case of transmission geometry the investigated tissue layer is between source and detector (e.g., mammography applications).

The light emitted from a source and scattered by the tissue is registered by the detector placed several centimeters apart. For more significant distances between source and detector, deeper layers of tissue may be penetrated (Figure 5.27); however, the attenuation of light limits the depth of penetration. In practice, a cortical activity more than 4 cm deep cannot be recorded due to the safety limitations in light emitter power.

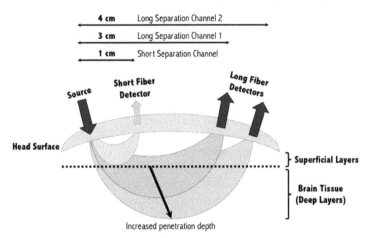

FIGURE 5.27
Banana shape profiles of the sampled functional near-infrared spectroscopy signal at multiple source-detector distances. Modified from [514].

There are similarities between fNIRS and fMRI since, in both techniques, neuronal activity through neuro-vascular coupling is related to changes in localized cerebral blood flow and hemoglobins concentrations. The advantages of fNIRS are low cost and portability. However, deeper structures cannot be penetrated by fNIRS, and its spatial resolution is much lower than fMRI.

The cortex's functional state in a specific location and time may be linked to the relative concentration of chromophores. Different studies show that neural activity and hemodynamic response maintain a linear relationship [14, 348], suggesting that changes in hemodynamic response could provide a useful marker for assessing neural activity. In the brain, during functional activation, oxygen consumption increases. The decreased tissue oxygenation is counteracted by the enhanced O_2 supply caused by increased cerebral blood flow (CBF) and cerebral blood volume. When hemoglobin transporting oxygen releases the oxygen, oxygenated hemoglobin (HbO) transforms into deoxyhemoglobin or reduced hemoglobin (HbR) via metabolic processes. Therefore, a cerebral region could be considered active when its CBF increases, producing changes in HbO and HbR concentrations [445].

Employing fNIRS tissue activity may be probed through differences in absorption spectra of HbO and HbR. Sometimes also, total hemoglobin changes—the sum of HbO and HbR concentration changes is considered. The optimal light spectrum window for studying brain functions ranges between 700–900 nm, which is regarded as the biological "optical window". Usually, two wavelengths are selected with one wavelength above and one below the

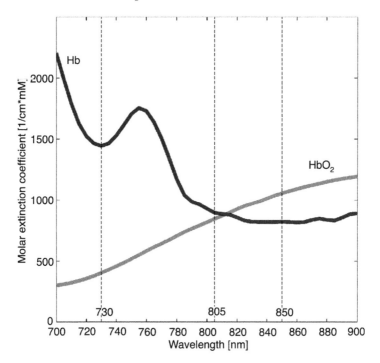

FIGURE 5.28
Absorption of light by oxygenated and deoxygenated hemoglobin (gray and black lines) and common NIRS light source wavelengths (730 and 850 nm) as well as Infrascanner wavelength (805 nm) marked. Modified from [21].

810 nm at which point HbO and HbR have identical absorption coefficients (Figure 5.28).

Brain fNIRS studies aim to assess neural activation. Other neuroimaging techniques have validated the application of fNIRS in cerebral functioning studies, showing that the NIRS signal correlates with: positon emission tomography (PET) changes in regional cerebral blood flow (rCBF), and the fMRI Blood Oxygen Level Dependent (BOLD) signal [226, 292, 618, 603, 361, 235].

The changes in concentration of the chromophores can be determined using the modified Beer-Lambert (B-L) law. According to B-L law the attenuation of the laser beam:

$$\Delta A_\lambda = -\log_{10} \frac{N_\lambda}{N_{0,\lambda}} \tag{5.43}$$

(where N_λ—intensity of the transmitted light beam, $N_{0,\lambda}$—intensity of the input light beam), is proportional to the the mean path lengths of photons and the sum of product of molar extinction coefficients and chromophores concentrations:

$$\Delta A_\lambda = \langle l \rangle \left(\Delta c_{\text{HbO}} \varepsilon_{\text{HbO}} + \Delta c_{\text{HbR}} \varepsilon_{\text{HbR}} \right) \tag{5.44}$$

where $\langle l \rangle$ mean path length of photons, ε_{HbO}, ε_{HbR}—molar extinction coefficients for HbO and HbR respectively, Δc_{HbO}, Δc_{HbR} changes of concentrations of HbO and HbR.

The mean path lengths of photons for the continuous wave technique are derived from Monte-Carlo simulations and physical models. When a time-resolved technique with picosecond light pulses is applied, $\langle l \rangle$ can be found from the distribution of times of flight of photons (DTOF).

From DTOF curve, mean time of flight of photons (MTF) can be determined and hence we get:

$$\langle l \rangle = \text{MTF}v \tag{5.45}$$

where v is the velocity of light.

From the measurements of the attenuation of the light beam for two wavelengths, λ_1 and λ_2, the relative concentrations of two chromophores may be determined:

$$\frac{\Delta A_{\lambda_1}}{v\text{MTF}_{\lambda_1}} = \Delta c_{\text{HbO}} \varepsilon_{\text{HbO}}(\lambda_1) + \Delta c_{\text{HbR}} \varepsilon_{\text{HbR}}(\lambda_1) \tag{5.46}$$

$$\frac{\Delta A_{\lambda_2}}{v\text{MTF}_{\lambda_2}} = \Delta c_{\text{HbO}} \varepsilon_{\text{HbO}}(\lambda_2) + \Delta c_{\text{HbR}} \varepsilon_{\text{HbR}}(\lambda_2) \tag{5.47}$$

The concentration changes Δc_{HbO} and Δc_{HbR} can be calculated by solving above equations since extinction coefficients are known [490] and N_λ and MTF_λ may be derived from DTOFs as its zeroth and first order moments.

The results of the fNIRS experiments yield the time evolution of HbO and HbR concentrations. Preprocessing of these signals involves removal of artifacts connected with instrumental noise, motion artifacts, and physiological noise due to heartbeat (1–1.5 Hz), respiration ($\sim$(0.2–0.5 Hz) and Mayer waves ($\sim$0.1 Hz) related to blood pressure fluctuations. The artifact removal methods involve band-pass filtering, adaptive filtering, auto-regression with exogenous inputs, PCA, and ICA. In general band-pass filter can remove a large part of instrumental noise and physiological noises. However, it cannot be used to filter noises, which overlap with the band of the hemodynamic response signal, for example, due to respiration. When necessary, more advanced preprocessing methods may be used; however performance of PCA and ICA depends on the number of available channels and is recommended only for a sufficient number of detectors.

Hemodynamic time series can be analyzed by the signal analysis methods described in previous Chapters, most commonly: spectral power, correlation, and coherence analysis are applied. The studies concerning spectral properties of HbO and HbR revealed the main frequency components: very low oscillations $\sim$0.04 Hz and low-frequency oscillations $\sim$0.1 Hz [444, 540, 659]. In chromophores concentrations, a peak around 1 Hz corresponding to the subject's heart rate and the peak at the respiratory frequency around 0.2–0.3 s were detected. The ∼0.1 Hz component is connected with the Mayer waves (MW), which appear in several physiological signals. The problem of MW will be touched in the Sect. 5.5 concerning the multimodal analysis of biomedical signals.

fNIRS derived measures of hemodynamic response have been used in numerous studies to assess cerebral functioning during resting state, e.g., [355, 527], and stimulation, e.g., visual [218], acoustic [660], also transcranial magnetic stimulation [562]. fNIRS technique was also applied in language studies [624] and motor action experiments e.g.: [443, 524, 651, 330, 319]. Usually, the brain activation is connected with an increase of HbO and decrease of HbR [445, 585]. The studies of Izzetoglu [243] indicated that human performance and cognitive activities such as attention, working memory, problem-solving, etc., can be assessed by fNIRS technology.

The resting-state connectivity analysis based on fNIRS signals was conducted, e.g., in [355], who calculated functional connectivity maps over the sensorimotor and auditory cortexes using seed-based correlation analysis and data-driven cluster analysis. Coherence analysis was used in [527] to assess more global connectivity patterns. The inter-hemispheric coherence was significant for the whole spectrum of HbO and HbR concentrations, while fronto-posterior coherence was lower and revealed a peak around 0.1 Hz. However, the study was based on a bi-variate connectivity measure, prone to the common drive effect, so the results were only approximate.

Several works aiming to understand neurovascular coupling involved combined fNIRS, and EEG techniques, e.g., [624, 562, 319], fNIRS and MEG [524]. In the above works, EEG or MEG signals were correlated with time changes of chromophores concentrations. Sood [562] used the ARX[9] model for continuous assessment of the transient coupling between the EEG and HbO signals representing spontaneous brain activity during transcranial direct current stimulation. In the motor experiment [319], the correlations between the evolution of EEG rhythms and concentrations of hemoglobins were investigated. The highly significant Person correlations between HbO and alpha and beta rhythms amplitude envelopes were (-0.69 ± 0.16 and -0.54 ± 0.32 respectively); for HbR, they were positive of slightly smaller values. The delays of changes in chromophore concentrations in respect to cue presentation were found as 2.7 ± 1 s for both HBO and HbR. These results were (within the error limits) in agreement with these found in [651]. In the above publication in motor imagery experiments, changes of blood chromophore concentrations were characterized by longer delays in respect to cue presentations than in real movements.

In recent years, the use of fNIRS has steadily increased. Due to its portability, small size, and reliability, it has become a valuable neuroimaging technique for brain research and medical applications. fNIRS clinical applications

---

[9]ARX is AR model with exogenous input.

multiply primarily in medical monitoring and diagnosis [167, 311, 629]. The fNIRS technique is ideal for studies in which subjects may have a difficult time with traditional neuroimaging techniques [242], namely in the case of:

- populations that may not be able to readily tolerate the confines of an fMRI magnet or be able to remain sufficiently still (young children, schizophrenics, people with dementia),

- patients that require the long-term monitoring of cerebral oxygenation, e.g., neonates or high-risk infants,

- bed-ridden patients,

- studies that require repeated, low-cost neuroimaging,

- measuring of brain activity during physical training.

A leading cause of death and disability and a major public health problem is traumatic brain injury (TBI). fNIRS can provide early detection of TBI, which is a crucial factor for successful therapy, e.g., [541]. The simultaneous study of EEG and fNIRS [431] for patients with temporal lobe epilepsy showed that seizure was accompanied by an initial increase in HbO concentration and decrease of HbR followed by the rise of HbR concentrations, which indicates the fNIRS potential for long-term monitoring of epileptic patients.

Among the signals used for brain-computer interfaces (BCI), fNIRS is relatively new. The topographical locations of fNIRS sensors are usually above the motor cortex or pre-frontal cortex. The second location has the advantage of a lack of artifacts due to hair. The paradigms are connected with motor action imagery or mental activity such as mental arithmetics, music, or landscape imagery. Signal features most frequently used for BCI applications are the mean, variance, peak value, slope, skewness, and kurtosis of the hemodynamic response in respect to rest. For classification, linear discriminant analysis, support vector machine, hidden Markov model, and artificial neural networks are used. The main drawbacks of fNIRS-BCI is slow information transfer rate and high error rate. The comprehensive review of fNIRS-based BCI can be found in [423].

Hybrid BCIs based on the information gained from EEG and NIRS were designed to enhance the performance of well-established BCI based on EEG. Both methods were applied simultaneously in a real-time sensory motor-rhythm-based BCI paradigm involving executed, and imaginary movements [152] to test how fNIRS data classification can complement ongoing real-time EEG classification. The results showed that simultaneous measurements of NIRS and EEG can significantly improve the classification accuracy of motor imagery in over 90% of considered subjects and increased performance by 5% on average ($p < 0.01$).

In hybrid BCI, EEG and fNIRS complemented each other in terms of information content, which not only increased performance for most subjects but also allowed meaningful classification rates for those who would otherwise not be able to operate a solely EEG-based BCI. However, the increased performance of hybrid BCI requires a more complicated measurement setup and decreases the information transfer rate.

5.2 Heart Signals

5.2.1 Electrocardiogram

Electrocardiogram (ECG) is a record of the electrical activity of a heart. The heart's contraction activity is driven by the electric pacemaker—the assembly of cardiac cells that possess the property of automatic generation of action potentials. Such cells are located in the sino-atrial node (SA), the atrioventricular node (AV), and in the atria and ventricles' specialized conduction systems. Myocardial cells have the unique property of transmitting action potentials from one cell to an adjacent cell through the direct current spread.

5.2.1.1 Measurement Standards

Einthoven—the pioneer in the field of electroradiology—proposed a two-dimensional dipole placed in a homogenous conducting volume of the thorax in the middle of a triangle with vertexes defined by the electrodes placed on the arms and left leg as a model of the electrical activity of the heart muscle. Einthoven standard leads I, II, III measured the potential differences: I-between the left arm (LA) and the right arm (RA), II—(between the left leg (LL) and the RA), III—(between the LL and the LA). The electrical activity propagation throughout the cardiac cycle can be represented by an instantaneous electrical heart vector (Figure 5.29). Despite the crude approximation of the body geometry, the projections of the heart vector on the Einthoven triangle's arms became a standard in electrocardiography.

The typical leads used in electrocardiography are shown in Figure 5.29. They include the three ECG standard leads: I, II, III, and the so-called augmented limb leads: aVR, aVL, aVF corresponding to the potential recorded between a given limb (correspondingly: right arm, left arm, left foot) with respect to the average of the potentials of the other two limbs. In standard clinical applications, signals from six leads placed on the torso are also used. The potentials measured between electrodes placed on the limbs correspond to the projections on the frontal plane. The electrodes, located on a torso, supply the information about ECG vector projection on the horizontal plane (Figure 5.29).

In a three-dimensional system proposed by Frank, electrodes are placed according to Cartesian XYZ axes (Figure 5.30). In principle, three orthogonal projections are sufficient to reconstruct the vector. However, in ECG, the information's redundancy is beneficial since the projection axes are non-orthogonal; the body is not a homogenous sphere and the vector changes in

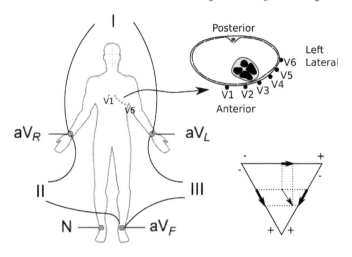

FIGURE 5.29
The typical leads used in electrocardiography. At left—limb leads; top right—chest leads. Bottom-right: the heart vector and its projection on the Einthoven triangle.

time. Besides, some particular features of ECG are more visible in specific derivations.

The sampling frequency of ECG has to be chosen according to the aim of the analysis. The commonly accepted range of ECG frequencies is 0.05–100 Hz. However, when late ventricular potentials are considered, the upper-limit reaches 500 Hz.

5.2.1.2 Physiological Background and Clinical Applications

The transmission of electrical heart activity is reflected in the features of the ECG evolution. When depolarization propagates toward the given electrode, the detected voltage is seen as positive and is represented by an upward deflection in the ECG. Let us consider the potential recorded by the lead II during the heart cycle. The activation starts in the SA node and propagates toward AV. The depolarization of atria follows with a heart vector pointing down and left[10] which is reflected in the positive P wave (Figure 5.31). Then there is a delay at the AV node corresponding to the flat portion of the ECG. The septal depolarization with the vector pointing to the right corresponds to the Q wave. The period between the P wave and the QRS complex corresponds to the time between the atrial and ventricular depolarization when ECG is on the isoelectric level. This period serves as a reference for ECG amplitude estimation. The prominent QRS peak corresponds to the ventricles'

---

[10]The left and right concerns subjects' left and right.

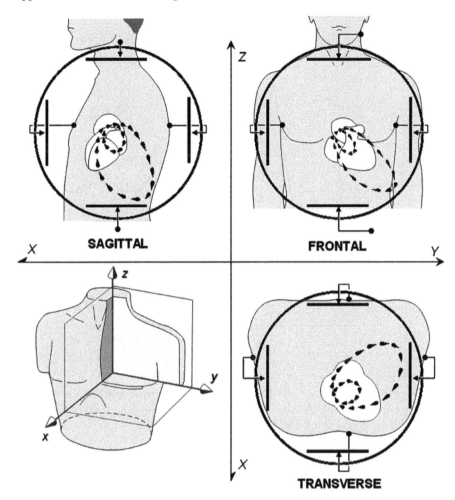

FIGURE 5.30
The 3-D Frank's electrode system. Black dots—the heart vector trajectory projections on sagittal, frontal, and transverse planes are marked. From [378].

depolarization with a heart vector pointing to the left and down. The width of the QRS complex depends on conduction speed through the ventricle. S-T segment corresponds to the pause between ventricular depolarization and repolarization. Next, a phase of repolarization follows, during which both the direction of propagation and the electrical polarity are changed, which results in a positive T wave. U wave, which can be observed using high-resolution electrocardiography, and which is also connected with repolarization, follows

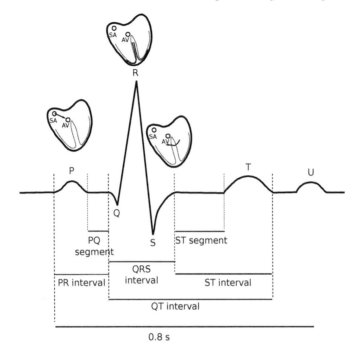

FIGURE 5.31
The potential recorded by lead II during the heart cycle. Main waves, intervals, and segments marked. Above schematic representations of electrical activity propagation during the cycle are marked.

the T wave. In the case of a short heart cycle, it might be impossible to register the U wave.

The heart pacemakers are influenced by the sympathetic and parasympathetic portions of the autonomic nervous system. The sympathetic system acts quickly to increase heart rate, blood pressure, and respiration; the parasympathetic system acts to decrease heart rate and slow down respiration and blood flow (e.g., during falling asleep). The disturbances in the generation or propagation of electrical activity of the heart are reflected in the ECG; hence it is a widely used clinical test for heart diseases. However, this doesn't mean that all abnormal electrical heart activity features will be observed in ECG. According to the Helmholtz law, already mentioned in the context of EEG, the same potential difference on the body surface can be due to different configurations of sources; hence abnormal activity may be masked. The information on the heart's electromagnetic activity, its generation, and measurement may be found in [378].

In the clinical environment, ECG recording is often accompanied by the measurement of respiration and blood flow since these signals are mutually dependent, e.g., respiration rhythm is visible in the heart rate signal.

The ECG is a clinical test in most heart diseases. Other medical conditions such as diabetes and metabolic disturbances may also have an impact on ECG. In ECG, the activity of the autonomous nervous system is reflected. Therefore ECG can be used as an indicator of activation of sympathetic and parasympathetic systems. Among the heart activity disturbances reflected in ECG are the arrhythmias, bradycardia (slowing of the rhythm), tachycardia (an increase of heart rate), ectopy, fibrillation, myocardial infarction, premature atrial/ventricular contraction, atrial/ventricular flutter, conduction blocks, cardiac ischemia, hyper- or hypokalemia (excess/deficiency of potassium), hyper-, hypocalcemia (excess/deficiency of calcium). The description of the influence of these and other pathological heart conditions on ECG may be found in [83].

5.2.1.3 Processing of ECG

In ECG, deterministic and stochastic features are manifested, and the repertoire of the methods applied for its analysis is very broad. It includes morphological features extraction, analysis in time and frequency domain, statistical methods, classification procedures. Another time series, namely HRV (heart rate variability), is derived from the ECG signal, and is a subject of further analysis.

In signal processing, before applying a new method to the given type of data or the differentiation of a particular pathological state, it is recommended to compare the results with other methods and test it on standard data. In the ECG case, fortunately, there is a PhysioNet database [188], which can be helpful in this respect. It is available at `http://www.physionet.org/physiobank/database/`. PhysioBank is an extensive and continuously updated archive of well-characterized digital recordings of physiologic signals and related data that may be downloaded at no cost. PhysioBank currently includes databases of multi-parameter cardiopulmonary, neural, and other biomedical signals from healthy subjects and patients with a variety of conditions with major public health implications, including sudden cardiac death, congestive heart failure, arrhythmia (MIT-BIH arrhythmia database). In PhysioNet, the software for analysis of ECG and HRV may be found as well.

Artifact Removal

The first step in ECG analysis is the elimination of the artifacts. ECG signals may be corrupted by technical artifacts such as power line interferences, artifacts due to bad electrode contacts, quantization or aliasing errors, noise generated by other medical equipment present in the patient care environment, and biological artifacts: patient-electrode motion artifacts, muscular activity, baseline drift—usually due to respiration. Technical artifacts may be avoided by designing proper measurement procedures. However, the elimination of biological artifacts is much more complicated and requires special signal analysis techniques. The filtering techniques used for denoising ECG involve linear and non-linear methods, model-based or model-free approaches.

For rejecting high-frequency noise, FIR filters or Butterworth 4-pole or 6-pole lowpass digital filters appear to be suitable [632]. The cut-off frequency is usually set around 40 Hz. For baseline wander, high-pass linear filters of cut-off frequency up to 0.8 Hz may be used. The cut-off frequency above 0.8 Hz would distort the ECG waveform. Among the linear, model-based approaches to denoising, Wiener filtering is an often-used method. Unfortunately, one of the Wiener filter assumptions is that both signal and noise are stochastic. However, ECG has a prominent deterministic character, which diminishes the performance of the Wiener filter. One of its drawbacks is the reduction of the signal amplitude.

Wavelet transform is a model-free method of ECG denoising. The discrete wavelet transform may be used to decompose ECG into time-frequency components (Sect. 3.4.2.2), and then the signal may be reconstructed only from presumably free of noise approximations. It is essential to choose the mother wavelet correctly; it must be maximally compatible with the ECG signal structures. In [88] biorthogonal spline wavelets were recommended for ECG reconstruction, which was performed by keeping only the first approximation $A_2^d(x)$. The procedure is equivalent in a way to lowpass filtering followed by resampling. However, the wavelet transform approach seems to have advantages over conventional FIR filters, which may produce oscillating structures in filtered signal (so-called Gibbs oscillations) [88, Fig 5.6]. The biorthogonal wavelets family is available in MATLAB Wavelet Toolbox. The wavelets in the family are named: biorJ.K, where J and K correspond to the number of vanishing moments in the lowpass and high-pass filters, respectively. The wavelets with higher J and K seem to better reproduce ECG signal structures.

Other methods useful in removing the noise components from ECG are PCA and ICA. It is an advantage of ICA that its subspace axes are not necessarily orthogonal, and the projections on them are maximally independent. ICA is a blind source separation method. The independent components for a set of signals can be found if the mixing matrix **D** is known (Sect. 4.6.2, equation 4.65). We can expect that specific ICA components will be connected with the noise/artifact sources. Since ICA components are transformed combinations of leads, to recover the information about the ECG signals, the back projection has to be performed (Sect. 4.6.2). During this procedure, the noise sources may be removed. Namely, the components corresponding to artifact/noise may be set to zero during reconstruction.

However, this procedure is not entirely straightforward since we have to determine which ICA components are noise. The system based upon kurtosis and variance has been devised to distinguish noise components automatically [217], but the quality of results may depend on the particular data set. The problem is complicated since the ICA mixing/demixing matrix must be tracked over time, and the filter response is continually evolving. A robust system of separating artifacts from ECG is still to be found.

Morphological ECG Features

Clinical assessment of ECG mostly relies on evaluating its time-domain morphological features such as positions, durations, amplitudes, and slopes of its complexes and segments (Figure 5.31). The morphological feature may be estimated using a sequence of individual heartbeats or using averaged heartbeat. The first step in the analysis is usually detecting the QRS complex; it serves as a marker for averaging heart cycles, evaluating heart rate, and finding the heart axis. Usually, the analyzed parameters include amplitudes of Q, R, S, T peaks, depression or elevation of the ST segment, durations of the QRS complex, QT interval, dispersion of QT (the difference between the longest and the shortest ST).

The algorithms for the determination of ECG morphological time features based on wave boundaries, positions, amplitudes, polarizations may be found, e.g., in [453, 454, 321].

Several approaches based on different formalisms have been proposed for feature extraction from ECG beats. Examples of such approaches include: Fourier transform [406], Hermite functions [320], wavelet transform [542, 657].

An interesting approach to ECG features quantification was proposed in [86]. Each training heartbeat was approximated with a small number of waveforms taken from a Wavelet Packet dictionary (Symlet 8). The matching pursuit algorithm was used as the approximation procedure. During the training procedure, each of the five classes of heartbeats was approximated by ten atoms, which were then used for the classification procedures. The results showed high classification accuracy. The matching pursuit was also used in the procedure of fitting the wavelets from the Daubechies family to the ECG structures to obtain sparse time-frequency representation of ECG [455].

At present, there are also ready-to-use systems for ECG feature extraction available on the Internet. The software for determination of the characteristic time points of ECG may be found in PhysioNet (`http://www.physionet.org/physiotools/ecgpuwave/`) as a routine `ecgpuwave`.

A classification system for the electrocardiogram features called the Minnesota Code [482] utilizes a defined set of measurement rules to assign specific numerical codes describing ECG morphology. It provides the ranges of normal values of ECG morphological features in the standard 12 leads. The system was developed in the late 1950s in response to the need for reporting ECG findings in uniform and objective terms and has been updated and improved since then. Minnesota Code incorporates ECG classification criteria that have been validated and accepted by clinicians [296].

ECG ST-segment analysis is of particular interest in detecting ischemia, heart rate changes, and other heart disturbances. The review of the ST segment analysis approaches, comparison of ST analyzers' performance, and algorithm for detection of transient ischemic heart rate related ST episodes may be found in [245].

Another feature of ECG important for clinical diagnosis is T-wave alternans (TWA). T wave is connected with the processes of repolarization of the heart muscle. Its character changes from beat to beat, and in particular, its amplitude has alternating higher and lower values in consecutive cycles. TWA is a valuable ECG index that indicates increased susceptibility for ventricular arrhythmia and the risk of cardiac arrest, one of the most frequent causes of sudden death. The parameter commonly used in diagnostics is T wave amplitude, although its phase and shape are also considered. Spectral analysis is usually applied to the series constructed from T wave amplitude values found in consecutive cycles to find subtle TWA fluctuations. The alternation of amplitude from beat to beat is reflected in the power spectrum of these signals as a peak at $0.5\,Hz$. The peak amplitude, normalized in respect to noise, gives the information about the T wave variability; its higher value reflects greater repolarization dispersion and indicates a risk of sudden cardiac arrest [422].

An important TWA index is its phase, which may be detected by quantifying sign-change between successive pairs of beats or using Fourier transform. TWA phase reversal is helpful in sudden cardiac arrest risk stratification.

Spatial Representation of ECG Activity; Body Surface Potential Mapping and Vectorcardiography

New diagnostic possibilities were opened by the methods involving mapping the spatial distribution of TWA on a thorax [294, 155]. It is known that sometimes there can be no changes in the standard ECG despite the evidence of coronary artery disease or myocardial infarction. Analysis of high-resolution ECG recorded from multiple electrodes placed on the torso enables detection of electric activity of individual fragments of the cardiac muscle, which opens up a possibility of developing a qualitatively new diagnostic method and a means to control the efficiency of treatment.

Figure 5.32 shows body surface maps of the QRST integral for a healthy volunteer and a patient after myocardial infarction (MI) with a documented episode of ventricular tachycardia (VT). The drastic change in the shape of isopotential boundaries may be easily observed.

The pioneer of body surface potential mapping (BSPM) is Taccardi [580, 581] who recorded the ECG from 200 electrodes. He described the time-varying potential distribution on the thorax, intending to relate its evolution to the propagation of the activity in the heart muscle. At present, the technique is used in many laboratories. The construction of maps involves triangulation of the body surface and interpolation, usually utilizing biharmonic splines [525]. The BSPM technique was used to quantify different phenomena occurring during the heart cycle, e.g., the spatial distribution of the late atrial potentials [259] and the distribution of electric field during the phase of repolarization

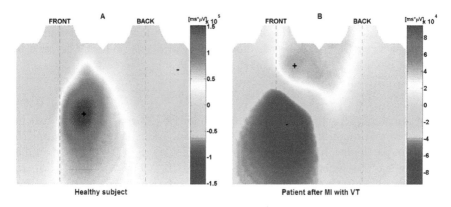

FIGURE 5.32
Body surface maps of QRST integral: A) healthy volunteer, B) patient after myocardial infarction (MI) with documented episode of ventricular tachycardia (VT). By courtesy of M. Fereniec from [154].

[287]. In the above work, the quantification of the electric field shape dispersion was proposed.

Diagnostic application of BSPM includes, e.g., detection of myocardial infarction, [108], left ventricular hypertrophy [95], and myocardial ischemia [212]. At present even portable devices for high resolution ECG mapping are available [512].

Vectorcardiography is a method of recording the magnitude and direction of the electrical forces generated by the heart using a continuous series of vectors that form curving lines around a central point. In vectorcardiography, the heart vector trajectory in 3D space is traced (Figure 5.30). The strength and direction of electric currents passing through the heart are represented as vector loops. The information about the directional evolution of the heart vector in time is found. This technique enables observation of the depolarization and repolarization in particular cardiac muscle fragments during its evolution, which means that it permits detection of even small changes in the electric activity of individual fragments of the cardiac muscle caused by ischemia or the applied treatment. Usually, the evolution of electrical activity connected with QRS complex, P-wave, or ST complex is followed. High-resolution vectorcardiography reveals changes related to an untypical propagation of the depolarization vector or distinctly weaker amplitude of depolarization from particular cardiac muscle fragments. It was reported that high-resolution vectorcardiography analysis permits a fast, non-invasive, and cheap confirmation of myocardial infarction and determination of its localization in persons whose standard and exercise ECG and echocardiography do not reveal changes. It also provides the recognition of changes in the depolarization amplitude related to, e.g., the effect of a drug on the cardiac muscle in ischemia [306]. At present, the on-line vectorcardiography is available and may be used for patient monitoring in hospital [357].

Statistical Methods and Models for ECG Analysis

Statistical methods and models are applied for ECG segmentation, feature extraction, and quantification. For ECG analysis, probabilistic models may be used; in particular, hidden Markov models (HMM) (Sect. 3.2.1) were applied for the signal segmentation. The first step is to associate each state in the model with a particular ECG feature, namely to connect the individual hidden states with structures such as P wave, QRS complex, etc. The next step involves the training of HMM, usually by supervised learning. The important aspect is the choice of observation model to capture the statistical characteristics of the signal samples from each hidden state. As the observation models, Gaussian density, autoregressive model, or wavelets may be used. The latter two methods were reported as performing better [232].

For ECG quantification and identification of the features helpful for the distinction of pathological changes in ECG, the analysis based on statistical procedures such as PCA and models such as AR may be applied without making explicit reference to time-amplitude features of ECG structures.

AR model technique was proposed to classify different types of cardiac arrhythmias [177]. The preprocessing involved removal of the noise, including respiration, baseline drift, and power line interference. The data window of 1.2 s encompassing R peak was used. AR model of order 4 was fitted to the data. Four model coefficients and two parameters characterizing noise level were used as input parameters to the classification procedure. For classification, a quadratic discriminant function was used. The classification performance averaged over 20 runs (different training and testing data sets), showed specificity and sensitivity above 90% for each of six classes of arrhythmias. These results demonstrated the usefulness of the AR model for the quantification of ECG. The model may also be used for ECG data compression and portable telemedicine ECG systems since the algorithms are relatively simple and work fast.

Wavelet transform was used to analyze high-resolution ECG for risk evaluation in tachycardia [336]. The time-frequency maps were obtained by a modified Morlet wavelet (Figure 5.33). For quantification of late potentials (appearing after QRS complex), the index called irregularity factor, quantifying the variation of energy, was proposed.

For ECG feature extraction and reduction of redundancy, the orthonormal function model may be used. It is based on the Karhunen-Loève transform, which provides signal representation through principal components (Sect. 4.6.1). Representing ECG morphology through PCA provides robust estimates in terms of few descriptive parameters and allows for effective comparison of

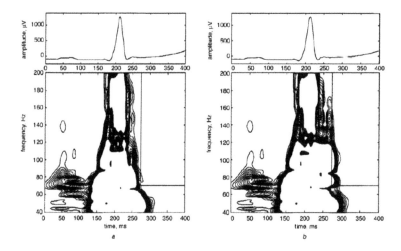

FIGURE 5.33
The time-frequency maps obtained by modified Morlet wavelet. a) ECG of healthy subject; b) the same signal with simulated late potentials. Time and frequency ranges for irregularity factor calculation are marked by straight lines. From [336, Fig. 2].

normal and deviating ECG pattern vectors exposing differences between them. In the procedure, the noisy outliers may be eliminated based on criteria concerning the normalized residual error.

The application of PCA in ECG signal analysis included compression, filtering and QRS complex classification [412], shape representation of ECG morphology [582], analysis of repolarization ST-T segment [322], detection of transient ST-segment episodes during ambulatory ECG monitoring [246].

ECG feature extraction serves diagnostic purposes and is usually followed by the classification procedures. Machine learning methods such as artificial neural networks are helpful in this respect. Different kinds of ANN including multilayer perceptrons and support vector machine classifiers have been used for ECG classification; training of networks may be performed by means of supervised or non-supervised learning. Introduction to supervised and unsupervised learning for ECG classification may be found in [449, 627].

5.2.2 Heart Rate Variability

The sequence of RR intervals (sometimes called normal-to-normal (NN) intervals)—that is, all intervals between adjacent QRS complexes resulting from sinus node depolarizations—forms the RR interval time series or RR tachogram. The sequence of successive times, t_i, $i \in \{1, 2, \ldots, n\}$, when the R wave appeared, e.g., identified by applying QRS detector to the ECG signal is:

$$RR_i = t_i - t_{i-1} \qquad (5.48)$$

Then, by means of interpolation the HRV signal is obtained. Heart rate variability (HRV) is a physiological phenomenon of the time interval variation between heartbeats. For the determination of the HRV series ECG should be appropriately sampled. A low sampling rate may produce jitter in the estimation of the R-wave fiducial points. The optimal recommended range is 250 to 500 Hz or perhaps even higher. In contrast, a lower sampling rate (in any case ≥ 100 Hz) may behave satisfactorily only if an adequately chosen interpolation algorithm is used to refine the R-wave fiducial point.

Since each RR interval is related to the previous one, the RR tachogram is unevenly sampled; therefore, it has to be resampled to produce evenly sampled time series. Typical resampling schemes involve linear or cubic spline interpolative sampling with resampling frequencies between 2 and 10 Hz. This procedure may introduce some bias in estimating power spectra. It was reported that the bias is lower for the procedure based on cubic splines than in the case of linear resampling [87].

The Task Force of the European Society of Cardiology and the North American Society of Pacing Electrophysiology [370] provided an extensive overview of measurement standards, physiological interpretation, and clinical HRV applications. It is believed that HRV is an indicator of sympathovagal interaction after acute myocardial infarction [350] and that it is helpful in the diagnosis of congestive heart failure and predicting survival in premature babies. HRV is also relevant to pathological conditions not necessarily connected with the heart action, e.g., diabetic neuropathy. Since the autonomous nervous system influences HRV, the phenomena related to its functioning impact HRV. An example may be the sleep-wake cycle.

5.2.2.1 Time-Domain Methods of HRV Analysis

Time-domain statistics are usually calculated on RR intervals without resampling. A review of the time domain methods of HRV analysis may be found in [370]. The primary indices involve:

- the standard deviation of RR intervals (which may be calculated over 24 hours, or short, usually 5 minutes, periods),

- the square root of the mean squared difference of successive RRs,

- the number of pairs of successive RRs that differ by more than 50 ms (so called NN50), the proportion of NN50 divided by total number of NNs (pNN50).

These markers and other tools helpful in HRV quantification may be found in PhysioNet [188].

The series of RR intervals can also be represented by density distribution of RR interval durations or density distribution of differences between adjacent RR intervals. In the density distribution construction, a reasonable number of RR intervals is needed. In practice, recordings of at least 20 minutes (but preferably 24 hours) are recommended to ensure correct performance. The histogram's discrete scale should be appropriate—not too fine or too coarse, permitting the construction of smoothed histograms.

The deviations from normal distribution may be quantified by fitting some geometrical shapes or using higher-order moments: skewness and kurtosis [88]. However, the higher the moment, the more sensitive it is to outliers and artifacts. Skewness—the indicator of asymmetry—is used to detect possible sudden acceleration followed by a more prolonged deceleration in the heart rate. This phenomenon may be connected with clinical problems. Another measure quantifying RR distribution is the HRV triangular index measurement; it is the integral of the density distribution (that is, the number of all RR intervals) divided by the maximum of the density distribution [371].

5.2.2.2 Frequency-Domain Methods of HRV Analysis

In HRV several rhythms can be distinguished, which are conventionally divided in four frequency bands:

- Ultra low frequency (ULF)): $0.0001\,\mathrm{Hz} \leq \mathrm{ULF} < 0.003\,\mathrm{Hz}$

- Very low frequency (VLF): $0.003\,\mathrm{Hz} \leq \mathrm{VLF} < 0.04\,\mathrm{Hz}$

- Low frequency (LF): $0.04\,\mathrm{Hz} \leq \mathrm{LF} < 0.15\,\mathrm{Hz}$

- High frequency (HF): $0.15\,\mathrm{Hz} \leq \mathrm{HF} < 0.4\,\mathrm{Hz}$

Fluctuations in VLF and ULF bands are thought to be due to long-term regulatory mechanisms such as thermoregulatory system, systems related to blood pressure, and chemical regulatory factors [82]. VLF appears to depend primarily on the parasympathetic system. HF is a measure of respiratory sinus arrhythmias and can be considered an index of vagal modulation [369]. Some studies suggest that LF, when expressed in normalized units, is a quantitative marker of sympathetic modulations; other studies view LF as reflecting sympathetic activity and vagal activity. Consequently, the LF/HF ratio is considered by some investigators to mirror sympathovagal balance or to reflect the sympathetic modulations [369]. It is important to note that HRV measures fluctuations in autonomic inputs to the heart rather than the mean level of autonomic inputs.

Three main spectral components can be distinguished in a spectrum calculated from short-term recordings of two to five minutes: VLF, LF, and HF. The ULF component may be found from long-term recording, usually 24 hours. In the case of long-term recordings, the problem of stationarity arises. If the modulations of frequencies are not stable, the interpretation of the results of frequency analysis is less well defined. In particular, physiological mechanisms of heart period modulations responsible for LF and HF power components cannot be considered stationary during the 24 hours; thus, they should be estimated based on shorter time epochs. To attribute individual spectral components to well-defined physiological mechanisms, such mechanisms modulating the heart rate should not change during the recording. Transient physiological phenomena should be analyzed by time-frequency methods. Traditional statistical tests may be used to check the signal's stability in terms of specific spectral components.

Both non-parametric and parametric methods are used for frequency analysis of HRV. The advantages of the non-parametric methods are the simplicity of the algorithm used (fast Fourier transform) and high processing speed. In comparison, the benefits of parametric methods are smoother spectral components that can be distinguished independent of pre-selected frequency bands, straightforward post-processing of the spectrum with an automatic calculation of low- and high-frequency power components (easy identification of the central frequency of each component), and an accurate estimation of PSD even on a small number of samples which is important for the quasi-stationary signal. A disadvantage of parametric methods may be the need to determine the model order, but this can be resolved using known criteria (Sect. 3.3.2.3).

For the determination of ECG-derived respiratory information, spectral analysis is usually applied. The respiratory activity is estimated as the HF component of the HRV signal. The respiratory rhythm may be determined as the central frequency of the HF peak of the power spectrum or, in the case of the AR model, directly as a frequency determined using FAD (Sect. 3.3.2.3). The time-varying AR model was used to analyze the coupling of respiratory and cardiac processes [403]. Respiratory activity may also be found based on beat morphology, namely from amplitude fluctuation of main complexes of ECG [30], or possibly both methods can be combined [327].

The time-varying features of HRV may be studied using wavelet analysis. Both, continuous, e.g., [601] and discrete wavelet analysis, e.g., [508] were applied in HRV analysis. The advantage of discrete wavelet transform is the parametric description of the signal, which is useful for further statistical analysis and classification.

HRV is related to other body signals e.g.: blood pressure and respiration. Interrelation between HRV and arterial blood pressure rhythms was shown by cross spectral analysis already in 1988 [37] and further studies indicated the possible clinical applications of simultaneous analysis of both signals [515]. The problem of coupling between cardiovascular, hemodynamic and brain signals will be considered in Sect. 5.5.

5.2.2.3 Non-Linear Methods of HRV Analysis

Given the complexity of the mechanisms regulating heart rate, it is reasonable to assume that applying HRV analysis based on non-linear dynamics methods will yield valuable information. The study concerning the properties of heart rate variability based on 24-hour recordings of 70 subjects was conducted in [611]. It was reported that variability due to the linearly correlated processes was dominant (in normal group 85%). But with the development of the disease and risk of cardiac arrest, the non-linear component increased. The contribution of the random noise variance was found at 5–15% of the overall signal variance. The exception was the case of atrial fibrillation, where the contribution of random noise achieved 60%. One of the first applications of the non-linear methods was assessing the correlation dimension of the HRV [25]. The most frequently used parameters applied to measure non-linear properties of HRV include the correlation dimension, Lyapunov exponents, and Kolmogorov entropy. Despite multiple attempts conducted using the above methods to demonstrate the prevalence of chaos in heart rate series, more rigorous testing based primarily on surrogate data techniques did not confirm this hypothesis. It revealed the weakness of classical non-linear methods connected with lack of sensitivity, specificity, and robustness to noise [487].

More recently, scientists have turned to the nonlinear methods which are less biased by noise and restricted length of the available data, namely, empirical mode decomposition, detrended fluctuation analysis, modified measures of entropy, Poincaré plots.

Empirical Mode Decomposition

Empirical mode decomposition (EMD) is based on the decomposition of the original signal into components called instantaneous mode functions (Sect. 3.4.2.2). The selection of modes corresponds to adaptive, signal-dependent, time-variant filtering. EMD found application in separation and tracking of rhythms present in HRV [136, 563].

In [136] a series of simulations was performed illustrating the behavior of the method for stationary, non-stationary, and chirp signals. The calculations were also performed for experimental data involving measurements during rhythmic breathing and HRV connected with the change of posture from seated to standing. First, the decomposition into instantaneous mode functions was performed, then the Hilbert transform was calculated for the consecutive components. Next, the instantaneous frequency was estimated from the phases of component time-series (Sect. 3.4.1). In Figure 5.34, the first four components, the difference between component one and four, and the instantaneous amplitude ratio of the third and first component (a_3/a_1) as a function of time are illustrated. The value of the component ratios may be considered as a convenient measure of the HRV rhythms evolution. One can observe the increase of the ratio a_3/a_1 after changing posture.

In Figure 5.35 instantaneous frequency as a function of time for first components presented in Figure 5.34 is shown. The instantaneous frequency (IF) seems to be a good index for tracking HRV rhythms changes. Still, in the IF determination procedure, appropriate boundary requirements have to be

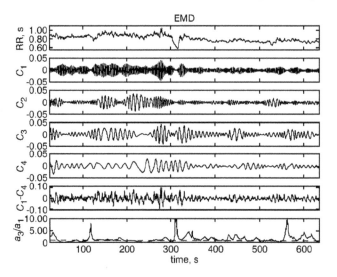

FIGURE 5.34
Empirical mode decomposition of a real, short-term HRV signal from a healthy young adult involving standing up movement at 300 s. Top graph presents original R-R interval series, and C1, C2, C3, C4 were used to describe the first four components obtained by EMD. Also shown is reconstructed (C1–C4) series obtained by first four components of decomposition. Plot in bottom graph represents instatenous amplitude ratio of third and first components. From [136].

applied, which is not a trivial problem [231]. Another problem in the application of EMD is the reproducibility of results. In the study [362] (described in more detail in Sect. 5.2.2.3) the reproducibility of EMD was found to be very poor. It seems that the EMD method has to be used with caution, especially in respect of fulfilling the assumptions of the method and in respect of statistical significance and reproducibility of results.

Entropy Measures

The indexes connected with information theory (Sect. 3.5.6): approximate entropy (ApEn) [484], conditional entropy [488], and sample entropy (SaEn) [503] were introduced and found applications in recognition of certain cardiovascular pathologies. For example in [469] a progressive decrease in ApEn indicating reduced complexity of heart rate variability was observed before the onset of atrial fibrillation. Conditional entropy was used for evaluation of regularity, synchronization, and coordination in cardiovascular beat-to-beat variability [489].

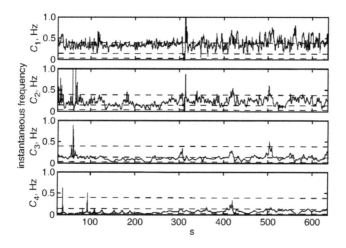

FIGURE 5.35
Instantaneous frequency as a function of time for the first four components presented in Figure 5.34. From [136].

SaEn was applied to study heart rate variability during obstructive sleep apnea episodes in [3]. Compared with spectral analysis in a minute-by-minute classification, sample entropy had an accuracy, sensitivity, and specificity slightly worse than the spectral analysis results. The combination of the two methods improved the results, but the improvement was not substantial. Nevertheless, SaEn has begun to be applied in practice, namely in wearable devices. It was reported that SaEn was used to assess respiratory biofeedback effect from HRV [346].

Detrended Fluctuation Analysis

Another measure of complexity is provided by detrended fluctuation analysis (DFA). In its framework, the fractal index of self-similarity is determined (Sect. 3.5.3).

The HRV scaling exponent value depends on the window length. The examples of DFA results illustrating the dependence of scaling exponents on data length are shown in Figure 5.36. The short-time fractal scaling exponent has a relatively good correlation with the low-to-high spectral component ratio in controlled recording conditions [469]. It was reported that short-term scaling exponents derived from the HRV series could predict vulnerability to ventricular tachycardia, ventricular fibrillation, arrhythmic death, and other heart conditions (review in [469]). The power-law relationship of heart rate variability was used as a predictor of mortality in the elderly [233]. However, the mono-fractal (scale-invariant) behavior need not testify to the chaotic

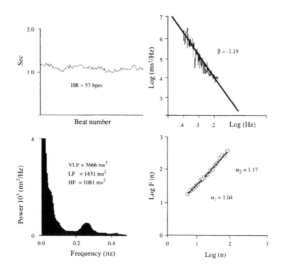

FIGURE 5.36
Scaling properties of 24 hour tachogram of a healthy subject. RR interval tachogram (upper left), power spectra (lower left), a power-law scaling slope of long-term fluctuations of heartbeats (upper right), and DFA results (lower right). β—long-term scaling slope, α_1 short-term fractal scaling exponent (4–11 beats), α_2 intermediate scaling exponent (> 11 bits). From [469].

character of the data. It was reported that the surrogate data test revealed the same α and β values for the original HRV series and for the ones with randomized phases [337].

In [465] the scaling exponent was determined for 24 h recordings of HRV of a healthy group and patients with congestive heart failure using DFA. Short-term and long-term scaling exponents α_1 and α_2 differed significantly for normal and pathological groups. Still, the separation was not complete— there was substantial overlap between both groups in parameter space (8 out of 12 healthy subjects and 11 out of 15 pathological subjects exhibited a "cross-over" between groups).

The same data (from PhysioNet database, Beth Israel Hospital) as those used in [465] were subjected to wavelet analysis and DFA in [598]. For classification, standard deviations of wavelet coefficients at scales 4–5 (16–32 heartbeats) were used, which yielded 100% classification (sensitivity and specificity) between the groups. In the above work, from the wavelet coefficients scaling exponents α were also determined. The discrimination between groups based on their values resulted in 87% sensitivity, and 100% specificity, which was better than in [465] but worse than for parametrization by wavelet standard

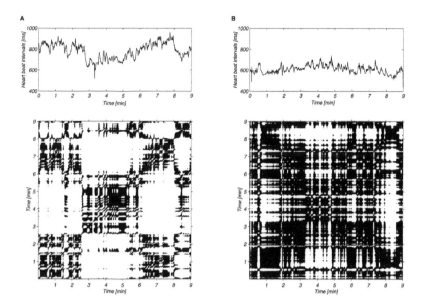

FIGURE 5.37
Recurrence plots of the heartbeat interval time series at a control time (a) and before a ventricular tachycardia (b) with an embedding dimension of 6. The RP before a life-threatening arrhythmia is characterized by big black rectangles, whereas RP from the control series shows only small rectangles. From [387].

deviation. It follows that wavelet analysis is more efficient than DFA in respect of differentiation between normal and pathological states.

Poincaré and Recurrence Plots

Other non-linear methods of analyzing heart rate variability are the Poincaré map (PM) and recurrence plot (RP). The advantage of these methods is that they can be applied to short and non-stationary data. Recurrence plots were applied to heart rate variability data in [387]. The examples of recurrence plots for HRV epochs before ventricular tachycardia (VT) in comparison to control epochs are shown in Figure 5.37. The RP for epoch before VT onset shows big rectangular structures. Usually, RP plots are characterized by specific shape parameters, among them diagonal line lengths and vertical line lengths, which are important for characterizing different conditions.

The study of HRV from 24 h ECG through entropy measures and Poincaré maps was conducted in [661]. The examples of three-dimensional RR interval return maps (Poincaré maps) are shown in Figure 5.38. One can observe different patterns for cases: before cardiac arrest and one year later after

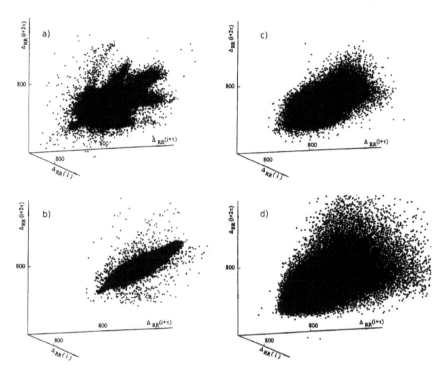

FIGURE 5.38
Three-dimensional images of RR interval return maps calculated from 24-h
Holter ECG recordings. a) and b) data from the same subject before cardiac
arrest and 1y after, respectively; plot b) is very similar to plots obtained for
healthy subjects; c) example plot for a ventricular arrhythmia; d) example
plot for atrial fibrillation. Adapted from [661].

medication. The authors point out that although pathological patterns differ
from the normal ones, it is difficult to distinguish between different pathological cases (Figure 5.38 c and d).

Effectiveness of Non-linear Methods

Different non-linear methods were compared in [362]. The authors computed
11 non-linear HRV indices found from detrended fluctuation analysis, entropy estimators including SaEn, empirical mode decomposition, and Poincaré
plots. The performance of these measures was evaluated by comparison of results of the two 5-minute sessions conducted on two consecutive days for a
group of 42 healthy adults. The reliability of measures was assessed using the
test of no difference between two measurements (paired *t*-test), and relative
reliability was based on the interclass correlation coefficient. The best results

were obtained for Poincaré plots, detrended fluctuation analysis, and entropy measures; other indices performed relatively poorly. These reliability measures were based on short-term measurements and two data sets only, so they must be considered with care. In assessing the utility of different methods, other measures such as sensitivity and specificity in distinguishing different conditions and robustness to noise have to be taken into account as well.

Although in principle, the techniques mentioned above are useful tools for the characterization of various complex systems, no major breakthrough has yet been achieved by their application to biomedical signals, including HRV analysis. No systematic study has been conducted using non-linear methods to investigate large patient populations. Therefore they were not recommended for diagnostic applications by the Task Force of the European Society of Cardiology and the North American Society of Electrophysiology [370]. At present, the standards concrning non-linear methods applications to HRV are still lacking, and none of these methods is in widespread clinical use. For validation of non-linear methods the studies involving both linear and non-linear approaches applied for the same data are needed.

5.2.3 Fetal ECG

The fetal electrocardiogram (fECG) is the electrical activity of a fetus' heart, which contains valuable information about the physiological state of a fetus and is an important tool in monitoring its well-being. For instance, pathological conditions like hypoxia or acidemia may be detected using fECG. The amplitude of fECG is in the range from 10 to $100\,\mu V$, whereas the maternal ECG (mECG) can be as high as $0.5\,mV$–$100\,mV$. When measured on the maternal abdomen, fECG is additionally disturbed by the muscular activity from abdomen and uterus, since the main frequency range of fetal and maternal ECG is in the frequency range 0.3–$100\,Hz$, while for the abdominal muscle, it is 10–$400\,Hz$ and for uterine contractions 0.01–$0.6\,Hz$. Therefore the problem of extraction of fECG from other disturbing bioelectric signals is not a simple one [279].

Several methods for extraction of fECG from maternal ECG (mECG) were devised including: adaptive filtering [384], singular value decomposition [277], fractals [504], neural networks [76], fuzzy logic [4], polynomial networks [17], combination of PCA and matched filtering [381], template construction [591]. Wavelet transform also found several applications to the fECG analysis, e.g., [279] and [288] who used biorthogonal quadratic spline wavelet. Two-level wavelet transform (Daubechie wavelets) followed by low-pass filtering was applied in [215] and also sequential source separation in the wavelet domain was used for fECG extraction [244].

The problem of fECG extraction from the signal registered on the abdomen can be formulated as a blind source separation problem (BSS) since it involves the separation of the source signals from a sensor array signals without knowledge of the transmission channels. BSS method involving singular value decomposition and ICA was applied to fECG in [112]. In the job of separation of fECG from mECG the independent component method performed better than PCA.

In solving the problem of fECG – mECG separation, especially by ICA-based BSS approaches, some theoretical limitations appear because the number of observations (registered signals) must be equal to or higher than the number of uncorrelated signals sources. However, in addition to heart signals, each electrode also picks up activity due to maternal myoelectric activity and other noises, which are the sources of partly uncorrelated noise different for each channel. Therefore BSS requires multielectrode arrays, usually including electrodes placed on the thorax, which is not always practical in the clinical environment.

The aim of several works concerning fECG was the extraction of a fetal heart rate variability (fHRV) signal, which can be used to find the synergetic control activity of the sympathetic and parasympathetic branches of the autonomous nervous system [80]. The fHRV estimation method based on the minimization of a cost function that measures the differences between the discrete Fourier transform (DFT) of the fetal ECG waveform and the DFTs of its circularly shifted forms was proposed in [520]. Using the DFT linear phase shift property, they showed that the minimization of this cost function is equivalent to finding the cosine waveform that matches best to the fECG power spectrum. The optimal cosine waveform was then used to estimate the fundamental period of fHRV.

The problem of separating the components of fHRV was also approached through empirical mode decomposition by Ortiz et al. [448]. High-frequency fHRV modes were obtained by application of EMD followed by the reconstruction of components above 0.3 Hz. The results showed differences in frequency components power for the episodes connected with breathing movements and body movements versus quiet epochs.

A three-stage method involving different signal analysis techniques for extracting fECG and determination of fHRV was proposed in [280]. The methodology can be applied to two or more channels. The first stage involved mECG elimination. After filtering the signals in the 4–20 Hz band, the parabolic fitting technique was used to detect maternal R-peaks. A phase space thresholding method based on Poincaré maps was used to determine maternal fiducial points. The axes in 3-D phase space were constructed from the signal, its first and second derivative. The thresholding method used for identifying the QRS complex was based on the observation that in the 3-D plot, the points in the center correspond mainly to P and T waves and low-frequency noise, while the points in the surrounding torus correspond to QRS and high-frequency noise. As a threshold for QRS separation, the 3-D ellipsoid was defined, making it possible to determine the fiducial points of QRS and perform the subtraction of mECG from the original (not filtered) signal. In the second stage, the denoising was performed through the PCA application to wavelet coefficients (6-levels Daubechies) followed by reconstruction. Then in the denoised signal,

R peaks of fECG not coincident with maternal R peaks were identified. For detection of the superimposed peaks, a histogram of R-R intervals was constructed, and the intervals of double-length were considered in the procedure of finding the lost fetal R peaks. This information was used to construct the final fHRV series.

The above-described method was tested on simulated signals with various signal-to-noise (SNR) ratios, generated according to [522] and on experimental signals from the University of Nottingham database [478]. Simulated data helped improve the ellipsoid parameters used for thresholding, which resulted in better sensitivity values. The results for the experimental time series also yielded high values (around 90%) of selectivity. The described approach was compared with other methods: time-frequency-based methodology [279], parabolic fitting [Zhang et al., 2006], template matching [182], fast ICA algorithm [238]. The accuracy for real data showed the best results for the above-described method and time-frequency method.

5.2.4 Magnetocardiogram and Fetal Magnetocardiogram

5.2.4.1 Magnetocardiogram

The magnetic field produced by a heart is around $50\,\text{pT}$, which is more than an order of magnitude more prominent than the one generated by the brain (about $1\,\text{pT}$), making MCG easier to record than MEG. The first magnetocardiogram (MCG) was measured almost 50 years ago by Baule and Mc Fee utilizing a gradiometer containing a set of copper coils with several millions of windings. The introduction of SQUID opened the era of magnetocardiology. At first, single sensors were used. At present, commercially available devices include multiple MCG leads, and the custom-made systems have up to 128 sensors (the description of the magnetic field sensors may be found in Sect. 5.1.4).

One of the advantages of MCG is the lack of direct contact of the sensors with the skin, which allows for fast screening of patients and is useful for people with burns. The same currents generate MCG and ECG signals, and their time evolution is similar but not identical. Tangential components of the fields are attenuated in the ECG. Contrary, the ideal magnetic lead is sensitive only to the tangential components of electric sources. It should be particularly responsive to abnormalities in the activation since normal activation sources are primarily radial. Another advantage of MCG stems from the fact that the tissue's magnetic permeability is that of free space. This feature allows recording activity of the posterior side of the heart. In ECG measurement, this activity is strongly attenuated by the resistivity of the lungs.

It was reported that MCG gives unique information on cardiac ischemia, arrhythmias, and fetal diagnosis [655]. MCG was successfully applied in the diagnosis of several pathological heart conditions. For example, inspecting ST shifts in MCG during an exercise test [89] demonstrated that injury currents

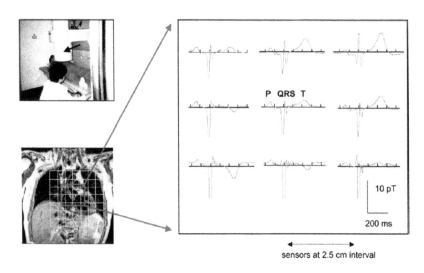

FIGURE 5.39
A 64-channel magnetocardiographic system. Overview of the system of 8 by 8 matrix of sensors (sensor interval: 2.5 cm, a measuring area: 17.5 by 17.5 cm) superimposed on magnetic resonance image; 9 out of 64 MCG signals. From [655].

resulting from ischemia, practically impossible to measure with ECG, may be detected by MCG. The usefulness of magnetic field maps to detect angina pectoris and myocardial infarction in cases when no ST complex elevation is present was reported in [340]. It was also reported that MCG makes it possible to distinguish the features characterizing coronary arterial disease and arrhythmia vulnerability [573].

The MCG time evolution and an example of a measurement system are shown in Figure 5.39. Construction of maps representing magnetic field distribution measured over the thorax is one of the methods useful for clinical diagnosis. An example of such an approach may be the work [185]. The MCG, recorded utilizing 36 sensors, was filtered in the frequency band 0.01 to 120 Hz and the artifacts were removed using the wavelet transform. The MCG evolutions synchronized in respect to the R peak were averaged. In Figure 5.40 the averaged traces together with a magnetic field map are displayed. The MCG shows the R waves' splitting in several positions (not visible in ECG), which is a symptom of the left bundle branch block. In the iso-magnetic field contour plot for R peak instant, the contributions of two equivalent current dipoles may be observed. In the same study, the fragmentation of the R peak of MCG and distinct contributions of more than one dipole in the contour maps were observed for several subjects, who didn't show specific changes in

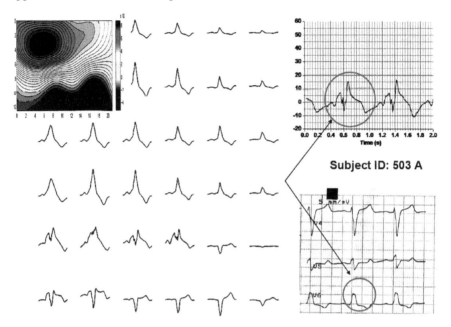

FIGURE 5.40
Averaged MCG at the spatial positions recorded over the chest. The top left panel is the corresponding magnetic field map. The ECG at lead V6 showing features of left bundle branch block (highlighted with a circle) is at bottom right. The inset at the top right shows the time trace of two cardiac cycles in MCG at a position above V4 of ECG. The split in the R wave is highlighted with a circle. From K. Gireesan, C. Sengottuvel, C. Parsakthi, P. Rajesh, M. Janawadkar, T. Radhakrishnan: Magnetocardiography study of cardiac anomalies, in Selma Supek, Ana Susac (Eds.): *Advances in Biomagnetism BIOMAG2010, IFMBE Proceedings* 28, 2010, pp. 431–435, Figure: 1, www.springerlink.com/ifmbe ©International Federation for Medical and Biological Engineering 2010. (This figure is courtesy of IFMBE.)

ECG. It was postulated that the fragmentation of the R peak of MCG was connected with altered ventricular depolarization. In the several investigated cases, the MCG has revealed more distinct features associated with the heart action disturbances than ECG, which opens the possibility to improve early detection of myocardial infarction using MCG.

The utility of the cardiac field mapping for the detection of coronary artery disease was studied in [195]. The comparison of the obtained maps to the ideal group mean maps was based on the three measures of similarity: Kullback-Leibler (KL) entropy (Sect. 4.4.1), normalized residual magnetic field strength and deviations in the magnetic field map orientation. The mean values of these

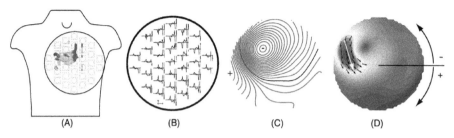

FIGURE 5.41

Recording and analysis of atrial magnetic fields. (A) The sensor arrangement of a 33 triple sensor (99-channel) magnetometer. Superimposed on sensor array is the antero-posterior view of the left atrium by electroanatomic mapping. (B) Signal—averaged magnetic field density on each magnetometer channel over cardiac cycle. The onset and end of atrial signal are determined automatically using filtering technique. (C) Spatial distribution of the magnetic field B_z component over the middle part of atrial complex interpolated from the measurement using multipole expansion. The blue color indicates flux out of the chest (-) and red color flux into the chest (+). The step between two consecutive lines is 200 fT. (D) Pseudocurrent map derived by rotating magnetic field gradients by 90°. The red-yellow color indicates the area of the top 30% of strongest currents, and the large arrow indicates their mean direction. Zero angle direction is pointing from subject's right to left and positive clockwise. From [268].

parameters during the depolarization and repolarization were used for the classification with logistic regression. The features set based on KL entropy demonstrated the best classification results, namely sensitivity/specificity of 85/80% was reported.

The possibility of identification of inter-atrial conduction pathways by MCG was reported in [268]. The experiment involved the patients undergoing catheter ablation of paroxysmal atrial fibrillation. The intra-cardiac electroanatomic mapping was compared with pseudocurrent maps obtained using MCG recorded by 33 triple sensors, namely two planar gradiometers (x-y plane) and a magnetometer oriented along the z-axis (axis orientation as in Figure 5.30). Figure 5.41 shows averaged magnetic field density at each z-magnetometer, the spatial distribution of B_z component, and pseudocurrent map. The pseudocurrent inversion is used to characterize the orientation of the magnetic field. The method is based on rotating the estimated planar gradients of B_z component by 90°:

$$b = \frac{\partial B_z}{\partial y}\bar{e}_x - \frac{\partial B_z}{\partial x}\bar{e}_y \qquad (5.49)$$

where $\bar{e}_x$, $\bar{e}_y$ are the perpendicular unit vectors on the sensor array plane. The resulting arrow map provides a zero-order approximation (pseudocurrent map) for the underlying electric current. The conclusions from the above studies were that by inspecting the pseudocurrent maps, the current propagation might be determined, and the damages in conducting pathways may be found. Overall findings imply that the non-invasive pseudocurrent mapping technique may help localizing the breakthroughs in conduction pathways and refining methods for the ablation treatment, if needed.

Biomagnetic methods are well suited to solve the problem of the localization of bioelectric sources in the body. Although magnetic fields are much less influenced by the body tissues than electric fields, MCG is to some extent modified by the anisotropic properties of cardiac and torso tissues. The best localization results were obtained when the activity started in a small, well-defined area. For the solution of inverse problems in magnetocardiography, both the equivalent current dipole approach and effective magnetic dipole models have been used. It was reported that they might provide localization with the accuracy of a few millimeters when realistic and patient-tailored torso models are used [470]. However, the approach is limited to localizing point-like focal sources such as those appearing, e.g., in focal arrhythmias or ventricular extrasystole [153].

In general, multiple current sources are simultaneously activated during the cardiac cycle; thus, the inverse solution based on current density distribution found several applications in magnetocardiography. For example, in [426] for equivalent current-density estimation in subjects with coronary artery disease, a patient-specific boundary-element torso model was used. Three different methods of regularization were confronted with the PET measurements. The results showed that non-invasive localization of ischemic areas in patients with coronary artery disease is possible when proper regularization is applied.

5.2.4.2 Fetal MCG

Magnetic field measurement is a promising technique for the evaluation of fetal well-being. Fetal biomagnetic signals are unaffected by the low electrical conductivity of the vernix caseosa. This waxy substance forms on the fetal skin at about 25 weeks' gestation and impedes fetal bioelectric signals transmission. Fetal magnetocardiography (fMCG) has the potential to provide beat-to-beat fetal heart rate in case of normal rhythm and its disturbances. Amplitude changes in fMCG and fetal heart rate acceleration make fetal movement assessment possible using a technique called actocardiography [668].

Various groups reported good quality fMCG recordings in which P, QRS, and T waves may be distinguished even without signal averaging. Usually, after bandpass filtering and maternal MCG removal, the averaged waveforms are computed after aligning fetal complexes, e.g., using autocorrelation. This kind of analysis was used for finding fetal cardiac repolarization abnormalities [667].

For determining fetal heart rate variability (fHRV), methods similar to the ones described in the section concerning fECG may be used. In [644] another approach was applied. After removal of maternal MCG, four approaches for fetal QRS identification were tested: Hilbert transform, ICA, ICA followed by Hilbert transform, and filtering. The Hilbert transform (HT) yields an analytic signal whose amplitude is always positive independently of signal polarity (Sect. 3.4.1). The value called the rate of change of HT amplitude (RHA) was defined as:

$$RHA(n) = \sqrt{(x_{n+1} - x_n)^2 + (y_{n+1} - y_n)^2} \qquad (5.50)$$

where x_n and y_n are real and imaginary parts of the transformed signal at point n. The series $RHA(n)$ is always positive, and signals from different channels may be easily averaged. The next step was the application of FastICA. FastICA (Sect. 4.6.2.3) finds the columns of separating matrix by maximizing the absolute value of kurtosis, which works particularly well for QRS extraction due to the high value of kurtosis for QRS complexes. The third approach used in [644] relied on the application of HT to ICA components. The fourth method involved the manual selection of QRS complexes in a signal filtered in the 1—100 Hz band. The next step in analysis involved correction for missed beats performed in a way similar to the method described in the section on fECG. In the quantitative analysis of results, the Hilbert method scored best, yielding the smallest number of total errors and efficiency of 99.6% in comparison to ICA (94.4%) and ICA-HT (96.4%).

Adaptive ICA was used in real-time fetal heart monitoring. The developed monitoring system [623] consisted of real-time access to fMCG data, an algorithm based on independent component analysis, and a graphical user interface. The authors reported that the algorithm extracts the current fetal and maternal heart signal from a noisy and artifact-contaminated data stream in real-time and can adapt automatically to varying environmental parameters.

Despite the numerous successful clinical applications based on the measurement of heart magnetic fields, the advantages and disadvantages of MCG in comparison with ECG are still a matter of debate [230, 572]. The advantage of ECG is lower cost and easier support of the devices. The long years of clinical experience in ECG have some meaning as well.

5.2.5   Ballistocardiogram, Seismocardiogram, Photoplethysmogram

Ballistocardiogram (BCG) is a signal generated by repetitive motions of the human body caused by blood ejection with each heartbeat. It is a well-accepted measure of the ejection force of the heart. The spectrum of BCG's possible clinical applications involves coronary heart disease, myocardial infarction, angina pectoris, monitoring cardiac reaction to exercise or stress. BCG was discovered in the 19th century and has been at the center of attention from the 1940s to the early 1980s. The decline of BCG was connected with the development of ultrasound and echocardiographic techniques. The scientific community's withdrawal from the BCG technique was associated with a lack of standard measurement techniques and a deeper understanding of the BCG's physiologic origin. The comeback of BCG may be noted with the present advance of unobtrusive recording devices, which may be integrated within ambient locations such as beds, garments, or can be placed on the human body. The wearable contemporary healthcare systems often involve, besides ECG and BCG, seismocardiogram (SCG) and photoplethysmogram (PPG).

SCG represents mechanical heart motion, which helps determine periods of cardiac quiescence within a cardiac cycle. SCG signal records cardiac movement transmitted to the chest wall utilizing one or more linear accelerometers placed on the chest. SCG time series are collected as acceleration versus time and have been shown as an accurate cardiac mechanical state indication.

Plethysmogram is a measure of blood volume changes in the organ. Photoplethysmogram (PPG) is an optically obtained plethysmogram, which shows blood volume variations in the microvascular bed of tissue. The pressure pulse causes a volume change; thus, each cardiac cycle appears as a peak. A PPG is often obtained using a pulse oximeter, which illuminates the skin with light-emitting diode and measures changes in light absorption.

BCG, SCG, and PPG, together with ECG, are used in wearable systems for monitoring patients with cardiovascular diseases and neurological disorders (especially epilepsy, Parkinson's disease, and multiple sclerosis) and for supervising the condition of people with health risk.

5.2.5.1 Wearable Devices

A wearable medical device (WMD) can be defined as an autonomous, small-sized non-invasive system that performs a specific medical function such as monitoring or support. Wearable devices can be classified into monitoring systems, rehabilitation assistance devices, and long-term medical aids or assistive technology devices, which allow their users to achieve some levels of independence. The range of applications of WMD in health care is broad and involves, among others:

- cardiovascular diseases

- neurological disorders (especially epilepsy, Parkinson's disease, multiple sclerosis, sleep apnea)

- diabetes

- assisting disabled and chronically ill

- helping the elderly suffering from memory loss and Alzheimer's disease

WMDs are also increasingly used by healthy people as health monitors or fitness assistants. The devices can be either supported directly on the human body as a patch, a piece of clothing, or a sort of a wristwatch. WMD should be easy to use, comfortable to wear, and have minimal size and weight, which is increasingly provided due to advancements in sensor and microprocessors technology.

WMD has a data input unit, a processing module, and an output unit as any typical device. The input mechanism involves the recording of the time series, typically: ECG, BCG, PPG, EMG, blood pressure, temperature, signals from accelerometers providing information about the posture or body movement, also EEG, EOG, SCG, ERD, respiration are included depending on the medical condition to be monitored. The processing unit's role is to handle the incoming information, often in real-time, to generate the appropriate feedback. This feedback is accessed directly by the user, may serve as a control for the system providing supporting functions, or may be transmitted to a remote monitoring unit. Output mechanisms can involve combinations of audio, visual, and electrical signals, including a telemetric service. The pre-processing involves removing artifacts due to, e.g., body motions, possible power line interference from subcomponents of the device, and the uncontrolled conditions of the measuring environment.

The incorporation of complex processing algorithms in wearable systems is limited by memory availability, processing capabilities, and power autonomy considerations. The methodologies vary from straightforward, e.g., comparing the detected signal with a pre-determined threshold based on prior knowledge, to more complicated processing involving rule-based, statistical, or machine learning techniques. A monitoring system's robustness and reliability may be increased by the fusion of information gained from multiple sensors. Signal fusion algorithms include probabilistic models (e.g., recursive operators), least-square techniques (e.g., Kalman filtering), intelligent fusion (e.g., fuzzy logic, neural networks, genetic algorithms).

If a system can be described with a linear model and both the system error and the sensor errors can be modeled as Gaussian noise, then the Kalman filter provides a statistically optimal estimate of the fused data [186]. The signal processing results, possibly combined with other information, are fed into a Bayesian inference system for complex interpretation and decision-making.

An example of a wearable medical device may be a system where BCG, SCG, ECG, body temperature and atmospheric pressure (to detect movements, e.g., climbing the stairs), were applied to monitor cardiovascular diseases [146]. The sensors were accommodated in a patch mounted on the chest. Miniature accelerometers measured whole-body movements and chest wall vibrations. All hardware was self-contained, including a microprocessor and associated circuitry required for storing the data locally on a micro secure digital card. The removal of artifacts relied on empirical mode decomposition and feature tracking algorithms. Through EMD, the SCG signal components related to the movement were sifted from the components related to heartbeats. EMD improved SCG waveform morphology, and the improvement was then quantified with dynamic time warping (DTW) [550]. DTW quantifies the similarity between two waveforms while allowing for time shift and temporal dilation/contraction. The system was robust against motion artifacts and sensor misplacement. It provided information on the cardiac output blood pressure, cardiac contractility, and key time intervals of cardiac cycle: pre-ejection period and pulse transit time.

Combined wearable and fixed component systems have been used to develop home monitoring platforms for the elderly, the critically ill, and the disabled. Health smart homes may encompass a wearable medical monitoring device with sensors accommodated, e.g., in beds and set of cameras monitoring users' movements.

Continuous progress in biomedical signal processing, novel human-computer interaction technologies, innovative sensors, machine learning, and wireless telemetry will possibly lead to an environment in which we will be surrounded by intelligent objects recognizing our presence and condition.

5.3 Electromyogram

Electromyogram (EMG) is a record of electrical muscle activity. In this section, only EMG of striated muscles will be discussed.

5.3.1 Measurement Techniques and Physiological Background

The EMG characteristics depend on the kind of applied electrodes. EMG may be recorded by means of intramuscular (needle or fine wire) or surface electrodes. The range of measured potential difference in case of intramuscular electrodes is 0.1–20 mV, and for surface electrodes 0.05–1 mV. In clinical practice EMG is used for routine diagnosis of neuromuscular disorders, based typically on single fiber (Figure 5.42 a) or concentric needle electrodes (Figure 5.42 b). Surface electrodes enable non-invasive EMG recording of the global muscle electrical activity. Earlier, surface electrodes were applied only in situations when the insertion of needle electrodes was difficult or not possible, e.g., for examination of children, long-term recordings, ergonomics, sports, or space medicine. Nowadays, with increasing demand concerning non-invasiveness, surface EMG electrode arrays (Figure 5.43) became popular, which was possible due to technical developments and progress in signal analysis.

The registered EMG signal is a superposition of activity of single motor units (MUs). A single MU consists of a motoneuron and the group of muscle fibers innervated by its axon (Figure 5.44 a). Fibers belonging to a given MU are distributed throughout a muscle (Figure 5.44 b). Every motoneuron discharge evokes contraction of all its muscle fibers, preceded by their depolarization, which is detected as a waveform called motor unit action potential (MUAP). The spike component of MUAP (and its peak-to-peak amplitude) is determined by a few fibers closest to the electrode; the more

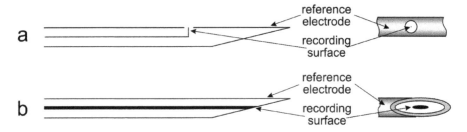

FIGURE 5.42
Types of EMG electrodes. a) Single wire electrode, b) needle electrode.

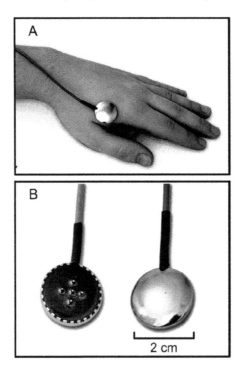

FIGURE 5.43
Surface electrodes array. From [424].

remote fibers influence mainly MUAP duration. MUAP shape depends also on the temporal dispersion between single fiber potentials. The temporal dispersion is determined mainly by the distribution of motor endplates along muscle fibers, resulting in the differences in arrival times to the electrode. The shapes of consecutive MUAPs are not identical due to the variability of these

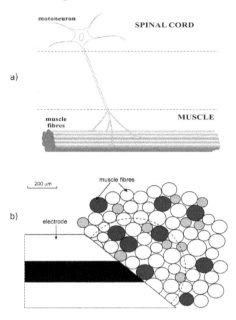

FIGURE 5.44
a) Motor unit, b) cross-section of a muscle and a needle electrode. Fibers belonging to the same MU are marked by the same shade.

arrival times caused by the small fluctuations in the synaptic transmission called jitter. In the healthy muscle, jitter is small (for biceps brachii around 20 µs) and for small force level the MUAPs generated by the given MU are easily identifiable by visual inspection.

A needle electrode typically detects the activity of several muscle fibers within its pick-up area, which belong to a few different MUs (Figure 5.44 b). Usually the shapes of MUAPs are different, since they depend on the geometrical arrangement of the fibers of given MU with respect to the electrode, thus at low force levels single MUAPs can be easily distinguished. During the steady muscle contraction MUs fire repetitively, generating MUAP trains with essentially constant mean firing rate. With the increasing force of muscle contraction, MU firing rate increases and additional MUs are recruited, so the probability of superposition of single MUAPs increases and the EMG shows a rise of amplitude and density (zero crossings of voltage). At strong muscle contractions a so-called interference pattern develops and EMG resembles stochastic signal. More detailed analysis of the relationship between EMG and muscle force can be found in [486].

In neuromuscular diseases, the shapes of MUAPs change (Figure 5.45), which makes their study useful for clinical diagnosis. Two main types of neuromuscular diseases are distinguished: neurogenic and myogenic. In

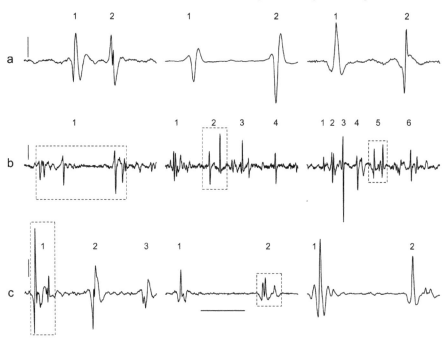

FIGURE 5.45

Examples of single MUAPs recorded during weak muscle contractions from normal and diseased muscles: a) control , b) Duchenne muscle dystrophy (myogenic disease), c) amyotrophic lateral sclerosis (neurogenic disease). For each type three 30-ms epochs from different subjects are shown, with a few single MUAPs marked by distinct numerals. Calibration bars: horizontal, 5 ms; vertical: 200 μV for a) and c), 100 μV for b). Note complex MUAPs in b) and c), indicated by broken line boxes. (By courtesy of M. Piotrkiewicz.)

neurogenic diseases the pathological process is related to the degeneration of the motoneuron; in myogenic ones the degeneration begins from the muscle fibers. In neurogenic processes some muscle fibers lose their innervation due to motoneuron death and are reinnervated by the surviving motoneurons; in effect the amplitude and duration of MUAP increases and the number of MUAPs in the pick-up area of electrode decreases. During the phase of acute reinnervation newly formed endplates are unstable and generate increased jitter. In myogenic disorders the amplitude and duration of MUAPs is reduced; they become polyphasic and contain satellite potentials. The typical examples of normal, neurogenic, and myogenic potentials are shown in Figure 5.45. More information about physiological and pathological processes in muscles and their influence on MUAP shapes may be found in [72].

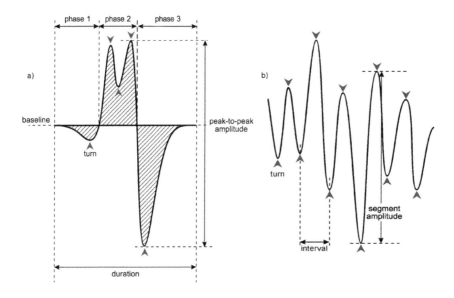

FIGURE 5.46
a) Parameters characterizing MUAP; b) parameters characterizing interference EMG. (By courtesy of M. Piotrkiewicz.)

5.3.2   Quantification of EMG Features

The first effort to quantify EMG was connected with the manual evaluation of mean MUAP parameters from a representative number of MUs recorded during a weak effort [71, 72]. The duration and the amplitude of MUAPs as well as the incidence of polyphasic potentials were evaluated from photographically recorded signals. The greater availability of computers in the 1970s and 1980s facilitated the development of computer-aided quantitative EMG processing methods. Automatic systems for EMG-based diagnosis including commercial systems evolved. These systems perform automatic identification of single MUAPs and extraction of their features, such as peak amplitude, rise time, duration, area, or number of phases (connected with baseline crossings). For interference patterns, usually the smaller fluctuations of amplitude are analyzed (so called turns-amplitude analysis). These parameters are illustrated in Figure 5.46.

Besides turns-amplitude analysis, the most popular automatic methods of interference pattern analysis include amplitude measurements, different spike counting methods, and power spectrum analyses. The methods of interference pattern analysis and their applications in diagnosis of myogenic disorders, in assessment of exerted force, muscle fatigue, and the condition of unused muscles and muscles in chronic pain syndromes are considered in [174].

In the system proposed by Pattichis and coworkers [462] for differentiation of myogenic and neurogenic signals from normal EMG, both time domain and frequency domain features of EMG were considered. The following time-domain parameters were automatically determined: 1) MUAP duration, 2) spike duration measured from the first to the last positive peak, 3) amplitude: difference between the minimum positive peak and maximum negative peak, 4) area: the rectified MUAP integrated over its duration, 5) spike area: the rectified MUAP integrated over spike duration, 6) phases, 7) turns. The frequency features were based on the spectra estimated by means of AR model of order 12. They included: 1) bandwidth, 2) quality factor—the ratio between the dominant peak frequency and bandwidth, 3) median frequency, 4) spectral moments describing the shape of the spectrum; additionally for classification AR coefficients were taken into account.

Univariate analysis was applied to find the coefficients which best separated the normal from pathological groups and multiple covariance analysis to select stepwise the best features and find the correlation between the parameters. For classification into three considered groups artificial neural networks (ANN) were applied. Three algorithms were used: back-propagation, the radial-basis function network, and self-organizing feature map. The univariate and multivariate analysis indicated as the best classifying parameters: among time domain parameters—duration, and among AR spectral measures—median frequency. Median frequency and central frequency for the neurogenic group were higher, and for the myogenic group these parameters were lower than for the normal EMG. In classification by ANN the highest diagnostic yield was obtained for time domain features and next for frequency domain features; AR coefficients gave the poorest results.

5.3.3　Decomposition of Needle EMG

The problem of automatic decomposition of an EMG signal into its constituent MUAP trains is important not only for medical diagnosis, but also for basic studies of the neuromuscular system. One of the first systems performing EMG decomposition was proposed by LeFever and de Luca [329]. The signal from the intramuscular needle electrode was sampled at 50 kHz and high-pass filtered in order to reduce the amplitude of slow rise-time MUAP waveforms recorded from fibers more distant from the electrode. The decomposition program was based on a template-matching algorithm and included a routine for continuous template updating, in which every consecutive MUAP could be classified as belonging to the existing template (then the template was updated), or used as the initial estimate of a new template, or discarded. The verification of the initial decomposition results was based on the a priori knowledge of firing statistics, i.e., inter-spike interval distribution within MUAP trains generated during isometric constant force contraction. The program might work automatically; however the reliability of the results was considerably higher if it was run interactively by an operator. To resolve parts of the EMG signal, presumably resulting from the overlapping MUAP waveforms, a two-step procedure was applied. The template fitted in the first step was subtracted from the signal. This produced a residue. In the second step, an attempt was made to fit another template to the residue.

In the relatively simple semi-automatic system proposed in [391] the first step of classification involved the criterion of similarity. Construction of a similarity vector was based on cross-correlation between the given MUAP and several templates indicated by an operator. In the next step the nearest neighbor classification algorithm was applied. The decomposition of complex potentials into superimposed MUAPs, based on consideration of the discharge rates of units, was performed interactively. The correct classification rate varied from 60% to 100% depending on the quality of the signal. The determined interspike intervals trains were inspected in the context of estimating afterhyperpolarization duration times of motoneurons. The results found application in the study of various neuromuscular disorders.

A program called ADEMG for extracting MUAPs from the EMG interference pattern for clinical diagnosis purposes was designed by McGill and coworkers [393]. The signal processing consisted of four steps: 1) EMG signal was digitally filtered to transform sharp rising edges into narrow spikes, 2) the spikes which exceeded a certain detection threshold were classified by a one-pass template matching method that automatically recognized and formed templates, 3) each tentatively identified train of MUAPs was verified by examining its firing rate, 4) the MUAPs corresponding to the verified spike trains were averaged from the raw signal using the identified spikes as triggers.

The conversion of MUAPs to spike trains was motivated by the fact that the spikes are better distinguishable than the MUAPs themselves, due to suppression of low-frequency noise, and can be reliably detected by a simple threshold crossing detector. ADMG in the last step of its operation adjusted the simple averages to cancel out possible interference from other MUAPs.

In the system designed by Stashuk [569] MUAPs were identified by a similar procedure as in [393]. MUAP detection was followed by clustering and supervised classification procedures in which the shape and firing pattern of motor units were considered. The fraction of correctly detected MUAPS was reported as 89% and the maximum false detections rate was 2.5%.

The ADMG system was further developed into an interactive EMG decomposition program EMGLAB [394], which can handle single- or multi-channel signals recorded by needle or fine-wire electrodes during low and moderate levels of muscle contraction. EMGLAB is a MATLAB program for viewing EMG signals, decomposing them into MUAP trains, and averaging MUAP waveforms. It provides a graphical interface for displaying and editing results and algorithms for template matching and resolving superpositions. The software may be found at: http://www.emglab.net. At this location also MTLEMG [158], a MATLAB function for multichannel decomposition of EMG, including a genetic algorithm for resolving superpositions may be found. EMG simulator available at http://www.emglab.net is a package for simulating normal and pathological EMG signals designed in [210]. The package contains executable code for PCs and Macintosh and a user interface written in MATLAB.

An automatic algorithm for decomposition of multichannel EMG recorded by wire electrodes (including signals from widely separated recording sites) into MUAPs was proposed by Florestal and coworkers [159]. The program uses the multichannel information, looking for the MUAPs with the largest single component first, and then applies matches obtained in one channel to guide the search in other channels. Each identification is confirmed or refuted on the grounds of information from other channels. The program identified 75% of 176 MUAPs trains with accuracy of 95%.

As was mentioned above, both time and frequency features are useful for EMG quantitative analysis [462]. Wavelet transform, which provides description of signals in time and frequency domain, was applied for analysis of EMG for normal, myogenic and neurogenic groups [463]. Four kinds of wavelets were tested including: Daubechies4, Daubechies20, Chui (linear spline), and Battle-Lemarie (cubic spline) wavelets [103]. The scalogram was inspected for different wavelet types and in respect of characteristics of investigated groups of patients. Scalograms obtained by means of Daubechies4 wavelets had the shortest time spread in each band and they detected sharp spikes with good accuracy. Daubechies20 wavelets had lower time resolution, however they provided higher frequency resolution. The linear-spline scalograms showed a large energy concentration in the lowest energy bands; on the contrary scalograms for cubic spline wavelet were spread in the upper frequency band.

As might have been expected, scalograms for the neurogenic group had larger time spreads and those for the myogenic group had shorter time-domain spreads in respect to the normal group. Also shifts in frequency similar to these found in [462] were reported in pathological groups. The reduction of information was provided, since MUAP signals were described by a small number of wavelet coefficients located around the main peak. High frequency coefficients were well localized in time and captured the MUAP spike changes, whereas low frequency coefficients described the time span and average behavior of the signal. Wavelet coefficients characterizing MUAPs were used as the input parameters in classification procedures performed by ANN. The best diagnostic yield was obtained for Daubechies4 wavelets. Although the 16 wavelet coefficients captured most of the energy of the MUAP signals the results of classification obtained in [463] were worse than those provided by the time domain parameters (listed above) defined in [462] (Sect. 5.3.2).

The EMG decomposition based on wavelet transform was proposed in [663] for analysis of long-term multichannel EMG. In the first step of analysis so-called active segments containing MUAPs were selected by thresholding based on an estimated signal to noise ratio. The Daubechies wavelets [103] were used and coefficients from lower frequency bands were applied for further steps of analysis. The high frequency coefficients were considered as more contaminated by noise and so called time-offset, which depended on the position of MUAP in the analysis window. The next steps of analysis included supervised classification and clustering based on the selected wavelet coefficients. The provisions were made to account for the possible changes of MUAP shape during the experimental session, by introducing adaptation of MUAPs templates. The achieved accuracy of classification was reported as 70%.

The problem inherent in application of discrete WT to EMG decomposition is connected with the fact that the description of the signal shape in terms of wavelet coefficients depends on the position of MUAP in the data window. This problem is usually alleviated by alignment of the main peaks of the analyzed potentials. Another approach which might be used for MUAP identification, which is free of restrictions imposed by the presence of the measurement window, is matching pursuit. To our knowledge such an attempt has not yet been undertaken.

5.3.4 Surface EMG

Because of its non-invasiveness, surface EMG (sEMG) has always attracted the attention of investigators seeking new methods for EMG analysis. One of the first applications of sEMG was studying muscle fatigue. With increasing fatigue, the sEMG spectrum shifts toward lower frequencies. It was suggested that the spectral shift was caused by the decrease of conduction velocity of action potential along the muscle fibers [342, 519]. However, according to [343] despite the definite and strong influence of the motor fiber conduction velocity on the power spectrum, the frequency shift cannot be explained by a change in propagation velocity alone. Nevertheless, reported changes of propagation velocity in some muscle diseases [671] indicate the usefulness of sEMG spectral analysis for medical diagnosis.

From sEMG spatial information concerning the MUAPs activity may also be gained. The analysis of surface EMG maps may provide indirect information on the spatial recruitment of motor units within a muscle [148].

The analysis of sEMG signals used for control of prostheses requires fast and efficient signal processing techniques. In this case the application of WT appeared quite successful [144]. The sEMG signals were recorded from biceps and triceps during elbow and forearm movements. The aim of the study was the best distinction of the four classes of EMG patterns. Time-frequency analysis was performed by means of short-time Fourier transform (STFT), wavelets, and wavelet packets (WP). Different kinds of wavelets were considered. The best results in case of WT were obtained for Coiflet-4 and for WP for Symmlet-5. The task of the reduction of dimensionality of feature sets was performed by Euclidean distance class separability (CS) criterion and principal component analysis (PCA). For classification linear discriminant analysis (LDA) and multilayer perceptron (MLP) were applied. The best performance was obtained for WP/PCA/LDA combination, yielding a classification error of 6%. This result was an improvement in respect to procedures based on time features of MUAPs, which for the same data gave an average error of 9%.

WT was also successfully applied for diagnosis and follow up of Parkinson disease [113]. The sEMG for ballistic movement was recorded from major pectoralis and posterior deltoid muscles. The Morlet complex wavelet was used and cross-correlation was computed between continuous wavelet transforms $W_f(a, \tau)$ and $W_g(a, \tau)$ of functions $f(t)$ and $g(t)$ describing signals from the considered muscles. The wavelet cross-scalogram is given by:

$$W_{fg}(a, \tau) = W_f^*(a, \tau)W_g(a, \tau) \tag{5.51}$$

The cross-correlation spectrum is defined as:

$$|W_{fg}(a, \tau)|^2 = |Re(W_{fg}(a, \tau))|^2 + |Im(W_{fg}(a, \tau))|^2 \tag{5.52}$$

The integration of the local wavelet cross-correlation spectrum over τ gives the global wavelet cross-spectrum. In distinction to FT here it is possible to select a proper integration time interval. In [113] the threshold was chosen as 5% above maximum wavelet power peak. The time-frequency distributions of cross-correlation power for the normal group was more concentrated in time and frequency than for the Parkinsonian group, where a large spread was observed, especially in time. In order to quantify these differences a parameter called global power was determined by integration of scalograms in frequency and time. This parameter differentiated the Parkinsonian group from normal subjects better than conventional medical measure. The global power index was proposed as a parameter useful in estimation of the disease progress.

5.3.4.1 Surface EMG Decomposition

Recently, the sEMG technique has drawn a lot of attention. One of the reasons is the fear of infection by needle electrodes. The problem inherent to sEMG technique is the influence of volume conduction and low-pass filtering effect of the tissues, which causes an increase in MUAPs duration, smoothing of their shapes, and a decrease of the differences between them. For these reasons sEMG was initially treated as an interference signal, from which only global properties of motor units and their firing rates could be assessed.

Nevertheless, the application of surface EMG for detection of single MU activity by means of small bipolar electrodes and their arrays was proposed by Gydikov and coworkers in the early 1970s [206, 204, 205, 301]. Further studies by these authors provided information on the velocity of the activity propagation along muscle fibers and on MU structure, in particular its size and location within the muscle, endplate distribution, muscle fibers topography [203].

The biggest challenge in sEMG processing—the extraction of individual MUAPs—became possible when the high-density grids, including sometimes hundreds of electrodes, became available. Figure 5.47 shows motor unit action potential recorded with a grid of 13×5 electrodes and illustrates tracking of

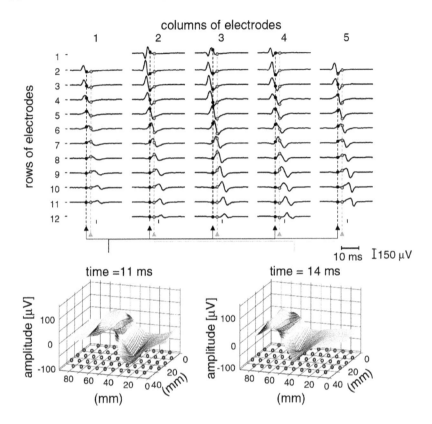

FIGURE 5.47

Motor unit action potentials recorded with a grid of 61 electrodes (13×5 electrodes, inter-electrode distance 5 mm) from the biceps brachii muscle during an isometric contraction at 10% of the maximal force. The bipolar signals derived along the direction of the muscle fibers are shown in the top panel. The electrode grid was placed distal with respect to the innervation zone of the motor unit. The signals detected along the rows show similar action potential shapes with a delay corresponding to the propagation along the muscle fibers. The multi-channel action potential is a three-dimensional signal in time and space. The two-dimensional spatial representations for two time instants (11 ms and 14 ms after the generation of the action potential) are shown in the bottom panel. The circles on the plane representing the spatial coordinates mark the locations of the electrodes of the grid. From [401].

the propagation of MUAP by multielectrode arrays. The process of the sEMG decomposition into MUAPs is shown in Figure 5.48.

The first step in analysis of sEMG is alleviation of the effect of volume conduction and suppression of signals from distant sources. This may be achieved

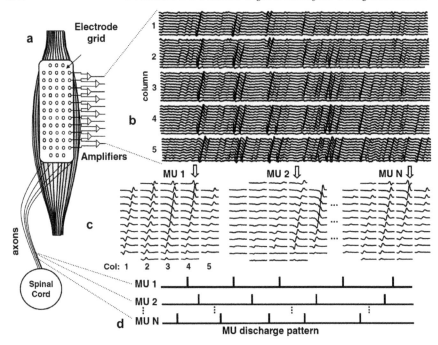

FIGURE 5.48
Representation of the process of decomposition of multi-channel surface EMG.
(a) Surface EMG was recorded from the biceps brachii muscle with a 13×5
electrode grid (corner electrodes are missing) with the columns parallel to the
fiber direction. (b) Segment of 500 ms duration of bipolar EMG detected by
each column of the grid. The action potentials propagation along the columns
may be noticed. (c) Multi-channel action potentials for three motor units
extracted from the interference signal with the decomposition algorithm described in [229]. (d) Estimated discharge patterns for the three motor units.
From [401].

by application of the Laplacian operator (Sect. 5.1.3). Spatial high-pass filtering effect may be achieved also by using bipolar montage [291]. The example
of spatially filtered signals is shown in Figure 5.49.

The procedure of MUAP identification is similar to these applied in case of
signals recorded by the needle electrodes. It includes setting the threshold for
candidate MUAPs, clustering, creating templates, template matching, decomposition. The serious problem is a small differentiation of the MUAP shapes
recorded by sEMG in comparison to signals registered by needle electrode.
This problem is now resolved by application of multielectrode grids, which
provide the two-dimensional spatial information. The differences in location
of single MUs with respect to the multielectrode and the information about

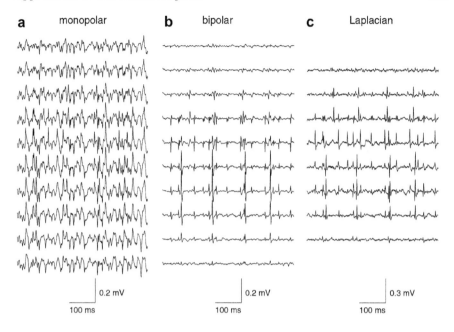

FIGURE 5.49
Example of spatially and temporally (high-pass at 15 Hz) filtered signals. Signal from the same 500 ms in monopolar (a), bipolar (b), and Laplacian (c) derivation. The upper traces correspond to the medial, the lower to the lateral electrode positions. From [291].

the delay of the given MUAP connected with propagation of MUAPs along the muscle fibers, create so-called MU signatures, which are utilized in the algorithms of sEMG decomposition.

The semi-automatic techniques for extracting single MUs from sEMG including operator-computer interaction were proposed in 1999, e.g., [123]. Later, the fully automatic system for extraction and classification of single MUAPs from sEMG was designed by Gazzoni et al. [176]. The segmentation phase was based on the matched continuous wavelet transform, and was performed in two steps. In the first step candidate MUAPs were detected by multi-scale matched filter. The second step was based on the properties of propagation of the action potentials along muscle fibers. Only MUAPs presenting clear propagation along the longitudinal fiber direction were selected. The clustering was performed by a set of neural networks, working in parallel, one for each available channel. They operated in an adaptive way updating the templates and creating new clusters for non-matching patterns. However, the system didn't offer the possibility of resolving the superimposed MUAPs.

In the system proposed by Kleine et al. [291] a 10 × 13 grid with inter-electrode distance 5 mm was used. Bipolar and Laplacian reference systems were considered (Figure 5.49). The MUAPs were isolated from the background activity by peak detection. In the clustering procedure both the spatial (waveform and amplitude difference between channels) and the temporal (time-course of the potential in each channel) information were taken into account. Additional information was provided by the firing statistics. The clusters were inspected and adjusted interactively; that is, the operator could split or merge clusters. Then the MUAP templates were subtracted from continuous sEMG. In the decomposition procedure a single row (perpendicular to muscle fibers) of bipolar derivations was taken into account. The template matching procedure was repeated with different subsets of available data. It was confirmed that two bipolar montages, separated by a few millimeters in the direction parallel to the fibers will record almost the same waveform, delayed by a few milliseconds. On the other hand in the direction perpendicular to the muscle fibers the amplitude of bipolar MUAP is largely attenuated (see Figure 5.47 and Figure 5.48). In this case the muscle fibers of a particular MU cover only a small part of muscle fibers from another MU, so a single row of channels gives a reasonable separation of MUAPs.

The limitation of the method is the restriction to low force contractions, since at higher forces two or more MUAPs may have almost identical shapes and also the superimposed potentials become a problem. It was suggested that sEMG may be useful for quantification of neurogenic diseases, where the number of remaining MUs and MUAP interference are low.

In general, sEMG may be helpful in case of disturbances of the neuromuscular system leading to decrease of firing rates of MUs. An example may be the automatic classification of MUAPs in sEMG recorded from muscles paralyzed by spinal cord injury, where involuntary EMG activity occurs [646]. However, it should be mentioned that changes in firing rates are not specific to any particular neuromuscular disorder.

The application of the blind source separation method for identification of motor unit potential trains in sEMG was proposed by Nakamura and coworkers [420]. The signals from eight electrodes placed perpendicularly to the muscle fibers were processed. Principal component analysis and independent component analysis were applied. The ICA approach performed better than PCA in separating groups of similar MUAP waveforms, however single MUAPs were not separated completely into independent components. In case of delays between potentials coming from the same motor unit ICA regards them as several different sources. The limitation of the method is connected also with the fact that the procedure does not take into account the firing rates of the motor units, which limits useful information available for classification. The BSS methods assume independence of the underlying sources which hampers their application for higher force level when the synchronization of the motor units is high. The advantages of the BSS are connected with the fact that the method does not rely on prior estimation of the shapes of MUAPs. It seems that the approach might be promising, for example, as a preprocessing step in the procedure of sEMG decomposition.

Blind source separation method was used in [227] where sEMG was measured from four muscles by means of 13x5 grid during low force contraction. The approach was based on the convolution kernel compensation (CKC) algorithm proposed in [229]. The signal was modeled as a convolutive mixture of sparse pulse trains, which carry information about the rising times of the detected symbols and the symbols themselves. The spatial and temporal statistics of symbols, i.e., convolution kernels, was combined with the information about their overlapping probability, in order to blindly reconstruct their pulse sequences. The model implied admixture of a white noise, however residual noise is dominated by the contributions of indecomposable MUAPs and does not meet the criterion of spatial and temporal independence. CKC assumed as well, that action potential of MU remains constant throughout a contraction. In case of one of the muscles the results give 98% of agreement in MUAPs identification with the intramuscular recordings. However, the method is limited to low force contractions (10% of maximal force) and biased toward the high-threshold MUs, which was admitted by the authors.

The performance of different methods of sEMG analysis is usually evaluated by comparison with the manual decomposition of a given signal. In view of the multitude of the approaches to the problem of EMG decomposition there emerged the need for a test tool to evaluate and compare different methods. Such a tool was proposed in [150]. The approach was based on the model for the generation of synthetic intra-muscular EMG signals. The library of 18 test signals was designed. The indexes estimating performance for segmentation and classification stages of decompositions and measure of association between model and detected classes of MUAPs were proposed. Additionally global indices were introduced, such as the difference between the number of model classes and the number of classes estimated by the algorithm, mean firing rate, and activation interval detection.

Yet another way of testing the decomposition algorithm was applied in [424]. It relied on reconstruction of the signal from the identified MUAPs and decomposition of this signal with some noise added. This procedure is illustrated in Figure 5.50. In the experiment a five-pin surface sensor was used (Figure 5.43) and the pair wise voltages between 5 pins were analyzed. The method of the decomposition was based on the algorithm proposed in [329]. The operation of the algorithm started by extracting as many MUAPs as possible from experimental sEMG action potential templates. Then it searched for signal regions where the extracted templates were in superposition with each other and with unidentified potentials. The algorithm required that the unidentified action potentials account for less than 25% of the signal energy. The assumption about inter-pulse intervals was only that they be less than 0.35 s. The algorithm allowed for changes in the action potential shape of each MUAP to evolve over the duration of the contraction. The authors considered

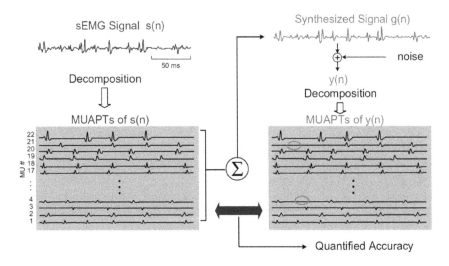

FIGURE 5.50
Illustration of "reconstruct-and-test" procedure for assessing the accuracy of the decomposition algorithm. An actual sEMG signal s(n) is decomposed to identify its MUAPTs (MUAP templates). Signal y(n) is synthesized by summing together the decomposed MUAPTs of s(n) and white Gaussian noise whose variance is set equal to that of the residual signal from the decomposition. The reconstituted y(n) signal is then decomposed and compared to the decomposed MUAPTs of s(n). The ellipses indicate discrepancies between the MUAPTs of y(n) and s(n). These are designated as errors. From [424].

22 signals recorded from 5 muscles for contractions reaching 100% force levels. They reported the accuracy estimated by the reconstruct-and-test procedure (Figure 5.50) ranging from 77–97%. Upon analysis of decompositions performed on these signals it was found that 92% of over 3000 firings from 11 decomposed MUAPs of the signal from one sensor were locked in with firings of 11 corresponding MUAPs decomposed from the other sensors' signals.

The method seems to be promising in respect to potential clinical applications, since the authors claimed that the system is able to detect morphology of the shapes of up to 40 concurrently active motor units without relocating the sensor and that the polyphasic MUAPs may be detected with the proposed technology.

The view on applicability of sEMG in clinical practice has changed over the years. In a technology assessment study issued by the American Association of Electrodiagnostic Medicine [207] it was concluded that there is no evidence to support the use of sEMG in the clinical diagnosis and management of nerve or muscle disease. Later study based on the literature covering the period 1994–2006 [399] stated that sEMG may be useful to detect the presence of some neuromuscular diseases and a study of fatigue, but there are insufficient data to support its utility for distinguishing between neuropathic and myopathic conditions. However, sEMG is able to provide additional, clinically valuable information, e.g., on MU size and topography or conduction velocity, which is complementary to the routine EMG diagnostic techniques [401].

The earlier opinions quoted above concerning the utility of sEMG were challenged by the recent studies. In the last years we can notice important advances in sEMG analysis associated with technical progress in construction of detection electrodes and developments in signal processing methods. Electrodes printed on soft films brought advantages connected with better signal transmission at the electrode-skin interface, electrode-skin stability, minimization of signal cross-talk between nearby electrodes, increase of user convenience and possibility of prolonged use [241, 662].

In parallel with progress in detection systems were developed: the algorithms of decomposition of the interference signals, the strategies used to edit the identified waveforms and techniques of discharge rate evaluation. Improvement in ICA based algorithms [99] and BSS techniques [425] contributed to advancement in solving the challenging problem of extracting motor unit potentials reliably from sEMG. The routines were designed to track MU longitudinally across experimental sessions, which allowed to identify changes in MU properties following training or during the progression of neuromuscular disorders [386].

Progress in MU discharge rate evaluation involved, e.g.: introduction of signal-based metric called pulse-to-noise-ratio [228] and novel editing routines based on decreasing shape variability, which made possible to estimate inter-discharge interval statistics [310].

Recent developments in sEMG analysis may be found in tutorial aiming to outline the best practices and to provide general guidelines for the proper signal evaluation [402]. Physiological validation of the decomposition of sEMG signals was considered in [145]. The author asserted that some sEMG decomposition methods are able to replicate many of the findings obtained from intramuscular recordings.

We can conclude that sEMG has the potential to provide the information on motor unit physiology and to offer useful approaches that may be translated into clinical practice. Indeed, Higashihara et al. [221] reported successful application of sEMG for diagnosis of neuromuscular disorders in children. High resolution sEMG may become in the near future widespread method for investigation of muscle function at the individual unit level and a useful tool in the diagnosis of neuromuscular conditions.

5.4   Acoustic Signals

5.4.1   Phonocardiogram

Phonocardiogram (PCG) is a recording of the acoustic signal produced by the heart's mechanical action. The main frequency range of normal PCG is in the range of 10–35 kHz; however, in the case of artificial heart valves, frequencies up to 50 kHz may be recorded. Heart sounds are generated by the vibration of the heart valves during their opening and closure and by the vibration of the myocardium and the associated structures. The sound generated by a human heart during a cardiac cycle consists of two dominant components called the first heart sound, S1, and the second heart sound, S2. The S1 sound corresponds to the QRS complex of the ECG, and the second heart sound, S2, follows the systolic pause in the normal cardiac cycle. In S1, two components, M1 and T1, may be distinguished. They are due to the closure of the mitral and the tricuspid valve, respectively. The components of S2 are due to the closure of the aortic valve (A2) and pulmonary valve (P2). Additionally, noise-like sounds called murmurs caused by the turbulence of the blood flow may be produced. They appear mostly during abnormal action of a heart. Examples of normal and pathological PCG are shown in Figure 5.51.

The technique of listening to the heart sounds, called auscultation, has been used for diagnostic purposes since the 16th century. However, the human ear is not very well suited to recognize the several short-duration events occurring in small intervals of time, especially since they occur in the low-frequency range, where the sensitivity of a human ear is not very good. Therefore signal processing techniques are beneficial in PCG analysis.

One of the first steps in PCG analysis is its segmentation. Usually, the ECG signal is used for this purpose; yet, the segmentation may be based on the PCG signal itself, taking into account its time-domain features and identification of S1 and S2 [10]. The recognition of S1 and S2 without referring to ECG was reported in [436], who used wavelet transform. PCG signal was found to be a convenient reference for establishing heart rate in fMRI studies since PCG, contrary to ECG, is not disturbed by the electromagnetic fields of the equipment [37].

The PCG signal is non-stationary and consists of short transients of changing frequency. Hence time-frequency methods are the appropriate tools in its analysis. The early works on PCG analysis were based mainly on spectrograms calculated by the Fourier transform. Later, the wavelet analysis was introduced to PCG analysis. The usefulness of different kinds of wavelets (Daubechies, Symlet, biorthogonal) for PCG reconstruction was tested in [115]. It was found that the best results were obtained using Daubechies db7 wavelet. The authors reported that the error of reconstruction could be used as a discriminatory parameter in classifying the pathological severity of the PCG. Daubechies wavelets were also used in the study [436] for finding the

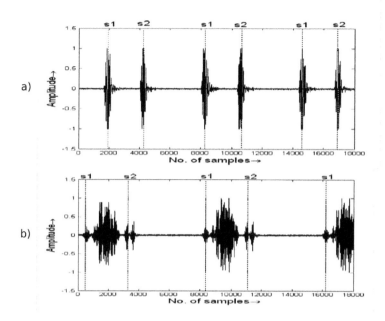

FIGURE 5.51
PCG signal a) for normal subject, b) for subject with pulmonary stenosis.
Sampling frequency 8012 Hz. Adapted from [10].

offset of the systolic murmur, which was then analyzed through the AR model
of the order 2. The approach proved effective in delineating a set of clinically
diagnosed systolic murmurs.

For evaluation of the condition of bioprosthetic valves, features of the
power spectrum such as the maximum peak location have been used to indicate whether the valve is normal or abnormal. For example, changes in the
frequency content of PCGs from bioprosthetic valves were described in [570],
who reported that the PCG of an abnormal valve has more high-frequency
content than that of a normal valve, and in patients with mitral bioprostheses,
implanted five years or longer, the higher frequencies have a greater proportion of sound energy compared to patients with implants of less than 1.5 years.
These findings were based on the power spectra computed using Fourier transform. Joo and coworkers [260] reported that a more accurate description of
heart sounds in patients with a bioprosthetic heart valve may be obtained by
a pole-zero model based on ARMA formalism.

In the study of the performance of a heart valve implanted in the mitral
position, the AR model was used in [303]. The authors applied an approach
similar to FAD [164] (Sect. 3.3.2.3). Namely, they decomposed the impulse
response function of the AR model into a sum of damped sinusoids. After
fitting the AR model, the very low level of residual signal suggested that the

recorded PCG is essentially composed of delayed and overlapped repetitions of the impulse response function of the AR model.

The matching pursuit approach is a high-resolution method of time-frequency analysis (Sect. 3.4.2.2). It was applied by Zhang et al. [664] for analysis-synthesis of PCG. The dyadic dictionary with certain limitations imposed on the scale was used. The results showed that PCG could be synthesized from a limited number of the highest energy atoms, and the method acts as a powerful de-noising filter. The separation of heart sounds from noise based on MP decomposition, and fuzzy detection was also applied in [590]. In a practical experiment, the heart sound signal was successfully separated from lung sounds and disturbances due to chest motion.

The usefulness of MP for PCG analysis was also confirmed by a study by Wang et al. [628], who reported that MP energy distribution provides a better time-frequency representation of PCG than spectrogram. The authors reported good results of mitral valve abnormality classification based on MP decomposition of the first heart sound. Figure 5.52 shows spectrograms and

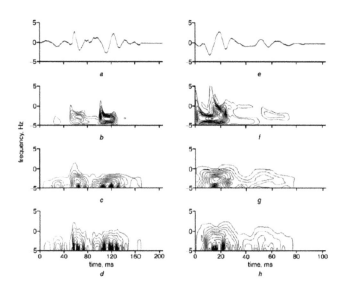

FIGURE 5.52
Comparison of time-frequency distributions obtained by means of MP and FT. (a) Averaged normal S1 of PCG and its contour plots using: (b) MP-based time-frequency distribution, (c) spectrogram with 64-point STFT window length, and (d) spectrogram with 32-point STFT window length. (e) Averaged abnormal S1 of PCG and its contour plots using: (f) MP-based time-frequency distribution, (g) spectrogram with 64-point STFT window length, and (h) spectrogram with 32-point STFT window length. From [628].

MP time-frequency maps of first heart sound for the normal and pathological case.

From the evidence gathered by the authors who applied different kinds of methods to PCG, it follows that the time-frequency methods such as wavelets and MP, and also parametric methods based on the autoregressive approach are suitable for PCG analysis and the choice of the method depends on the problem to be solved.

5.4.2 Otoacoustic Emissions

Otoacoustic emissions (OAEs) are weak acoustic signals generated by the inner ear in response to the stimulation and sometimes also spontaneously. OAEs are measured by a sensitive microphone placed in the ear canal. Their typical frequencies range from 0.5 to 6 kHz [492]. Responses evoked by short acoustic stimuli are referred to as transiently evoked otoacoustic emissions (TEOAEs). They are usually measured in 20 ms window from stimulus onset. Long-lasting OAEs recorded in 80 ms windows in response to broadband click stimulus are called synchronized spontaneous OAEs (SSOAEs). They are considered to represent spontaneous otoacoustic emissions (SOAEs). Distortion product otoacoustic emissions (DPOAEs) are the emissions which are generated by two tones f_1 and f_2, where f_1/f_2 is typically around $1.2f_1$. The emitted otoacoustic signal has the largest amplitude at $2f_1 - f_2$ frequency.

OAEs were first registered by Kemp [285] and quickly became an important tool in diagnosing hearing impairment. However, the mechanisms of their generation, which are closely related to the hearing process, are still a matter of debate between two theories, namely traveling wave [672, 546, 588] and resonance theory [554, 39, 40]. OAEs are especially useful as a screening hearing test in neonates and small children. The review of OAE techniques and clinical applications may be found in [507].

During the recording procedure, consecutive OAEs are stored in two buffers. The correlation between the averages from these two buffers serves as an index of the reproducibility and hence, the signal's quality. Usually, the discrimination between TEOAE responses from normal-hearing subjects and patients with sensorineural hearing losses is based on the reproducibility of the TEOAE responses obtained from two buffers, the signal to noise ratio (S/N) and overall TEOAE response level. Sometimes, the signal's frequency characteristics are considered; OAE measurement devices usually permit the calculation of Fourier spectra.

More recently, other TEOAE parameters have proven to be good signal descriptors, and among these are the latencies of the TEOAE components. It was demonstrated in several papers that there are differences in latencies of

OAEs between subjects with normal hearing and subjects exposed to noise, e.g., [599, 555, 249]. OAE signal is a basic test of hearing impairment in small children. Much attention has been devoted to clinical significance and determination of latencies in neonates, e.g., [600, 252].

In the research concerning the mechanisms of OAE generation, the relation between frequency and latency of components is one of the important tests of the models. For example, the scale-invariance hypothesis [588] led to the prediction of the inverse relationship between frequency and latency. However, this kind of relationship was not confirmed experimentally, and exponential dependence with different exponents' values was reported [555, 599, 251].

The non-stationary character of OAE and rising interest in frequency-latency dependencies promoted the application of time-frequency methods to this signal. Several methods have been devised to estimate the time-frequency distributions of OAEs, including short-time Fourier transform [216], minimum variance spectral estimation [666] methods based on the Wigner-Ville transform [85], or Choi-Williams transform [451]. However, the last two methods are biased by the presence of the cross-terms (Sect. 3.4.2).

One of the first applications of wavelets to the analysis of OAE concerned time-frequency decomposition of click-evoked and synthesized emissions [649]. Both continuous [599] and discrete [555] WT were used. The method which combined signal decomposition by discrete WT, non-linear denoising and scale-dependent time windowing was used to improve pass/fail separation during transient evoked otoacoustic emission (TEOAE) hearing screening[247].

WT was also applied for extraction of instantaneous frequencies of OAE [119] and for construction of multiscale detector of TEOAE [379]. In the above-quoted contribution, the detector performed adaptive splitting of the signal into different frequency bands using either wavelet or wavelet packet decomposition. The authors reported that the method performed significantly better than existing TEOAE detectors based on wave reproducibility or the modified variance ratio.

OAEs are a superposition of components characterized by specific frequencies and latencies. The dependence between their frequency and latency has been extensively studied in the context of verifying hypotheses concerning the mechanisms of OAE generation. To this avail, the latency has to be determined. The problem of the identification of the OAE latencies was approached by discrete and continuous wavelet transform. Sisto and Moleti [555] proposed the method based on the combination of spectral analysis and WT. In this respect, the limitation of WT is octave-band resolution, which influences the accuracy, especially for high frequencies. To surmount this difficulty, the authors proposed a method relying on a visual comparison between wavelet data and TEOAEs spectra, followed by identifying the wavelet contribution to a given spectral line. However, as was

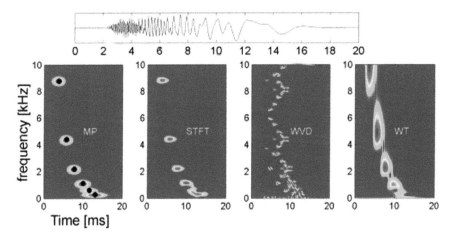

FIGURE 5.53
Time-frequency energy distributions obtained by means of different methods (named on pictures) for simulated signal (shown above) consisting of six gamma tones of frequencies 280, 550, 1100, 2200, 4400, 8800 Hz.

pointed out in [408] wavelet transform tends to underestimate the latency-frequency relation.

Substantial progress in the OAE field was achieved by the introduction to its analysis matching pursuit [251]. Matching pursuit decomposes signals into components of specific amplitudes, frequencies, latencies, and time spans, so these parameters are given explicitly. A comparison of different methods used for OAE analysis is shown in Figure 5.53. The simulated signal used for testing the methods was constructed from the so-called gammatones—functions resembling the shape of click evoked OAE at single resonant frequency. The gammatone is expressed by a function, whose envelope rises as t^3 and decays exponentially with a constant Γ:

$$\gamma(t) = \gamma_0 t^3 e^{-2\pi\Gamma t} \sin(2\pi f t) \tag{5.53}$$

The test signal was constructed from six gammatones of different frequencies (given in the legend to Figure 5.53). From Figure 5.53 it is easy to see that the components are best recognized by the MP method. The performance of a spectrogram is also quite good, but two components of the lowest frequencies are not distinguished. The additional advantage of MP is the parametric description of the components.

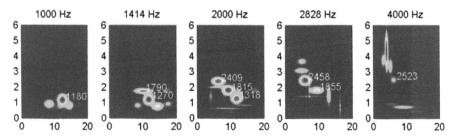

FIGURE 5.54
Time-frequency energy distribution of OAE obtained for tonal stimulation of frequencies shown above the pictures. The same resonance modes are excited by tonal stimuli of different frequencies.

The MP method allowed for the identification of the resonant modes of the OAE, which are characteristic for each ear [251, 250]. In the case of tonal stimulation, the ear does not respond in the frequency of the stimulus, but several resonant modes are excited (different in the left and right ear). Those closest to the stimulation frequency have the highest amplitude (Figure 5.54). The explicit identification of the latency of components allowed for the determination of the frequency-latency dependence [251] and found the application for evaluation of the influence of an exposition of subjects to noise [249].

One of the parameters returned by MP is a time-span of a component. It was observed that the histogram of time-spans of TEOA components is bimodal [251] and that long-lasting components have a very narrow frequency band and do not obey exponential frequency-latency dependence. This kind of observation indicated that they are connected with SOAEs. The application of MP allowed for differentiation of TEOAE components into short and long-lasting ones connected with SOAEs [248] and explanation of paradoxically longer latencies found for pre-term neonates [252].

Closer inspection of OAE components revealed that some of them, especially the long-lasting ones, have an asymmetric shape. To better approximate the signal, basic asymmetric functions were introduced to the MP dictionary [253]. These functions are composed of two parts; the first is based on Gabor, and the second on an exponential function. This kind of waveform can have different rise and fall times for the same frequency. Such a waveform can be described by the formula:

$$\Lambda\left(t; \mu, \sigma, \omega, \phi, T_f\right) = N \left\{ \begin{array}{ll} e^{-\frac{(t-\mu)^2}{2\sigma^2}} & \text{for } \quad t \le T_f \\ e^{-\alpha(t-\tau)} & \text{for } \quad t > T_f \end{array} \right\} \cos(\omega t + \phi) \qquad (5.54)$$

where $\alpha = \frac{T_f - \mu}{\sigma^2}$ and $\tau = \frac{T_f + \mu}{2}$. The additional parameter $T_f > \mu$ determines the asymmetry of the atom. T_f describes the point at which the

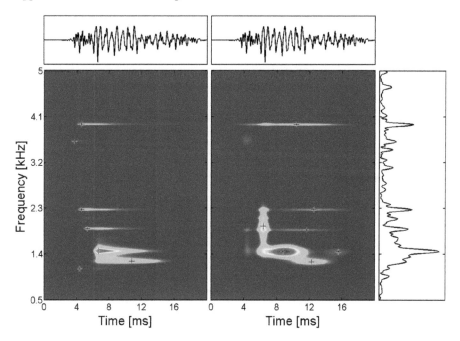

FIGURE 5.55
Time-frequency distributions of OAE signal obtained by means of MP with dictionary containing asymmetric functions (left) and dictionary of Gabor atoms only (right). At the far right, Fourier spectrum of the signal is shown.

Gaussian envelope changes into an exponential decay function. The function thus obtained is continuous up to the first-order derivative.

The introduction of an enriched dictionary encompassing asymmetric functions results in a better time resolution of time-frequency distributions of signal energy. The comparison of TOAE decomposition using enriched and Gabor dictionaries is shown in Figure 5.55. One can see that in some cases, two Gabor atoms are needed to describe one component. The enriched dictionary provides a more sparse representation. Furthermore, the new approach gets rid of the "pre-echo" effect, i.e., presence of false energy traces before the start of the signal visible on TF maps created from MP with Gabor dictionary. Analyses of simulated and real signals demonstrated that enriched dictionary provides correct estimation of latencies for components of short and long decay time.

A hybrid matching pursuit algorithm that included Fourier spectral information was developed in [439] to speed up computation times. Additionally, the procedure was capable of identifying atoms whose latency-frequency relation was not compatible with the frequency-latency dependence. These atoms could be associated with several known phenomena, either intrinsic, such as intermodulation distortion, spontaneous emissions, and multiple internal reflections, or extrinsic, such as instrumental noise, linear ringing, and the acquisition window onset.

The inspection of the development of the techniques used for OAE analysis and the concrete results provided by them indicates that the matching pursuit is the most appropriate method for OAE analysis.

5.5 Multimodal Analysis of Biomedical Signals

Simultaneous recording of biomedical signals has been conducted since the second half of the 20th century. It became a routine measurement for sleep analysis since 1968 when Rechtschaffen and Kales [502] published the first guidelines for determining sleep stages (see Sect. 5.1.6.6). Multimodal recordings are routinely used for monitoring anesthesia. Nowadays, batteries of medical signals are recorded for monitoring and support of body functions, including rehabilitation assistance devices and long-term medical aids (see Sect. 5.2.5.1). However, in the above applications, the intrinsic relations between the registered signals are not investigated.

In the organism, neural, cardiovascular, hemodynamic, and respiration systems are mutually related. Understanding of these relations and, in effect, understanding the mechanisms of control in the human autonomous system requires multimodal and simultaneous study of body signals. Biomedical signals have different spectral characteristics, and their simultaneous analysis requires adequate preprocessing. EEG and ECG have frequency bands from about single Hz to tenths Hz and even higher in the case of intracranial EEG, contrary to hemodynamic variables whose spectral power is concentrated in the frequency range of about 0.01–0.5 Hz. In the cardiac activity, the heart rate variability signal (HRV) is usually used since it has a frequency band similar to hemodynamic signals, and besides, it correlates with them. A similar frequency band can be observed for the amplitude modulation of the EEG rhythms, which can be evaluated as the instantaneous amplitudes (see Sect. 3.4.1). The instantaneous amplitude is estimated as the absolute value of the analytical signal using the Hilbert transform. The signals to be analyzed simultaneously have to be resampled to the same sampling frequency.

An interesting phenomenon, which can be an example of mutual dependencies between electrophysiological and hemodynamic signals, are so-called Mayer waves (MW)—oscillations of frequency centered around ∼0.1 Hz. They were first discovered by Sigmund Mayer [390] in blood pressure (BP) signal—hence their name. The oscillations of ∼0.1 Hz were later identified in HRV, hemoglobin concentration changes (HbO, HbR), BOLD signals, and EEG rhythms amplitudes fluctuations. Here, we shall call ∼0.1 Hz oscillations

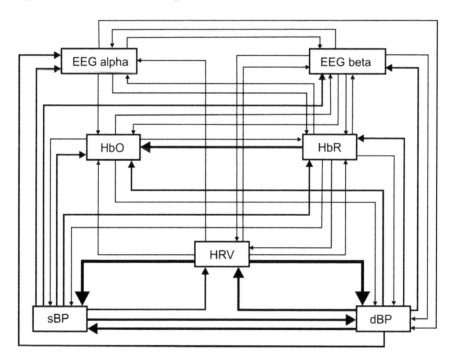

FIGURE 5.56
Scheme showing significant couplings between electrophysiological and vascular signals in the ~0.1 Hz frequency range for spontaneous activity. The thickness of the arrows represents the strength of connections. The thickest lines (relative thickness 3) denote the coupling values of 1.09–1.42, the lines of relative thickness 2 denote the values of 1.08–0.75, the lines of relative thickness 1.2 indicate the values below 0.75. From [318]

Mayer waves. The origin and mechanisms of MW generation are still enigmatic. One theory assumes the existence of a pacemaker in the brain stem or spinal cord. Another one connects their generation with a baroreflex loop [266].

The relations between MW occurring in different physiological signals were first investigated using bivariate methods such as coherence, correlation, and phase-locking index. These studies indicated the leading role of BP since MW fluctuations in this signal usually preceded the ones in the other time series. However, the results were sometimes ambiguous. Besides, bivariate methods cannot identify reciprocal connections, which may link the considered variables. As an example of a multivariate approach involving signals of different origins, we shall report a study in which: diastolic (dBP) and systolic(sBP) blood pressure, HRV, HbO, HbR, and alpha and beta rhythms envelopes were analyzed [319]. All these signals were resampled to 2 Hz. Directed Transfer Function (DTF) was used for the simultaneous analysis of all seven signals in the resting state and during voluntary finger movement. The obtained DTF functions showed a prominent peak at ~0.1 Hz. The integrated power around 0.1 Hz peak represented the measure of causal coupling between the signals.

The scheme of interactions is shown in Figure 5.56. We can observe directed influences of sBP and dBP on all the other variables and strong interaction of HbR on HbO. The prominent feature of the scheme is the presence of reciprocal interactions between HRV, sBP, and dBP. During movement, some of the connections visible in Figure 5.56 are weakened; however, the strong relationships between HRV, SBP, and dBP persisted. This result supported the baroreflex hypothesis since the loop involving these signals can generate ~0.1 Hz oscillations. Nevertheless, this conclusion is not decisive since the battery of investigated signals did not involve subcortical brain structures capable of generating MW. In the studies involving BOLD signal, some evidence of MW driving by subcortical structures was reported [476]. However, this finding requires further confirmation.

The multimodal studies [319, 57] supplied the information concerning the common mechanism of MW generation and their possible synchronization role serving the control in cardiovascular, hemodynamic, and nervous systems. We may postulate that multimodal studies have the potential of unraveling the mechanisms of steering and operation of body systems.

Bibliography

[1] R. Abreu, A. Leal, and P. Figueiredo. EEG-informed fMRI: A review of data analysis methods. *Frontiers in Human Neuroscience*, 12:29, 2018.

[2] P. Achermann, R. Hartmann, A. Gunzinger, W. Guggenbühl, and A.A. Borbély. All night sleep and artificial stochastic control signals have similar correloation dimension. *Electroencephalography and Clinical Neurophysiology*, 90:384–387, 1994.

[3] H.M. Al-Angari and A.V. Sahakian. Use of sample entropy approach to study heart rate variability in obstructive sleep apnea syndrome. *IEEE Transactions on Biomedical Engineering*, 54:1900–1904, Oct 2007.

[4] A.Al-Zaben and A.Al-Smadi. Extraction of foetal ECG by combination of singular value decomposition and neuro-fuzzy inference system. *Physics in Medicine & Biology*, 51:137–143, Jan 2006.

[5] P.J.Allen, O. Josephs, and R. Turner. A method for removing imaging artifact from continuous EEG recorded during functional mri. *NeuroImage*, 12(2):230–239, 2000.

[6] M. Amiri, B. Frauscher, and J. Gotman. Phase-amplitude coupling is elevated in deep sleep and in the onset zone of focal epileptic seizures. *Frontiers in Human Neuroscience*, 10(August):12, 2016.

[7] P. Anderer, S. Roberts, A. Schlogl, G. Gruber, G. Klosch, W. Herrmann, P. Rappelsberger, O. Filz, M.J. Barbanoj, G. Dorffner, and B. Saletu. Artifact processing in computerized analysis of sleep EEG–a review. *Neuropsychobiology*, 40:150–157, Sep 1999.

[8] R.G. Andrzejak, A. Kraskov, H. Stogbauer, F. Mormann, and T. Kreuz. Bivariate surrogate techniques: necessity, strengths, and caveats. *Physical review. E*, 68:066202, Dec 2003.

[9] R.G. Andrzejak, F. Mormann, T. Kreuz, C. Rieke, A. Kraskov, C.E. Elger, and K. Lehnertz. Testing the null hypothesis of the nonexistence of a preseizure state. *Physical review. E*, 67(1):010901, 2003.

[10] S. Ari, P. Kumar, and G. Saha. A robust heart sound segmentation algorithm for commonly occurring heart valve diseases. *Journal of Medical Engineering and Technology*, 32:456–465, 2008.

[11] P. Armitage, G. Berry, and J.N.S. Matthews JNS. *Statistical Methods in Medical Research.* Blackwell Science, Oxford, 4th edition, 2002.

[12] J. Arnhold, P. Grassberger, K. Lehnertz, and C.E. Elger. A robust method for detecting interdependences: application to intracranially recorded EEG. *Physica D: Nonlinear Phenomena,* 134(4):419–430, 1999.

[13] M. Arnold, W.H. Miltner, H. Witte, R. Bauer, and C. Braun. Adaptive AR modeling of nonstationary time series by means of Kalman filtering. *IEEE Transactions on Biomedical Engineering,* 45:553–562, May 1998.

[14] O.J. Arthurs and S. Boniface. How well do we understand the neural origins of the fMRI bold signal? *Trends in Neurosciences,* 25(1):27–31, 2002.

[15] J. Aru, J. Aru, V. Priesemann, M. Wibral, L. Lana, G. Pipa, W. Singer, and R. Vicente. Untangling cross-frequency coupling in neuroscience. *Current Opinion in Neurobiology,* 31:51–61, Apr 2015.

[16] R. Aschenbrenner-Scheibe, T. Maiwald, M. Winterhalder, H.U. Voss, J. Timmer, and A. Schulze-Bonhage. How well can epileptic seizures be predicted? An evaluation of a nonlinear method. *Brain,* 126:2616–2626, Dec 2003.

[17] K. Assaleh and H. Al-Nashash. A novel technique for the extraction of fetal ECG using polynomial networks. *IEEE Transactions on Biomedical Engineering,* 52:1148–1152, Jun 2005.

[18] L. Astolfi, F. Cincotti, D. Mattia, F. De Vico Fallani, A. Tocci, A. Colosimo, S. Salinari, M.G. Marciani, W. Hesse, H. Witte, M. Ursino, M. Zavaglia, and F. Babiloni. Tracking the time-varying cortical connectivity patterns by adaptive multivariate estimators. *IEEE Transactions on Biomedical Engineering,* 55:902–913, Mar 2008.

[19] L. Astolfi, F. Cincotti, D. Mattia, MG Marciani, LA Baccala, F de Vico Fallani, S Salinari, M Ursino, M Zavaglia, and F. Babiloni. Assessing cortical functional connectivity by partial directed coherence: simulations and application to real data. *IEEE Transactions on Biomedical Engineering,* 53(9):1802–1812, 2006.

[20] N. Axmacher, M.M. Henseler, O. Jensen, I. Weinreich, C.E. Elger, and J. Fell. Cross-frequency coupling supports multi-item working memory in the human hippocampus. *PNAS,* 107(7):3228–3233, 2010.

[21] H. Ayaz, M. Izzetoglu, K. Izzetoglu, B. Onaral, and B.B. Dor. Early diagnosis of traumatic intracranial hematomas. *Journal of Biomedical Optics,* 24(5):1–10, Feb 2019.

[22] H. Azami and J. Escudero. Refined composite multivariate generalized multiscale fuzzy entropy: A tool for complexity analysis of multichannel signals. *Physica A: Statistical Mechanics and its Applications*, 465:261–276, 2017.

[23] F. Babiloni, C. Babiloni, L. Locche, F. Cincotti, P.M. Rossini, and F. Carducci. High-resolution electro-encephalogram: source estimates of Laplacian-transformed somatosensory-evoked potentials using a realistic subject head model constructed from magnetic resonance images. *Medical & Biological Engineering & Computing*, 38:512–519, Sep 2000.

[24] F. Babiloni, D. Mattia, C. Babiloni, L. Astolfi, S. Salinari, A. Basilisco, P.M. Rossini, M.G. Marciani, and F. Cincotti. Multimodal integration of EEG, MEG and fMRI data for the solution of the neuroimage puzzle. *Magnetic Resonance Imaging*, 22(10):1471–1476, Dec 2004.

[25] A. Babloyantz and A. Destexhe. Is the normal heart a periodic oscillator? *Biological Cybernetics*, 58:203–211, 1988.

[26] L.A. Baccala, K. Sameshima, G. Ballester, and A. Do Valleand C. Timo-Laria. Studying the interaction between brain structures via directed coherence and granger causality. *Applied Signal Processing*, 5(5):40–48, 1998.

[27] L.A. Baccala, K. Sameshima, and D.Y. Takahashi. Generalized partial directed coherence. In *2007 15th International Conference on Digital Signal Processing*, pages 163–166, 2007.

[28] L.A. Baccala, M.Y. Alvarenga, K. Sameshima, C.L. Jorge, and L.H. Castro. Graph theoretical characterization and tracking of the effective neural connectivity during episodes of mesial temporal epileptic seizure. *Journal of Integrative Neuroscience*, 3(4):379–395, 2004.

[29] L.A. Baccala and K. Sameshima. Partial directed coherence: a new concept in neural structure determination. *Biological Cybernetics*, 84:463–74, 2001.

[30] R. Bailon, L. Sornmo, and P. Laguna. ECG derived respiratory frequency estimation. In G.D. Clifford, F. Azuaje, and P.E. McSharry, editors, *Advanced Methods and Tools for ECG Data Analysis*, pages 215–244. Artech House Inc, Norwood, MA, 2006.

[31] T. Ball, E. Demandt, I. Mutschler, E. Neitzel, C. Mehring, K. Vogt, A. Aertsen, and A. Schulze-Bonhage. Movement related activity in the high gamma range of the human EEG. *Neuroimage*, 41:302–310, Jun 2008.

[32] A. Barachant, S. Bonnet, M. Congedo, and C. Jutten. Multiclass brain–computer interface classification by riemannian geometry. *IEEE Transactions on Biomedical Engineering*, 59(4):920–928, 2012.

[33] L. Barnett, A.B. Barrett, and A.K. Seth. Granger causality and transfer entropy are equivalent for gaussian variables. *Physical Review Letters*, 103:238701, Dec 2009.

[34] E.A. Bartnik, K.J. Blinowska, and P.J. Durka. Single evoked potential reconstruction by means of wavelet transform. *Biological Cybernetics*, 67:175–181, 1992.

[35] D.S. Bassett, A. Meyer-Lindenberg, S. Achard, T. Duke, and E. Bullmore. Adaptive reconfiguration of fractal small-world human brain functional networks. *PNAS USA*, 103(51):19518–19523, Dec 2006.

[36] L.E. Baum, T. Petrie, G. Soules, and N. Weiss. A maximization technique occurring in the statistical analysis of probabilistic functions of Markov chains. *The Annals of Mathematical Statistics*, 41(1):164–171, 1970.

[37] M. Becker, T. Frauenrath, F. Hezel, G.A. Krombach, U. Kremer, B. Koppers, C. Butenweg, A. Goemmel, J.F. Utting, J. Schulz-Menger, and T. Niendorf. Comparison of left ventricular function assessment using phonocardiogram- and electrocardiogram-triggered 2D SSFP CINE MR imaging at 1.5 T and 3.0 T. *European Radiology*, 20:1344–1355, Jun 2010.

[38] C.F Beckmann, M. DeLuca, J.T Devlin, and S.M Smith. Investigations into resting-state connectivity using independent component analysis. *Philosophical Transactions of the Royal Society B: Biological Sciences*, 360(1457):1001–1013, 2005.

[39] A. Bell. Musical ratios in geometrical spacing of outer hair cells in colchea: strings of an underwater piano? In C.C. Stevens, D. Burnham, G. Mc Pherson, E. Schubert, and J. Renwick, editors, *Proc. of the Intern. Conference on Music Perception and Cognition*, 2002.

[40] A. Bell and N.H. Fletcher. The cochlear amplifier as a standing wave: "squirting" waves between rows of outer hair cells? *Journal of the Acoustical Society of America*, 116(2):1016–1024, Aug 2004.

[41] C.G. Bénar, T. Papadopoulo, B. Torrésani, and M. Clerc. Consensus Matching Pursuit for multi-trial EEG signals. *Journal of Neuroscience Methods*, 180:161–170, May 2009.

[42] J.S. Bendat and A.G. Piersol. *Random data: Analysis and measurement procedures*. Wiley&Sons, New York, London, Sydney, Toronto, 1971.

[43] Y. Benjamini and Y. Hochberg. Controlling the false discovery rate: a practical and powerful approach to multiple testing. *Journal of the Royal Statistical Society*, 57(1):289–300, 1995.

[44] Y. Benjamini and D. Yekutieli. The control of the false discovery rate in multiple testing under dependency. *Annals of Statistics*, 29(4):1165–1188, 2001.

[45] G.K. Bergey and P.J. Franaszczuk. Epileptic seizures are characterized by changing signal complexity. *Clinical Neurophysiology*, 112:241–249, Feb 2001.

[46] J.I. Berman, S. Liu, L. Bloy, L. Blaskey, T.P. L. Roberts, and J.C. Edgar. Alpha-to-Gamma phase-amplitude coupling methods and application to Autism Spectrum Disorder. *Brain Connectivity*, 5(2):80–90, Mar 2015.

[47] M. Bertini, M. Ferrara, L. De Gennaro, G. Curcio, F. Moroni, F. Vecchio, M. De Gasperis, P.M. Rossini, and C. Babiloni. Directional information flows between brain hemispheres during presleep wake and early sleep stages. *Cerebral Cortex*, 17(8):1970–1978, 2007.

[48] A. Bezerianos, S. Tong, and N. Thakor. Time-dependent entropy estimation of EEG rhythm changes following brain ischemia. *Annals of Biomedical Engineering*, 31:221–232, Feb 2003.

[49] BB Biswal, J Van Kylen, and JS Hyde. Simultaneous assessment of flow and bold signals in resting-state functional connectivity maps. *NMR in Biomedicine*, 10(4–5):165–170, 1997.

[50] R.B. Blackman and J.W. Tukey. *The Measurment of Power Spectra*. Dover Publications, Dover, New York, 1958.

[51] C. Bledowski, D. Prvulovic, K. Hoechstetter, M. Scherg, M. Wibral, R. Goebel, and D. E J Linden. Localizing p300 generators in visual target and distractor processing: a combined event-related potential and functional magnetic resonance imaging study. *The Journal of Neuroscience: the Official Journal of the Society for Neuroscience*, 24(42):9353–9360, Oct 2004.

[52] K. Blinowska, R. Kus, M. Kaminski, and J. Janiszewska. Transmission of brain activity during cognitive task. *Brain Topography*, 23:205–213, Jun 2010.

[53] K. Blinowska, G. Müller-Putz, V. Kaiser, L. Astolfi, K. Vanderperren, S. Van Huffel, and L. Lemieux. Multi-modal imaging of human brain activity: rational, biophysical aspects and modes of integration. *Computational Intelligence and Neuroscience*, 2009, 2009. Article ID 813607, 10 pages.

[54] K.J. Blinowska, P.J. Durka, and J. Zygierewicz. Time-frequency analysis of brain electrical activity–adaptive approximations. *Methods of Information in Medicine*, 43:70–73, 2004.

[55] K.J. Blinowska, M. Kamiński, A. Brzezicka, and J. Kamiński. Application of directed transfer function and network formalism for the assessment of functional connectivity in working memory task. *Philosophical Transactions of The Royal Society A Mathematical Physical and Engineering Sciences*, 371(1997):20110614, Aug 2013.

[56] K.J. Blinowska, M. Kaminski, J. Kaminski, and A. Brzezicka. Information processing in brain and dynamic patterns of transmission during working memory task by the SDTF function. *Conference proceedings - IEEE engineering in medicine and biology society*, 2010:1722–1725, 2010.

[57] K.J. Blinowska, P. Lachert, J. Zygierewicz, D. Janusek, P. Sawosz, M. Kacprzak, and A. Liebert. Characteristic of mayer waves in electrophysiological, hemodynamic and vascular signals. *International Journal of Neural Systems*, 30(3):2050003, Mar 2020.

[58] K.J. Blinowska, F. Rakowski, M. Kaminski, F. De Vico Fallani, C. Del Percio, R. Lizio, and C. Babiloni. Functional and effective brain connectivity for discrimination between Alzheimer's patients and healthy individuals: A study on resting state EEG rhythms. *Clinical Neurophysiology*, 128(4):667–680, Apr 2017.

[59] K.J. Blinowska and M. Kaminski. Functional brain networks: Random, "small world" or deterministic? *PLOS ONE*, 8(10):1–9, Oct 2013.

[60] K.J. Blinowska and M. Kaminski. Functional brain networks: Random, "small world" or deterministic? *PLOS ONE*, 8(10):1–9, Oct 2013.

[61] K.J. Blinowska. Non-linear analysis of physiological signals—highlights and pitfalls. In W. Klonowski, editor, *Attractors, Signals and Synergetics. Proceedings of EUROATTRACTOR 2000*, pages 341–350. Pabst Science Publishers, Lengerich, Berlin, Bremen, 2002.

[62] K.J. Blinowska and P.J. Durka. *The application of wavelet transform and matching pursuit to the time varying EEG signals*, volume 4, chapter Intelligent Engineering Systems through Artificial Neural Networks, pages 535–540. ASME Press, 1994.

[63] K.J. Blinowska and P.J. Franaszczuk. A model of the generation of electrocortical rhythms. In *Brain Dynamics*, Springer Series in Brain Dynamics 2, pages 192–201. Springer Verlag, 1989.

[64] K.J. Blinowska, R. Kus, and M. Kaminski. Granger causality and information flow in multivariate processes. *Physical Review E*, 2004.

[65] K.J. Blinowska and M. Malinowski. Non-linear and linear forecasting of the EEG time series. *Biological Cybernetics*, 66:159–165, 1991.

[66] G.E.P. Box and D.R. Cox. An analysis of transformations. *Journal of the Royal Statistical Society, Series B*, 26(2):211–252, 1964.

[67] M.A.B. Brazier and J.U. Casby. An application of the M.I.T. digital electronic correlator to a problem in EEG. *Electroencephalography*, 8:32, 1956.

[68] E.N. Brown, R.E. Kass, and P.P. Mitra. Multiple neural spike train data analysis: state-of-the-art and future challenges. *Nature Neuroscience*, 7:456–461, May 2004.

[69] Bryant and Yarnold. Principal components analysis and exploratory and confirmatory factor analysis. In *Reading and understanding multivariate analysis*. American Psychological Association Books, 1995.

[70] A. Brzezicka, G. Sedek, A. Marchewka, M. Gola, K. Jednoróg, L. Królicki, and A. Wróbel. A role for the right prefrontal and bilateral parietal cortex in four-term transitive reasoning: an fMRI study with abstract linear syllogism tasks. *Acta Neurobiologiae Experimentalis*, 71(4):479–495, 2011.

[71] F. Buchthal. Studies on muscular action potentials in the normal and diseased muscle. *Deutsche Zeitschrift für Nervenheilkunde*, 173:448–454, 1955.

[72] F. Buchthal. Electromyography in the evaluation of muscle diseases. *Neurologic Clinics*, 3:573–598, Aug 1985.

[73] T.H. Bullock, M.C. McClune, J.Z. Achimowicz, V.J. Iragui-Madoz, R. B. Duckrow, and S.S. Spencer. EEG coherence has structure in the millimeter domain: subdural and hippocampal recordings from epileptic patients. *Electroencephalography and Clinical Neurophysiology*, 95:161–177, Sep 1995.

[74] RB Buxton, EC Wong, and LR Frank. Dynamics of blood flow and oxygenation changes during brain activation: the balloon model. *Magnetic Resonance in Medicine*, 39(6):855–864, June 1998.

[75] G. Buzsaki and A. Draguhn. Neuronal oscillations in cortical networks. *Science*, 304:1926–1929, Jun 2004.

[76] G. Camps, M. Martinez, and E. Soria. Fetal ECG extraction using an FIR neural network. In *Proc. Computers in Cardiology*, volume 28, pages 249–252. IEEE Press, 2001.

[77] R.T. Canolty, E. Edwards, S.S. Dalal, M. Soltani, S.S. Nagarajan, H.E. Kirsch, M.S. Berger, N.M. Barbaro, and R.T Knight. High gamma power is phase-locked to theta oscillations in human neocortex. *Science (New York, N.Y.)*, 313(5793):1626–1628, 2006.

[78] J.-F. Cardoso. High-order contrasts for independent component analysis. *Neural Computation*, 11(1):157–192, 1999.

[79] F. De Carli, L. Nobili, F. Ferrillo, P. Gelcichand, and F. Ferrillo. A method for the automatic detection of arousals during sleep. *Sleep*, 22:561–572, 1999.

[80] S. Cerutti, G. Baselli, S. Civardi, E. Ferrazzi, A.M. Marconi, M. Pagani, and G. Pardi. Variability analysis of fetal heart rate signals as obtained from abdominal electrocardiographic recordings. *Journal of Perinatal Medicine*, 14:445–452, 1986.

[81] S. Cerutti, G. Baselli, D. Liberati, and G. Pavesi. Single sweep analysis of visual evoked potentials through a model of parametric identification. *Biological Cybernetics*, 56:111–120, 1987.

[82] S. Cerutti, A.M. Bianchi, and L.T. Mainardi. Spectral analysis of heart rate variability signal. In M. Malik and A.J. Camm, editors, *Heart Rate Variability*. Futura Publishing, Armonk, New York, 1995.

[83] J. Chee and S-C. Seow. Electrocardiogram. In U.R. Acharya, J.S. Suri, J.A.E. Spaan, and S.M. Krishnan, editors, *Advances in Cardiac Signal Processing*. Springer-Verlag, Berlin, Heidelberg, New York, 2007.

[84] Y. Chen, G. Rangarajan, J. Feng, and M. Ding. Analyzing multiple nonlinear time series with extended Granger causality. *Physics Letters A*, 324:26–35, 2004.

[85] J. Cheng. Time-frequency analysis of transient evoked otoacoustic emissions via smoothed pseudo Wigner distribution. *Scandinavian Audiology*, 24:91–96, 1995.

[86] I. Christov, G. Gomez-Herrero, V. Krasteva, I. Jekova, A. Gotchev, and K. Egiazarian. Comparative study of morphological and time-frequency ECG descriptors for heartbeat classification. *Medical Engineering & Physics*, 28:876–887, Nov 2006.

[87] G.D. Clifford and L. Tarassenko. Segmenting cardiac-related data using sleep stages increases separation between normal subjects and apnoeic patients. *Physiological Measurement*, 25:27–35, Dec 2004.

[88] G.D. Clifford. Linear filtering methods. In G.D. Clifford, F. Azuaje, and P.E. McSharry, editors, *Advanced Methods and Tools for ECG Data Analysis*, pages 135–170. Artech House Inc, Norwood, MA, 2006.

[89] D. Cohen, P. Savard, R.D. Rifkin, E. Lepeschkin, and W.E. Strauss. Magnetic measurement of S-T and T-Q segment shifts in humans. Part II: Exercise-induced S-T segment depression. *Circulation Research*, 53:274–279, Aug 1983.

[90] RR Coifman and MV Wickerhauser. Entropy-based algorithms for best basis selection. *IEEE Transactions on Information Theory*, 38(2):713–718, 1992.

[91] D. Cole, S. Smith, and C. Beckmann. Advances and pitfalls in the analysis and interpretation of resting-state fMRI data. *Frontiers in Systems Neuroscience*, 4:8, 2010.

[92] M. Congedo, A. Barachant, and R. Bhatia. Riemannian geometry for EEG-based brain-computer interfaces; a primer and a review. *Brain-Computer Interfaces*, 4(3):155–174, 2017.

[93] I A. Cook, A.M. Hunter, A. Korb, H. Farahbod, and A.F. Leuchter. *Quantitative EEG analysis methods and clinical applications*, chapter EEG signals in psychiatry: Biomarkers for Depression Management. Artech House, 2009.

[94] D Cordes, V M Haughton, K Arfanakis, J D Carew, P A Turski, C H Moritz, M A Quigley, and M E Meyerand. Frequencies contributing to functional connectivity in the cerebral cortex in "resting-state" data. *American Journal of Neuroradiology*, 22(7):1326–1333, Aug 2001.

[95] A.D. Corlan and L. De Ambroggi. New quantitative methods of ventricular repolarization analysis in patients with left ventricular hypertrophy. *Italian heart journal*, 1(8):542–548, Aug 2000.

[96] M. Costa, AL Goldberger, and C-K. Peng. Multiscale entropy analysis of complex physiologic time series. *Physical Review Letters*, 89:068102, 2002.

[97] R. Cox and J. Fell. Analyzing human sleep EEG: A methodological primer with code implementation. *Sleep Medicine Reviews*, 54:101353, 2020.

[98] N.E. Crone, D.L. Miglioretti, B. Gordon, and R.P. Lesser. Functional mapping of human sensorimotor cortex with electrocorticographic spectral analysis. II. Event-related synchronization in the gamma band. *Brain*, 121 (Pt 12):2301–2315, Dec 1998.

[99] C. Dai and X. Hu. Independent component analysis based algorithms for high-density electromyogram decomposition: Systematic evaluation through simulation. *Computers in Biology and Medicine*, 109:171–181, 2019.

[100] S.S. Dalal, A.G. Guggisberg, E. Edwards, K. Sekihara, A.M. Findlay, R. T. Canolty, M.S. Berger, R.T. Knight, N.M. Barbaro, H.E. Kirsch, and S. S. Nagarajan. Five-dimensional neuroimaging: localization of the time-frequency dynamics of cortical activity. *Neuroimage*, 40:1686–1700, May 2008.

[101] AM Dale, AK Liu, BR Fischl, RL Buckner, JW Belliveau, JD Lewine, and E Halgren. Dynamic statistical parametric mapping: combining fMRI and MEG for high-resolution imaging of cortical activity. *Neuron*, 26(1):55–67, Apr 2000.

[102] F. Darvas, R. Scherer, J.G. Ojemann, R.P. Rao, K.J. Miller, and L.B. Sorensen. High gamma mapping using EEG. *Neuroimage*, 49:930–938, Jan 2010.

[103] I. Daubechies. *Ten Lectures on Wavelets*. SIAM, 1992.

[104] O. David, D. Cosmelli, and Friston K.J. Evaluation of different measures of functional connectivity using a neural mass model. *Neuroimage*, 21:659–673, 2004.

[105] G. Davis. *Adaptive Nonlinear Approximations*. PhD thesis, New York University, 1994.

[106] T. Davis, K. Kwong, R. Weisskoff, and B. Rosen. Calibrated functional mri: mapping the dynamics of oxidative metabolism. *Proceedings of the National Academy of Sciences of the United States of America*, 95(4):1834–1839, 1998.

[107] G.D. Dawson. Cerebral responses to electrical stimulation of peripheral nerve in man. *Journal of Neurology, Neurosurgery, and Psychiatry*, 10:134–140, 1947.

[108] L. De Ambroggi, T. Bertoni, M.L. Breghi, M. Marconi, and M. Mosca. Diagnostic value of body surface potential mapping in old anterior non-Q myocardial infarction. *Journal of Electrocardiology*, 21:321–329, Nov 1988.

[109] C. de Hemptinne, E.S. Ryapolova-Webb, E.L. Air, P.A. Garcia, K.J. Miller, J.G. Ojemann, J.L. Ostrem, N.B. Galifianakis, and P.A. Starr. Exaggerated phase-amplitude coupling in the primary motor cortex in Parkinson disease. *Proceedings of the National Academy of Sciences*, 110(12):4780–4785, Mar 2013.

[110] C. de Hemptinne, E.S. Ryapolova-Webb, E.L. Air, P.A. Garcia, K.J. Miller, J.G. Ojemann, J.L. Ostrem, N.B. Galianakis, and P.A. Starr Exaggerated phase-amplitude coupling in the primary motor cortex in parkinson disease. *Proceedings of the National Academy of Sciences U S A*, 110(12):4780–4785, Mar 2013.

[111] C. De Hemptinne, N.C. Swann, Jill L. Ostrem, E.S. Ryapolova-Webb, M.S. Luciano, N.B. Galifianakis, and P.A. Starr. Therapeutic deep brain stimulation reduces cortical phase-amplitude coupling in Parkinson's disease. *Nature Neuroscience*, 18(5):779–786, 2015.

[112] L. De Lathauwer, B. De Moor, and J. Vandewalle. Fetal electrocardiogram extraction by blind source subspace separation. *IEEE Transactions on Biomedical Engineering*, 47:567–572, May 2000.

[113] G. De Michele, S. Sello, M.C. Carboncini, B. Rossi, and S.K. Strambi. Cross-correlation time-frequency analysis for multiple EMG signals in Parkinson's disease: a wavelet approach. *Medical Engineering & Physics*, 25:361–369, Jun 2003.

[114] F. De Vico Fallani, L. Astolfi, F. Cincotti, D. Mattia, M.G. Marciani, A. Tocci, S. Salinari, H. Witte, W. Hesse, S. Gao, A. Colosimo, and F. Babiloni. Cortical network dynamics during foot movements. *Neuroinformatics*, 6:23–34, 2008.

[115] S.M. Debbal and F. Bereksi-Reguig. Filtering and classification of phonocardiogram signals using wavelet transform. *Journal of Medical Engineering & Technology*, 32:53–65, 2008.

[116] Fani Deligianni, Maria Centeno, David W. Carmichael, and Jonathan D. Clayden. Relating resting-state fMRI and EEG whole-brain connectomes across frequency bands. *Frontiers in Neuroscience*, 8:258, 2014.

[117] A. Delorme and S. Makeig. EEGLAB: an open source toolbox for analysis of single-trial EEG dynamics including independent component analysis. *Journal of Neuroscience Methods*, 134:9–21, Mar 2004.

[118] A. Delorme, T. Sejnowski, and S. Makeig. Enhanced detection of artifacts in EEG data using higher-order statistics and independent component analysis. *Neuroimage*, 34:1443–1449, Feb 2007.

[119] N. Delprat, B. Escudié, P. Guillemain, R. Kronland-Martinet, P. Tchamitchian, and B. Torrésani. Asymptotic wavelet and Gabor analysis: extraction of instantaneous frequencies. *IEEE Transactions on Information Theory*, 38:644–664, 1992.

[120] G. Deshpande, S. LaConte, S. Peltier, and X. Hu. Directed transfer function analysis of fMRI data to investigate network dynamics. In *2006 International Conference of the IEEE Engineering in Medicine and Biology Society*, pages 671–674, 2006.

[121] M. Dhamala, G. Rangarajan, and M. Ding. Analyzing information flow in brain networks with nonparametric Granger causality. *Neuroimage*, 41(2):354–362, Jun 2008.

[122] G. Dietch. Fourier analyse von elektroenzephalogramen des menschen. *Pflügers Archiv European Journal of Physiology*, 230:106–112, 1932.

[123] C. Disselhorst-Klug, G. Rau, A. Schmeer, and J. Silny. Non-invasive detection of the single motor unit action potential by averaging the

spatial potential distribution triggered on a spatially filtered motor unit action potential. *Journal of Electromyography & Kinesiology*, 9:67–72, Feb 1999.

[124] M. Dovgialo, A. Chabuda, A. Duszyk, M. Zieleniewska, Marcin Pietrzak, Piotr Rózański, and Piotr Durka. Assessment of Statistically Significant Command-Following in Pediatric Patients with Disorders of Consciousness, Based on Visual, Auditory and Tactile Event-Related Potentials. *International Journal of Neural Systems*, 29(3):1–14, 2019.

[125] D.J. Doyle. Some comments on the use of Wiener filtering in the estimation of evoked potentials. *Electroencephalography and Clinical Neurophysiology*, 38:533–534, 1975.

[126] C.C. Duncan, R.J. Barry, J.F. Connolly, C. Fischer, P.T. Michie, R. Naatanen, J. Polich, I. Reinvang, and C. Van Petten. Event-related potentials in clinical research: guidelines for eliciting, recording, and quantifying mismatch negativity, P300, and N400. *Clinical Neurophysiology*, 120:1883–1908, Nov 2009.

[127] P.J. Durka. From wavelets to adaptive approximations: time-frequency parametrization of EEG. *BioMedical Engineering OnLine*, 2:1, Jan 2003. © 2003 Durka; licensee BioMed Central Ltd. This is an Open Access article: verbatim copying and redistribution of this article are permitted in all media for any purpose, provided this notice is preserved along with the article's original URL.

[128] P.J. Durka, D. Ircha, and K.J. Blinowska. Stochastic time-frequency dictionaries for matching pursuit. *IEEE Transactions on Signal Processing*, 49(3):507–510, Mar 2001.

[129] P.J. Durka, D. Ircha, Ch. Neuper, and G. Pfurtscheller. Time-frequency microstructure of event-related desynchronization and synchronization. *Medical & Biological Engineering & Computing*, 39(3):315–321, May 2001.

[130] P.J. Durka, U. Malinowska, W. Szelenberger, A Wakarow, and K.J. Blinowska. High resolution parametric description of slow wave sleep. *Journal of Neuroscience Methods*, 147(1):15–21, 2005.

[131] P.J. Durka, A. Matysiak, E.M. Montes, P. Valdés Sosa, and K. J. Blinowska. Multichannel matching pursuit and EEG inverse solutions. *Journal of Neuroscience Methods*, 148(1):49–59, 2005.

[132] P.J. Durka, J. Zygierewicz, H. Klekowicz, J. Ginter, and K.J. Blinowska. On the statistical significance of event-related EEG desynchronization and synchronization in the time-frequency plane. *IEEE Transactions on Biomedical Engineering*, 51:1167–1175, Jul 2004.

[133] P.J. Durka, A. Matysiak, E. Martínez Montes, P. Valdés Sosa, and K. J. Blinowska. Multichannel matching pursuit and EEG inverse solutions. *Journal of Neuroscience Methods*, 148(1):49–59, 2005.

[134] P.J. Durka. Adaptive time-frequency parametrization of epileptic EEG spikes. *Physical Review E*, 69:051914, 2004.

[135] P.J. Durka, W. Szelenberger, K.J. Blinowska, W. Androsiuk, and M. Myszka. Adaptive time-frequency parametrization in pharmaco EEG. *Journal of Neuroscience Methods*, 117:65–71, 2002.

[136] J.C. Echeverria, J.A. Crowe, M.S. Woolfson, and B.R. Hayes-Gill. Application of empirical mode decomposition to heart rate variability analysis. *Medical & Biological Engineering & Computing*, 39:471–479, Jul 2001.

[137] J.-P. Eckmann, S. Oliffson Kamphorst, and D. Ruelle. Recurrence plots of dynamical systems. *Europhysics Letters*, 4(9):973–977, 1987.

[138] B. Efron and R.J. Tibshirani. *An Introduction to the Bootstrap*. Chapman & Hall, New York, 1993.

[139] T. Eichele, V.D. Calhoun, M. Moosmann, K. Specht, M.L. Jongsma, R. Q. Quiroga, H. Nordby, and K. Hugdahl. Unmixing concurrent EEG-fMRI with parallel independent component analysis. *International Journal of Psychophysiology*, 67:222–234, Mar 2008.

[140] T. Eichele, K. Specht, M. Moosmann, M.L. Jongsma, R.Q. Quiroga, H. Nordby, and K. Hugdahl. Assessing the spatiotemporal evolution of neuronal activation with single-trial event-related potentials and functional MRI. *Proceedings of the National Academy of Sciences USA*, 102:17798–17803, Dec 2005.

[141] M. Eichler. On the evaluation of the information flow in multivariate systems by the Direct Transfer Function. *Biological Cybernetics*, 94:469–482, 2006.

[142] C.E. Elger and K. Lehnertz. Seizure prediction by non-linear time series analysis of brain electrical activity. *Eur J Neurosci*, 10:786–789, Feb 1998.

[143] A.K. Engel and W. Singer. Temporal binding and the neural correlates of sensory awareness. *Trends in Cognitive Sciences (Regul Ed)*, 5:16–25, Jan 2001.

[144] K. Englehart, B. Hudgins, P.A. Parker, and M. Stevenson. Classification of the myoelectric signal using time-frequency based representations. *Medical Engineering & Physics*, 21:431–438, 1999.

[145] R. Enoka. Physiological validation of the decomposition of surface emg signals. *Journal of electromyography and kinesiology : official journal of the International Society of Electrophysiological Kinesiology*, 46:70–83, 2019.

[146] M. Etemadi and O.T. Inan. Wearable ballistocardiogram and seismocardiogram systems for health and performance. *Journal of Applied Physiology (1985)*, 124(2):452–461, Feb 2018.

[147] A.C. Evans, D.L. Collins, S.R. Mills, E.D. Brown, R.L. Kelly, and T.M. Peters. 3D statistical neuroanatomical models from 305 MRI volumes. In *1993 IEEE Conference Record Nuclear Science Symposium and Medical Imaging Conference*, pages 1813–1817. volume 3, 1993.

[148] D. Falla and D. Farina. Non-uniform adaptation of motor unit discharge rates during sustained static contraction of the upper trapezius muscle. *Experimental Brain Research*, 191:363–370, Nov 2008.

[149] E.E. Fanselow, K. Sameshima, L.A. Baccala, and M.A.L. Nicolelis. Thalamic bursting in rats during different awake behavioral states. *Proceedings of the National Academy of Sciences of the United States of America*, 98(26):15330–15335, 2001.

[150] D. Farina, R. Colombo, R. Merletti, and H.B. Olsen. Evaluation of intramuscular emg signal decomposition algorithms. *J Electromyography and Kinesiology*, 11:175–187, 2001.

[151] A. Fasoula, Y. Attal, and D. Schwartz. Comparative performance evaluation of data-driven causality measures applied to brain networks. *Journal of Neuroscience Methods*, 215(2):170–189, 2013.

[152] S. Fazli, J. Mehnert, J. Steinbrink, G. Curio, A. Villringer, K.R. Müller, and B. Blankertz. Enhanced performance by a hybrid NIRS-EEG brain computer interface. *Neuroimage*, 59(1):519–529, Jan 2012.

[153] R. Fenici and D. Brisinda. Magnetocardiography provides non-invasive three-dimensional electroanatomical imaging of cardiac electrophysiology. *International Journal of Cardiovascular Imaging*, 22:595–597, 2006.

[154] M. Fereniec. *The Evaluation of Space Variability of the High-Resolution ECG During Atrial Heart Activity*. PhD thesis, IBIB PAN, Warsaw, 2008.

[155] M. Fereniec and G. Karpinski. Analysis of T-wave shape variability in hr ECG mapping. *Technology and Health Care*, 12(2):125–127, 2004.

[156] R. Ferri, F. Rundo, O. Bruni, M.G. Terzano, and C.J. Stam. Dynamics of the EEG slow-wave synchronization during sleep. *Clinical Neurophysiology*, 116(12):2783–2795, Dec 2005.

[157] M.J. Fischer, G. Scheler, and H. Stefan. Utilization of magnetoencephalography results to obtain favourable outcomes in epilepsy surgery. *Brain*, 128:153–157, Jan 2005.

[158] J.R. Florestal, P.A. Mathieu, and A. Malanda. Automated decomposition of intramuscular electromyographic signals. *IEEE Transactions on Biomedical Engineering*, 53:832–839, May 2006.

[159] J.R. Florestal, P.A. Mathieu, and K.C. McGill. Automatic decomposition of multichannel intramuscular EMG signals. *Journal of Electromyography and Kinesiology*, 19:1–9, Feb 2009.

[160] E. Florin and S. Baillet. The brain's resting-state activity is shaped by synchronized cross-frequency coupling of neural oscillations. *NeuroImage*, 111:26–35, 2015.

[161] G. Folland and A. Sitaram. The uncertainty principle: A mathematical survey. *Journal of Fourier Analysis and Applications*, 3(3):207–238, 1997.

[162] P.J. Franaszczuk and G.K. Bergey. An autoregressive method for the measurement of synchronization of interictal and ictal EEG signals. *Biological Cybernetics*, 81:3–9, Jul 1999.

[163] P.J. Franaszczuk, G.K. Bergey, and M.J. Kaminski. Analysis of mesial temporal seizure onset and propagation using the directed transfer function method. *Electroencephalography and Clinical Neurophysiology*, 91:413–427, Dec 1994.

[164] P.J. Franaszczuk, P. Mitraszewski, and K. Blinowska. FAD-parametric description of EEG time series. *Acta physiologica Polonica*, 40:418–422, 1989.

[165] P.J. Franaszczuk and K.J. Blinowska. Linear model of brain electrical activity–EEG as a superposition of damped oscillatory modes. *Biological Cybernetics*, 53:19–25, 1985.

[166] P.J. Franaszczuk, K.J. Blinowska, and M. Kowalczyk. The application of parametric multichannel spectral estimates in the study of electrical brain activity. *Biological Cybernetics*, 51:239–247, 1985.

[167] J.V. Frangioni. New technologies for human cancer imaging. *Journal of Clinical Oncology*, 26(24):4012–21, 2008.

[168] W.J. Freeman. *Mass Action in the Nervous System*. Academic Press, New York, San Francisco, London, 1975.

[169] F. Freyer, R. Becker, K. Anami, G. Curio, A. Villringer, and P. Ritter. Ultrahigh-frequency EEG during fMRI: Pushing the limits of imaging-artifact correction. *NeuroImage*, 48:94–108, 2009.

[170] K.J. Friston. Functional and effective connectivity: A review. *Brain Connectivity*, 1(1):13–36, 2011.

[171] K.J. Friston, J. Ashburner, S.J. Kiebel, T.E. Nichols, and W.D. Penny, editors. *Statistical Parametric Mapping: The Analysis of Functional Brain Images*. Academic Press, 2007.

[172] K.J. Friston, L. Harrison, and W. Penny. Dynamic causal modelling. *NeuroImage*, 19(4):1273–1302, 2003.

[173] K.J. Friston, Katrin H. Preller, Chris Mathys, Hayriye Cagnan, Jakob Heinzle, Adeel Razi, and Peter Zeidman. Dynamic causal modelling revisited. *NeuroImage*, 199:730–744, 2019.

[174] A. Fuglsang-Frederiksen. The utility of interference pattern analysis. *Muscle Nerve*, 23:18–36, Jan 2000.

[175] J.A. Gaxiola-Tirado, R. Salazar-Varas, and D. Gutiérrez. Using the partial directed coherence to assess functional connectivity in electroencephalography data for brain–computer interfaces. *IEEE Transactions on Cognitive and Developmental Systems*, 10(3):776–783, 2018.

[176] M. Gazzoni, D. Farina, and R. Merletti. A new method for the extraction and classification of single motor unit action potentials from surface EMG signals. *Journal of Neuroscience Methods*, 136:165–177, Jul 2004.

[177] D. Ge, N. Srinivassan, and S.M. Krishnan. The application of autoregressive modeling in cardiac arrhythmia classification. In U.R. Acharya, J.S. Suri, J.A.E. Spaan, and S.M. Krishnan, editors, *Advances in Cardiac Signal Processing*. Springer- Verlag, Berlin, Heidelberg, New York, 2007.

[178] E.M. Gerber, B. Sadeh, A, Ward, R.T. Knight, and Leon Y. Deouell. Non-sinusoidal activity can produce cross-frequency coupling in cortical signals in the absence of functional interaction between neural sources. *PLOS ONE*, 11(12):e0167351, Dec 2016.

[179] J. Geweke. Measurement of linear dependence and feedback between multiple time series. *Journal of the American Statistical Association*, 77(378):304–324, 1982.

[180] J.F. Geweke. Measures of conditional linear dependence and feedback between time series. *Journal of the American Statistical Association*, 79(388):907–915, 1984.

[181] M.H. Giard, J. Lavikahen, K. Reinikainen, F. Perrin, O. Bertrand, J. Pernier, and R. Näätänen. Separate representation of stimulus frequency, intensity, and duration in auditory sensory memory: An event-related potential and dipole-model analysis. *J Cognitive Neuroscience*, 7(2):133–143, 1995.

[182] N.M. Gibson, M.S. Woolfson, and J.A. Crowe. Detection of fetal electrocardiogram signals using matched filters with adaptive normalisation. *Medical & Biological Engineering & Computing*, 35:216–222, May 1997.

[183] J. Ginter, K.J. Blinowska, M. Kaminski, and P.J. Durka. Phase and amplitude analysis in time-frequency space–application to voluntary finger movement. *Journal of Neuroscience Methods*, 110:113–124, Sep 2001.

[184] J. Ginter, K.J. Blinowska, M. Kaminski, P.J. Durka, G. Pfurtscheller, and C. Neuper. Propagation of EEG activity in the beta and gamma band during movement imagery in humans. *Methods of Information in Medicine*, 44:106–113, 2005.

[185] K. Gireesan, C. Sengottuvel, C. Parsakthi, P. Rajesh, M.P. Janawadkar, and T.S. Radhakrishnan. Magnetocardiography study of cardiac anomalies. In S. Supek and A. Susac, editors, *Advances in Biomagnetism BIOMAG2010, IFMBE Proceedings*, volume 28, pages 431–435. Springer, 2010.

[186] C. Glaros and D.I. Fotiadis. *Wearable Devices in Healthcare*, pages 237–264. Springer, Berlin, Heidelberg, 2005.

[187] J. Goense, H. Merkle, and N. Logothetis. High-resolution fMRI reveals laminar differences in neurovascular coupling between positive and negative bold responses. *Neuron*, 76:629–639, 2012.

[188] A.L. Goldberger, L.A. Amaral, L. Glass, J.M. Hausdorff, P.C. Ivanov, R. G. Mark, J.E. Mietus, G.B. Moody, C.K. Peng, and H.E. Stanley. PhysioBank, PhysioToolkit, and PhysioNet: components of a new research resource for complex physiologic signals. *Circulation*, 101:E215–220, Jun 2000.

[189] R. Goldman, J. Stern, J. Engel, and M. Cohen. Simultaneous EEG and fMRI of the alpha rhythm. *NeuroReport*, 13:2487–2492, 2002.

[190] C.J. Gonsalvez and J. Polich. P300 amplitude is determined by target-to-target interval. *Psychophysiology*, 39(3):388–396, 2002.

[191] J. Gotman and M.G. Marciani. Electroencephalographic spiking activity, drug levels, and seizure occurrence in epileptic patients. *Annals of Neurology*, 17:597–603, Jun 1985.

[192] R.B. Govindan, J.D. Wilson, H. Eswaran, C.L. Lowery, and H. Preißl. Revisiting sample entropy analysis. *Physica A: Statistical Mechanics and its Applications*, 376:158 –164, 2007.

[193] B. Graimann, J.E. Huggins, S.P. Levine, and G. Pfurtscheller. Visualization of significant ERD/ERS patterns in multichannel EEG and ECoG data. *Clinical Neurophysiology*, 113:43–47, 2002.

[194] C.W.J. Granger. Investigating causal relations in by econometric models and cross-spectral methods. *Econometrica*, 37:424–38, 1969.

[195] A. Grapelyuk, A. Schirdewan, R. Fischer, and N. Wessel. Cardiac magnetic field mapping quantified by Kullback-Leibler entropy detects patients with coronary artery disease. *Physiological Measurement*, 31(10):1345–1354, 2010.

[196] P. Grassberger and I. Procaccia. Measuring the strangeness of strange attractors. *Physica D*, 9:189–208, 1983.

[197] C. Grau, L. Fuentemilla, and J. Marco-Pallarés. Functional neural dynamics underlying auditory event-related n1 and n1 suppression response. *Neuroimage*, 36(6):522–31, 2007.

[198] R. Grave de Peralta Mendez and S.L. Gonzales Andino. Distributed source models: standard solutions and new developments. In C. Uhl, editor, *Analysis of Neurophysiological Brain Functioning*, pages 176–291. Springer, Berlin, Heidelberg, New York, 1999.

[199] F. Grouiller, L. Vercueil, A. Krainik, C. Segebarth, P. Kahane, and O. David. A comparative study of different artefact removal algorithms for EEG signals acquired during functional mri. *Neuroimage*, 38(1):124–137, Oct 2007.

[200] P. Groves and R. Thompson. Habituation: a dual-process theory. *Psychological Review*, 77:419–450, 1970.

[201] M. Guirgis, Y. Chinvarun, M. Del Campo, P.L. Carlen, and Berj L. Bardakjian. Defining regions of interest using cross-frequency coupling in extratemporal lobe epilepsy patients. *Journal of Neural Engineering*, 12(2), 2015.

[202] A. Gunji, R. Ishii, W. Chau, R. Kakigi, and C. Pantev. Rhythmic brain activities related to singing in humans. *Neuroimage*, 34:426–434, Jan 2007.

[203] A. Gydikov and N. Gantchev. Velocity of spreading of the excitation along the muscle fibres of human motor units measured by superficial electrodes. *Electromyography and clinical neurophysiology*, 29:131–138, Apr 1989.

[204] A. Gydikov and P. Gatev. Human single muscle fiber potentials at different radial distances from the fibers determined by a method of location. *Experimental Neurology*, 76:25–34, Apr 1982.

[205] A. Gydikov, L. Gerilovsky, N. Radicheva, and N. Trayanova. Influence of the muscle fibre end geometry on the extracellular potentials. *Biological Cybernetics*, 54:1–8, 1986.

[206] A. Gydikov and D. Kosarov. Studies of the activity of alpha motoneurons in man by means of a new electromyographic method. In G. Somjem, editor, *Proceedings of Conference Neurophysiology Studies in Man*, pages 219–227, 1972.

[207] A.J. Haig, J.B. Gelblum, J.J. Rechtien, and A.J. Gitter. Technology assessment: the use of surface EMG in the diagnosis and treatment of nerve and muscle disorders. *Muscle Nerve*, 19:392–395, Mar 1996.

[208] M.S. Hämäläinen and R.J. Ilmoniemi. Interpreting magnetic fields of the brain: minimum norm estimates. *Medical & Biological Engineering & Computing*, 32(1):35–42, January 1994.

[209] M. Hämäläinen, R. Hari, R.J. Ilmoniemi, J. Knuutila, and O.V. Lounasmaa. Magnetoencephalography—theory, instrumentation, and applications to noninvasive studies of the working human brain. *Reviews of Modern Physics*, 65(2):413–497, Apr 1993.

[210] A. Hamilton-Wright and D.W. Stashuk. Physiologically based simulation of clinical EMG signals. *IEEE Transactions on Biomedical Engineering*, 52:171–183, Feb 2005.

[211] B. Händel and T. Haarmeier. Cross-frequency coupling of brain oscillations indicates the success in visual motion discrimination. *NeuroImage*, 45(3):1040–1046, April 2009.

[212] H. Hanninen, P. Takala, M. Makijarvi, J. Montonen, P. Korhonen, L. Oikarinen, K. Simelius, J. Nenonen, T. Katila, and L. Toivonen. Recording locations in multichannel magnetocardiography and body surface potential mapping sensitive for regional exercise-induced myocardial ischemia. *Basic Research in Cardiology*, 96:405–414, Jul 2001.

[213] F.J. Harris. On the use of windows for harmonic analysis with the Discrete Fourier Transform. *Proceedings of the IEEE*, 66(1):51–83, 1978.

[214] M. A. F. Harrison, I. Osorio, M.G. Frei, S. Asuri, and Y.-Ch. Lai. Correlation dimension and integral do not predict epileptic seizures. *Chaos: An Interdisciplinary Journal of Nonlinear Science*, 15(3):033106, 2005.

[215] H. Hassanpour and A. Parsaei. Fetal ECG extraction using wavelet transform. In *International Conference on Computational Intelligence for Modeling, Control and Automation, 2006 and International Conference on Intelligent Agents, Web Technologies and Internet Commerce*, page 179, 2006.

[216] S. Hatzopoulos, J. Cheng, A. Grzanka, and A. Martini. Time-frequency analyses of TEOAE recordings from normals and SNHL patients. *Audiology*, 39:1–12, 2000.

[217] T. He, G.D. Clifford, and L. Tarassenko. Application of ICA in removing artifacts from the ECG. *Neural Computing & Applications*, 15(2):105–116, 2006.

[218] M.J. Herrmann, T. Huter, M.M. Plichta, A. Christine Ehlis, G.W. Alpers, A. Mühlberger, and A.J. Fallgatter. Enhancement of activity of the primary visual cortex during processing of emotional stimuli as measured with event-related functional near-infrared spectroscopy and event-related potentials. *Human Brain Mapping*, 29(1):28–35, 2008.

[219] W. Hesse, E. Moller, M. Arnold, and B. Schack. The use of time-variant EEG Granger causality for inspecting directed interdependencies of neural assemblies. *Journal of Neuroscience Methods*, 124:27–44, Mar 2003.

[220] A.C. Heusser, D. Poeppel, Y. Ezzyat, and L. Davachi. Episodic sequence memory is supported by a theta-gamma phase code. *Nature Neuroscience*, 19(10):1374–1380, 2016.

[221] M. Higashihara, M. Sonoo, A. Ishiyama, Y. Nagashima, K. Matsumoto, H. Uesugi, M. Mori-Yoshimura, M. Murata, S. Murayama, and H. Komaki. Quantitative analysis of surface electromyography for pediatric neuromuscular disorders. *Muscle Nerve*, 58(6):824–827, 2018.

[222] C. C Hilgetag and A. Goulas. Is the brain really a small-world network? *Brain Structure & Function*, 221(4):2361–2366, May 2016. Edition: 2015/04/18. Publisher: Springer, Berlin, Heidelberg.

[223] B. Hjorth. An on-line transformation of EEG scalp potentials into orthogonal source derivations. *Electroencephalography and Clinical Neurophysiology*, 39:526–530, 1975.

[224] J. Hlinka, D. Hartman, N. Jajcay, D. Tomeček, J. Tintěra, and M. Paluš Small-world bias of correlation networks: From brain to climate. *Chaos*, 27(3):035812, Mar 2017.

[225] Y. Hochberg and A.C. Tamhane. *Multiple Comparison Procedures*. Wiley, New York, 1987.

[226] C. Hock, K. Villringer, F. Müller-Spahn, R. Wenzel, H. Heekeren, S. Schuh-Hofer, M. Hofmann, S. Minoshima, M. Schwaiger, U. Dirnagl, and A. Villringer. Decrease in parietal cerebral hemoglobin oxygenation during performance of a verbal fluency task in patients with alzheimer's disease monitored by means of near-infrared spectroscopy (nirs) — correlation with simultaneous rcbf-pet measurements. *Brain Research*, 755(2):293–303, 1997.

[227] A. Holobar, D. Farina, M. Gazzoni, R. Merletti, and D. Zazula. Estimating motor unit discharge patterns from high-density surface electromyogram. *Electroencephalography and Clinical Neurophysiology*, 120:551–562, Mar 2009.

[228] A Holobar, MA Minetto, and D Farina. Accurate identification of motor unit discharge patterns from high-density surface emg and validation with a novel signal-based performance metric. *Journal of Neural Engineering*, 11(1):016008, February 2014.

[229] A. Holobar and D. Zazula. Multichannel blind source separation using convolution kernel compensation. *IEEE Transactions on Signal Processing*, 55:4487–4496, 2007.

[230] R. Hren, G. Stroink, and B.M. Horacek. Spatial resolution of body surface potential maps and magnetic field maps: a simulation study applied to the identification of ventricular pre-excitation sites. *Medical & Biological Engineering & Computing*, 36:145–157, Mar 1998.

[231] NE Huang, Z Shen, SR Long, MC Wu, HH Shih, Q Zheng, NC Yen, Ch Tung, and HH Liu. The empirical mode decomposition and the Hilbert spectrum for nonlinear and non-stationary time series analysis. *Proceedings of the Royal Society of London*, 454:903–995, 1998.

[232] N.P. Hughes. Probabilistic approaches to ECG segmentation and feature extraction. In G.D. Clifford, F. Azuaje, and P.E. McSharry, editors, *Advanced Methods and Tools for ECG Data Analysis*, pages 291–317. Artech House Inc, Norwood, MA, 2006.

[233] H.V. Huikuri, T.H. Makikallio, K.E. Airaksinen, T. Seppanen, P. Puukka, I.J. Raiha, and L.B. Sourander. Power-law relationship of heart rate variability as a predictor of mortality in the elderly. *Circulation*, 97:2031–2036, May 1998.

[234] M.D. Humphries and K. Gurney. Network 'small-world-ness': a quantitative method for determining canonical network equivalence. *PLoS ONE*, 3(4):e0002051, Apr 2008.

[235] T.J. Huppert, R.D. Hoge, S.G. Diamond, M.A. Franceschini, and D.A. Boas. A temporal comparison of bold, asl, and nirs hemodynamic responses to motor stimuli in adult humans. *NeuroImage*, 29(2):368–382, 2006.

[236] R.J. Huster, S. Debener, T. Eichele, and C.S. Herrmann. Methods for simultaneous EEG-fMRI: An introductory review. *Journal of Neuroscience*, 32(18):6053–6060, 2012.

[237] S. Hyde. Likelihood based inference on the Box-Cox family of transformations: Sas and Matlab programs. Technical report, Mathematical Sciences, Montana State University, 1999. Avaliable online at: http://www.math.montana.edu/hyde/papers/mspaper99.pdf.

[238] A. Hyvarinen. Fast and robust fixed-point algorithms for independent component analysis. *IEEE Transactions on Neural Networks*, 10:626–634, 1999.

[239] A. Hyvärinen and E. Oja. Independent component analysis: Algorithms and applications. *Neural Networks*, 13(4-5):411–430, 2000.

[240] L.D. Iasemidis, D.S. Shiau, W. Chaovalitwongse, J.C. Sackellares, P. M. Pardalos, J.C. Principe, P.R. Carney, A. Prasad, B. Veeramani, and K. Tsakalis. Adaptive epileptic seizure prediction system. *IEEE Transactions on Biomedical Engineering*, 50:616–627, May 2003.

[241] L. Inzelberg and Y. Hanein. Electrophysiology meets printed electronics: The beginning of a beautiful friendship. *Frontiers in Neuroscience*, 12:992, 2019.

[242] K. Izzetoglu, S. Bunce, M. Izzetoglu, B. Onaral, and K. Pourrezaei. Functional near-infrared neuroimaging. In *The 26th Annual International Conference of the IEEE Engineering in Medicine and Biology Society*, volume 2, pages 5333–5336, 2004.

[243] M. Izzetoglu, S. Bunce, K. Izzetoglu, B. Onaral, and A. Pourrezaei. Functional brain imaging using near-infrared technology. *IEEE Engineering in Medicine and Biology Magazine*, 26(4):38–46, 2007.

[244] M.G. Jafari and J.A. Chambers. Fetal electrocardiogram extraction by sequential source separation in the wavelet domain. *IEEE Transactions on Biomedical Engineering*, 52:390–400, Mar 2005.

[245] F. Jager. ST analysis. In G.D. Clifford, F. Azuaje, and P.E. McSharry, editors, *Advanced Methods and Tools for ECG Data Analysis*, pages 269–290. Artech House Inc, Norwood, MA, 2006.

[246] F. Jager, G.B. Moody, and R.G. Mark. Detection of transient ST segment episodes during ambulatory ECG monitoring. *Computers and Biomedical Research*, 31:305–322, Oct 1998.

[247] A. Janušauskas, V. Marozas, B. Engdahl, H.J. Hoffman, O. Svensson, and L. Sornmo. Otoacoustic emissions and improved pass/fail separation using wavelet analysis and time windowing. *Medical & Biological Engineering & Computing*, 39:134–139, Jan 2001.

[248] W.W. Jedrzejczak, K.J. Blinowska, K. Kochanek, and H. Skarzynski. Synchronized spontaneous otoacoustic emissions analyzed in a time-frequency domain. *Journal of the Acoustical Society of America*, 124:3720–3729, Dec 2008.

[249] W.W. Jedrzejczak, K.J. Blinowska, and W. Konopka. Time-frequency analysis of transiently evoked otoacoustic emissions of subjects exposed to noise. *Hear Res*, 205:249–255, Jul 2005.

[250] W.W. Jedrzejczak, K.J. Blinowska, and W. Konopka. Resonant modes in transiently evoked otoacoustic emissions and asymmetries between left and right ear. *J Acoust Soc Am*, 119:2226–2231, Apr 2006.

[251] W.W. Jedrzejczak, K.J. Blinowska, W. Konopka, A. Grzanka, and P.J. Durka. Identification of otoacoustic emissions components by means of adaptive approximations. *J Acoust Soc Am*, 115:2148–2158, May 2004.

[252] W.W. Jedrzejczak, S. Hatzopoulos, A. Martini, and K.J. Blinowska. Otoacoustic emissions latency difference between full-term and preterm neonates. *Hear Res*, 231:54–62, Sep 2007.

[253] W.W. Jedrzejczak, K. Kwaskiewicz, K.J. Blinowska, K. Kochanek, and H. Skarzynski. Use of the matching pursuit algorithm with a dictionary of asymmetric waveforms in the analysis of transient evoked otoacoustic emissions. *J Acoust Soc Am*, 126:3137–3146, Dec 2009.

[254] K.K. Jerger, S.L. Weinstein, T. Sauer, and S.J. Schiff. Multivariate linear discrimination of seizures. *Clinical Neurophysiology*, 116:545–551, 2005.

[255] P. Jezzard, P.M. Matthews, and S.M. Smith, editors. *Functional MRI: An Introduction to Methods*. Oxford University Press, Oxford, New York, 2001.

[256] M. Jobert. Pattern recognition by matched filtering; an analysis of sleep spindle and K-complex density under the influence of lormetazepam and zopiclone. *Electroencephalography and Clinical Neurophysiology*, 26:100–107, 1992.

[257] E.R. John, D.S. Ruchkin, and J.J. Vidal. *Event Related Brain Potentials in Man*, chapter Measurment of event related potentials, pages 99–138. Academic Press, 1978.

[258] E.R. John and R.W. Thatcher. *Neurometrics: Clinical Applications of Quantitative Electrophysiology*. Functional Neuroscience. Lawrence Erlbaum Associates, 1977.

[259] T. Jokiniemi, K. Simelius, J. Nenonen, I. Tierala, L. Toivonen, and T. Katilal. Baseline reconstruction for localization of rapid ventricular tachycardia from body surface potential maps. *Physiological Measurement*, 24(3):641–651, Aug 2003.

[260] T.H. Joo, J.H. McClellan, R.A. Foale, G.S. Myers, and R.S. Lees. Pole-zero modeling and classification of phonocardiograms. *IEEE Transactions on Biomedical Engineering*, 30:110–118, Feb 1983.

[261] J. Lee, S. Nemati, I. Silva, B. A Edwards, J. P Butler, and A. Malhotra. Transfer Entropy Estimation and Directional Coupling Change Detection in Biomedical Time Series. *BioMedical Engineering OnLine*, 11(19):1–17, 2012.

[262] J. Jorge, W. van der Zwaag, and P. Figueiredo. EEG-fMRI integration for the study of human brain function. *NeuroImage*, 102 Pt 1:24–34, Nov 2014.

[263] C.C. Jouny, P.J. Franaszczuk, and G.K. Bergey. Characterization of epileptic seizure dynamics using Gabor atom density. *Clinical Neurophysiology*, 114:426–437, Mar 2003.

[264] C.C. Jouny, P.J. Franaszczuk, and G.K. Bergey. Signal complexity and synchrony of epileptic seizures: is there an identifiable preictal period? *Clin Neurophysiol*, 116:552–558, Mar 2005.

[265] C.A. Joyce, I.F. Gorodnitsky, and M. Kutas. Automatic removal of eye movement and blink artifacts from EEG data using blind component separation. *Psychophysiology*, 41:313–325, Mar 2004.

[266] C. Julien. The enigma of mayer waves: Facts and models. *Cardiovascular Research*, 70(1):12–21, Apr 2006.

[267] G.J. Jurkiewicz, M.J. Hunt, and J. Żygierewicz. Addressing pitfalls in phase-amplitude coupling analysis with an extended modulation index toolbox. *Neuroinformatics*, Aug 2020.

[268] R. Jurkko, V. Mantynen, J.M. Tapanainen, J. Montonen, H. Vaananen, H. Parikka, and L. Toivonen. Non-invasive detection of conduction pathways to left atrium using magnetocardiography: validation by intracardiac electroanatomic mapping. *Europace*, 11:169–177, Feb 2009.

[269] S. Kalitzin, J. Parra, D.N. Velis, and F.H. Lopes da Silva. Enhancement of phase clustering in the EEG/MEG gamma frequency band anticipates transitions to paroxysmal epileptiform activity in epileptic patients with known visual sensitivity. *IEEE Transactions on Biomedical Engineering*, 49:1279–1286, Nov 2002.

[270] S. Kalitzin, D. Velis, P. Suffczynski, J. Parra, and F.L. da Silva. Electrical brain-stimulation paradigm for estimating the seizure onset site and the time to ictal transition in temporal lobe epilepsy. *Clin Neurophysiol*, 116:718–728, Mar 2005.

[271] M. Kaminski and K.J. Blinowska. Is Graph Theoretical Analysis a Useful Tool for Quantification of Connectivity Obtained by Means of EEG/MEG Techniques? *Frontiers in Neural Circuits*, 12:76, 2018.

[272] M. Kaminski and K.J. Blinowska. A new method of the description of the information flow in brain structures. *Biol Cybern*, 65:203–210, 1991.

[273] M. Kaminski, K.J. Blinowska, and W. Szelenberger. Topographic analysis of coherence and propagation of EEG activity during sleep and wakefulness. *Electroencephalography and Clinical Neurophysiology*, 102:216–227, 1997.

[274] M. Kaminski, A. Brzezicka, J. Kaminski, and K.J. Blinowska. Coupling Between Brain Structures During Visual and Auditory Working Memory Tasks. *International Journal of Neural Systems*, 29(3):1850046, Apr 2019.

[275] M. Kaminski, M. Ding, W. Truccolo, and S. Bressler. Evaluating causal relations in neural systems: Granger causality, directed transfer function and statistical assessment of significance. *Biol Cybern*, 85:145–157, 2001.

[276] M. Kaminski, P. Szerling, and K.J. Blinowska. Comparison of methods for estimation of time-varying transmission in multichannel data. In *Proc. 10th IEEE International Conference on Information Technology and Applications in Biomedicine*, Korfu, 2010.

[277] P.P. Kanjilal, S. Palit, and G. Saha. Fetal ECG extraction from single-channel maternal ECG using singular value decomposition. *IEEE Trans Biomed Eng*, 44:51–59, Jan 1997.

[278] H. Kantz and T. Schreiber. *Nonlinear Time Series Analysis*. Cambridge University Press, Cambridge, 2000.

[279] E.C. Karvounis, M.G. Tsipouras, D.I. Fotiadis, and K.K. Naka. An automated methodology for fetal heart rate extraction from the abdominal electrocardiogram. *IEEE Transactions on Information Technology*, 11:628–638, Nov 2007.

[280] EC Karvounis, MG Tsipouras, and DI Fotiadis. Detection of fetal heart rate through 3-D phase space analysis from multivariate abdominal recordings. *IEEE Transactions on Biomedical Engineering*, 56(5):1394–1406, 2009.

[281] A. Katz, D.A. Marks, G. McCarthy, and S.S. Spencer. Does interictal spiking change prior to seizures? *Electroencephalography and Clinical Neurophysiology*, 79(2):153–156, 1991.

[282] S.M. Kay. *Modern Spectral Estimation: Theory and Application*. Prentice-Hall, Englewood Cliffs, New Jersey, USA, 1988.

[283] S. Keilholz. The neural basis of time-varying resting-state functional connectivity. *Brain Connectivity*, 4(10):769–779, 2014.

[284] B. Kemp, P. Jaspers, J.M. Franzen, and A.J.M.W. Janssen. An optimal monitor of the electroencephalographic sigma sleep state. *Biological Cybernetics*, 51:263–270, 1985.

[285] D.T. Kemp. Stimulated acoustic emissions from within the human auditory system. *Journal of the Acoustical Society of America*, 64:1386–1391, Nov 1978.

[286] B. Kerous, F. Skola, and F. Liarokapis. EEG-based BCI and video games: a progress report. *Virtual Reality*, 22(2):119–135, 2018.

[287] B. Khaddoumi, H. Rix, O. Meste, M. Fereniec, and R. Maniewski. Body surface ECG signal shape dispersion. *IEEE Trans Biomed Eng*, 53(12):2491–2500, Dec 2006.

[288] A. Khamene and S. Negahdaripour. A new method for the extraction of fetal ECG from the composite abdominal signal. *IEEE Transactions on Biomedical Engineering.*, 47:507–516, Apr 2000.

[289] JM Kilner, J Mattout, R Henson, and KJ Friston. Hemodynamic correlates of EEG: a heuristic. *NeuroImage*, 28(1):280–286, Oct 2005.

[290] V. Kiviniemi, Juha-Heikki Kantola, J. Jauhiainen, A. Hyvärinen, and O. Tervonen. Independent component analysis of nondeterministic fMRI signal sources. *NeuroImage*, 19:253–260, 2003.

[291] B.U. Kleine, J.P. van Dijk, B.G. Lapatki, M.J. Zwarts, and D.F. Stegeman. Using two-dimensional spatial information in decomposition of surface EMG signals. *Journal of Electromyography & Kinesiology*, 17(5):535–548, Oct 2007.

[292] A. Kleinschmidt, H. Obrig, M. Requardt, Kl-D. Merboldt, U. Dirnagl, A. Villringer, and J. Frahm. Simultaneous recording of cerebral blood oxygenation changes during human brain activation by magnetic resonance imaging and near-infrared spectroscopy. *Journal of Cerebral Blood Flow & Metabolism*, 16(5):817–826, 1996.

[293] K. Klimaszewska and J.J. Zebrowski. Detection of the type of intermittency using characteristic patterns in recurrence plots. *Physical review. E*, 80:026214, Aug 2009.

[294] T. Klingenheben, P. Ptaszynski, and S.H. Hohnloser. Quantitative assessment of microvolt T-wave alternans in patients with congestive heart failure. *J Cardiovasc Electrophysiol*, 16:620–624, Jun 2005.

[295] B. Kocsis and M. Kaminski. Dynamic changes in the direction of the

theta rhythmic drive between supramammillary nucleus and the septo-hippocampal system. *Hippocampus*, 16:531–540, 2006.

[296] J.A. Kors, G. van Herpen, J. Wu, Z. Zhang, R.J. Prineas, and J.H. van Bemmel. Validation of a new computer program for Minnesota coding. *Journal of Electrocardiology*, 29(1):83–88, 1996.

[297] A. Korzeniewska, M. Cervenka, Christophe C. Jouny, J.R. Perilla, J. Harezlak, G. Bergey, P. Franaszczuk, and N. Crone. Ictal propagation of high frequency activity is recapitulated in interictal recordings: Effective connectivity of epileptogenic networks recorded with intracranial EEG. *NeuroImage*, 101:96–113, 2014.

[298] A. Korzeniewska, C.M. Crainiceanu, R. Kuś, P.J. Franaszczuk, and N. E. Crone. Dynamics of event-related causality in brain electrical activity. *Human Brain Mapping*, 29:1170–1192, Oct 2008.

[299] A. Korzeniewska, S. Kasicki, M. Kaminski, and K.J. Blinowska. Information flow between hippocampus and related structures during various types of rat's behavior. *Journal of Neuroscience Methods*, 73:49–60, 1997.

[300] A. Korzeniewska, M. Manczak, M. Kaminski, K.J. Blinowska, and S. Kasicki. Determination of information flow direction among brain structures by a modified directed transfer function (dDTF) method. *Journal of Neuroscience Methods*, 125:195–207, May 2003.

[301] D. Kosarov. Vectorelectromiographic control on the position of surface electrodes in a relation to active motor units in the human muscles. *Acta Physiological et Pharmacologica Bulgarica*, 1:85–93, 1974.

[302] E. Koutsoukos, E. Angelopoulos, A. Maillis, G.N. Papadimitriou, and C. Stefanis. Indication of increased phase coupling between theta and gamma EEG rhythms associated with the experience of auditory verbal hallucinations. *Neuroscience Letters*, 534:242–245, Feb 2013.

[303] H. Koymen, B.K. Altay, and Y.Z. Ider. A study of prosthetic heart valve sounds. *IEEE Transactions on Biomedical Engineering*, 34:853–863, Nov 1987.

[304] M.A. Kramer, A. B.L. Tort, and N.J. Kopell. Sharp edge artifacts and spurious coupling in EEG frequency comodulation measures. *Journal of Neuroscience Methods*, 170(2):352–357, May 2008.

[305] T. Kreuz, R.G. Andrzejak, F. Mormann, A. Kraskov, H. Stögbauer, C.E. Elger, K. Lehnertz, and P. Grassberger. Measure profile surrogates: A method to validate the performance of epileptic seizure prediction algorithms. *Physical Review E*, 69(6):061915, 2004.

[306] R. Krzyminiewski, G. Panek, and R. Stepien. Correlation of results of coronarographic, SPECT examination and high-resolution vectorcardiography. *European Medical & Biological Engineering & Computing*, 37(2):514–515, 1999.

[307] M. Kubat, G. Pfurtscheller, and D. Flotzinger. AI-based approach to automatic sleep classification. *Biological Cybernetics*, 70(5):443–448, 1994.

[308] S. Kubicki, L. Holler, I. Berg, C. Pastelak-Price, and R. Dorow. Sleep EEG evaluation: a comparison of results obtained by visual scoring and automatic analysis with the Oxford sleep stager. *Sleep*, 12:140–149, Apr 1989.

[309] S. Kullback and R.A. Leibler. On information and sufficiency. *Annals of Mathematical Statistics*, 22(1):79–86, 1951.

[310] R. I Kumar, M. M Mallette, S. S Cheung, D. W Stashuk, and D. A Gabriel. A method for editing motor unit potential trains obtained by decomposition of surface electromyographic signals. *Journal of electromyography and kinesiology : official journal of the International Society of Electrophysiological Kinesiology*, 50:102383, Feb 2020.

[311] C.D. Kurth, J.M. Steven, and S.C. Nicolson. Cerebral oxygenation during pediatric cardiac surgery using deep hypothermic circulatory arrest. *Anesthesiology*, 82(1):74–82, Jan 1995.

[312] R. Kus, J.S. Ginter, and K.J. Blinowska. Propagation of EEG activity during finger movement and its imagination. *Acta Neurobiologiae Experimentalis (Wars)*, 66:195–206, 2006.

[313] R. Kus, M. Kaminski, and K.J. Blinowska. Determination of EEG activity propagation: pair-wise versus multichannel estimate. *IEEE Trans Biomed Eng*, 51:1501–1510, 2004.

[314] R. Kus, D. Valbuena, J. Zygierewicz, T. Malechka, A. Graeser, and P. Durka. Asynchronous BCI based on motor imagery with automated calibration and neurofeedback training. *IEEE Transactions on Neural Systems and Rehabilitation Engineering*, 20(6), 2012.

[315] R. Kuś, P. T. Różański, and P. J. Durka. Multivariate matching pursuit in optimal Gabor dictionaries: theory and software with interface for EEG/MEG via Svarog. *BioMedical Engineering OnLine*, 12(1):94, September 2013.

[316] J. Lachaux, P.E. Rodriguez, J. Martinerie, and F.J. Varela. Measuring phase synchrony in brain signals. *Human brain mapping*, 8(4):194–208, 1999.

[317] J.-P. Lachaux, E. Rodriguez, J. Martinerie, C. Adam, D. Hasboun, and F. J. Varela. A quantitative study of gamma-band activity in human intracranial recordings triggered by visual simuli. *European Journal of Neuroscience*, 12:2608–2622, 2000.

[318] P Lachert, J Zygierewicz, D Janusek, P Pulawski, P Sawosz, M Kacprzak, A Liebert, and K J Blinowska. Causal coupling between electrophysiological signals, cerebral hemodynamics and systemic blood supply oscillations in mayer wave frequency range. *International Journal of Neural Systems*, 29(5):1850033, Jun 2019.

[319] P. Lachert, D. Janusek, P. Pulawski, A. Liebert, D. Milej, and K.J. Blinowska. Coupling of oxy- and deoxyhemoglobin concentrations with EEG rhythms during motor task. *Scientific Reports*, 7(1), nov 2017.

[320] M. Lagerholm, C. Peterson, G. Braccini, L. Edenbrandt, and L. Sornmo. Clustering ECG complexes using Hermite functions and self-organizing maps. *IEEE Trans Biomed Eng*, 47:838–848, Jul 2000.

[321] P. Laguna, R. Jane, and P. Caminal. Automatic detection of wave boundaries in multilead ECG signals: validation with the CSE database. *Computers and Biomedical Research*, 27:45–60, Feb 1994.

[322] P. Laguna, G.B. Moody, J. Garcia, A.L. Goldberger, R.G. Mark, and A. L. Goldberger. Analysis of the ST-T complex of the electrocardiogram using the Karhunen–Loève transform: adaptive monitoring and alternans detection. *Med Biol Eng Comput*, 37:175–189, Mar 1999.

[323] Y.C. Lai, M.A. Harrison, M.G. Frei, and I. Osorio. Inability of Lyapunov exponents to predict epileptic seizures. *Phys Rev Lett*, 91:068102, Aug 2003.

[324] R. Lamothe and G. Stroink. Orthogonal expansions: their applicability to signal extraction in electrophysiological mapping data. *Medical and Biological Engineering and Computing*, 29(5):522–528, 1991.

[325] H.H. Lange, J.P. Lieb, J. Engel, and P.H. Crandall. Temporo-spatial patterns of pre-ictal spike activity in human temporal lobe epilepsy. *Electroencephalography and Clinical Neurophysiology*, 56:543–555, Dec 1983.

[326] H. Laufs. A personalized history of EEG-fMRI integration. *Neuroimage*, 62(2):1056–1067, Aug 2012.

[327] S. Leanderson, P. Laguna, and L. Sornmo. Estimation of the respiratory frequency using spatial information in the VCG. *Medical Engineering & Physics*, 25:501–507, Jul 2003.

[328] J. Lee and A.S. Gevins. Method to reduce blur distortion from EEGs using a realistic head model. *IEEE Trans Biomed Eng*, 6:517–528, 1993.

[329] R.S. LeFever and C.J. De Luca. A procedure for decomposing the myoelectric signal into its constituent action potentials – Part I: Technique, theory, and implementation. *IEEE Trans Biomed Eng*, 29:149–157, Mar 1982.

[330] D.R. Leff, F. Orihuela-Espina, C.E. Elwell, T. Athanasiou, D.T. Delpy, A. W. Darzi, and G.Z. Yang. Assessment of the cerebral cortex during motor task behaviours in adults: a systematic review of functional near infrared spectroscopy (fNIRS) studies. *Neuroimage*, 54(4):2922–2936, Feb 2011.

[331] B. Lega, J. Burke, J. Jacobs, and M.J. Kahana. slow-theta-to-gamma phase-amplitude coupling in human hippocampus supports the formation of new episodic memories. *Cerebral Cortex*, 26(1):268–278, 2016.

[332] D. Lehmann and W. Skrandies. Reference-free identification of components of checkerboard-evoked multichannel potential fields. *Electroencephalography and Clinical Neurophysiology*, 48:609–621, Jun 1980.

[333] S.J. Leistedt, N. Coumans, M. Dumont, J.P. Lanquart, C.J. Stam, and P. Linkowski. Altered sleep brain functional connectivity in acutely depressed patients. *Hum Brain Mapp*, 30(7):2207–2219, Jul 2009.

[334] L. Leistritz, B. Pester, A. Doering, K. Schiecke, F. Babiloni, L. Astolfi, and H. Witte. Time-variant partial directed coherence for analysing connectivity: a methodological study. *Philosophical Transactions of the Royal Society A: Mathematical, Physical and Engineering Sciences*, 371(1997):20110616, 2013.

[335] L. Lemieux, P.J. Allen, F. Franconi, M. Symms, and D. Fish. Recording of EEG during fMRI experiments: Patient safety. *Medical & Biological Engineering & Computing*, 38, 1997.

[336] P. Lewandowski, O. Meste, R. Maniewski, T. Mroczka, K. Steinbach, and H. Rix. Risk evaluation of ventricular tachycardia using wavelet transform irregularity of the high-resolution electrocardiogram. *Medical & Biological Engineering & Computing*, 38:666–673, Nov 2000.

[337] C. Li, D.K. Tang, D.A. Zheng, G.H. Ding, C.S. Poon, and G.Q. Wu. Comparison of nonlinear indices in analyses of heart rate variability. In *Proc. 39th Annual International IEEE EMBES Conference*, pages 2145–2148, 2008.

[338] T. S Ligeza, M. Wyczesany, A. D Tymorek, and M. Kamiński. Interactions Between the Prefrontal Cortex and Attentional Systems During Volitional Affective Regulation: An Effective Connectivity Reappraisal Study. *Brain topography*, 29(2):253–261, 2016.

[339] H.W. Lilliefors. On the Kolmogorov-Smirnov test for normality with mean and variance unknown. *Journal of the American Statistical Association*, 62:399–402, 1967.

[340] H.K. Lim, H. Kwon, N. Chung, Y.G. Ko, J.M. Kim, I.S. Kim, and Y.K. Park. Usefulness of magnetocardiogram to detect unstable angina pectoris and non-ST elevation myocardial infarction. *Am J Cardiol*, 103:448–454, Feb 2009.

[341] M. Lindquist, J.M. Loh, L. Atlas, and T. Wager. Modeling the hemodynamic response function in fMRI: Efficiency, bias and mis-modeling. *NeuroImage*, 45:s187–s198, 2009.

[342] L. Lindstrom, R. Magnusson, and I. Petersen. Muscular fatigue and action potential conduction velocity changes studied with frequency analysis of EMG signals. *Electromyography*, 10:341–356, 1970.

[343] W.H. Linssen, D.F. Stegeman, E.M. Joosten, M.A. van't Hof, R.A. Binkhorst, and S.L. Notermans. Variability and interrelationships of surface EMG parameters during local muscle fatigue. *Muscle Nerve*, 16(8):849–856, Aug 1993.

[344] J. Lisman. The theta/gamma discrete phase code occuring during the hippocampal phase precession may be a more general brain coding scheme. *Hippocampus*, 15(7):913–922, 2005.

[345] A.D. Liston, J.C. De Munck, K. Hamandi, H. Laufs, P. Ossenblok, J.S. Duncan, and L. Lemieux. Analysis of EEG-fMRI data in focal epilepsy based on automated spike classification and Signal Space Projection. *Neuroimage*, 31:1015–1024, Jul 2006.

[346] G.Z. Liu, D. Wu, G.R. Zhao, B-Y Huang, Z-Y Mei, Guo Y-W, and L. Wang. Use of refined sample entropy and heart rate variability to assess the effects of wearable respiratory biofeedback. In *3rd International Conference on Biomedical Engineering and Informatics (BMEI), 2010*, pages 1915–1919, Oct. 2010.

[347] N. Logothetis. What we can do and what we cannot do with fMRI. *Nature*, 453:869–878, 2008.

[348] N.K. Logothetis, J. Pauls, M. Augath, T. Trinath, and A. Oeltermann. Neurophysiological investigation of the basis of the fMRI signal. *Nature*, 412(6843):150–157, Jul 2001.

[349] NK Logothetis, J Pauls, M Augath, T Trinath, and A Oeltermann. Neurophysiological investigation of the basis of the fMRI signal. *Nature*, 412(6843):150–157, July 2001.

[350] F. Lombardi, G. Sandrone, S. Pernpruner, R. Sala, M. Garimoldi, S. Cerutti, G. Baselli, M. Pagani, and A. Malliani. Heart rate variability as an index of sympathovagal interaction after acute myocardial infarction. *Am J Cardiol*, 60:1239–1245, Dec 1987.

[351] F.H. Lopes da Silva, W. Blanes, S.N. Kalitzin, J. Parra, P. Suffczynski, and D.N. Velis. Dynamical diseases of brain systems: different routes to epileptic seizures. *IEEE Trans Biomed Eng*, 50:540–548, May 2003.

[352] F.H. Lopes da Silva. *Comprehensive Human Physiology*, chapter The Generation of Electric and Magnetic Signals of the Brain by Local Networks. Springer-Verlag, 1996.

[353] F.H. Lopes da Silva, W. Blanes, S.N. Kalitzin, J. Parra, and D.N. Suffczynski, P. Velis. Epilepsies as dynamical diseases of brain systems: Basic models of the transition between normal and epileptic activity. *Epilepsia*, 44(s12):72–83, 2003.

[354] F. Lotte, L. Bougrain, A. Cichocki, M. Clerc, M. Congedo, A. Rakotomamonjy, and F. Yger. A review of classification algorithms for EEG-based brain-computer interfaces: A 10 year update. *Journal of Neural Engineering*, 15(3), 2018.

[355] C.M. Lu, Y.J. Zhang, B.B. Biswal, Y.F. Zang, D.L. Peng, and C.Z. Zhu. Use of fNIRS to assess resting state functional connectivity. *Journal of Neuroscience Methods*, 186(2):242–249, Feb 2010.

[356] S.J. Luck. *An Introduction to Event-Related Potentials Technique*. MIT Press, Cambridge, MA, 2005.

[357] P. Lundin, S.V. Eriksson, L.E. Strandberg, and N. Rehnqvist. Prognostic information from on-line vectorcardiography in acute myocardial infarction. *Am J Cardiol*, 74:1103–1108, Dec 1994.

[358] H. Lutkepohl. *Introduction to Multiple Time Series Analysis*. Springer Verlag, Berlin, Heidelberg, 1993.

[359] M. Kaminski, A. Brzezicka, J. Kaminski, and K.J. Blinowska Information transfer during auditory working memory task. In Kyriacou E., Christofides S., and Pattichis C., editors, *XIV Mediterranean Conference on Medical and Biological Engineering and Computing 2016*, volume 57. Springer, Cham., 2016.

[360] M. Kaminski and K. J. Blinowska The influence of volume conduction on dtf estimate and the problem of its mitigation. *Frontiers in Computational Neuroscience*, 11:36, 2017.

[361] B. J MacIntosh, L. M. Klassen, and R. S Menon. Transient hemodynamics during a breath hold challenge in a two part functional imaging
study with simultaneous near-infrared spectroscopy in adult humans.
NeuroImage, 20(2):1246–1252, 2003.

[362] R. Maestri, G.D. Pinna, A. Porta, R. Balocchi, R. Sassi, M.G. Signorini,
M. Dudziak, and G. Raczak. Assessing nonlinear properties of heart rate
variability from short-term recordings: are these measurements reliable?
Physiological Measurement, 28:1067–1077, Sep 2007.

[363] J. N. Mak and J. R. Wolpaw. Clinical Applications of Brain—Computer
Interfaces: Current State and Future Prospects. *IEEE Reviews in
Biomedical Engineering*, 2:187–199, 2009.

[364] S. Makeig. Auditory event-related dynamics of the EEG spectrum and
effects of exposure to tones. *Electroencephalography and Clinical Neurophysiology*, 86:283–293, 1993.

[365] S. Makeig. EEGLAB: ICA toolbox for psychophysiological research.
World Wide Web Publication, https://sccn.ucsd.edu/~scott/ica.html,
2000. WWW Site, Swartz Center for Computational Neuroscience, Institute of Neural Computation, University of San Diego, California.

[366] S. Makeig, A.J. Bell, T-P Jung, and T.J. Sejnowski. Independent
component analysis of electroencephalographic data. In D. Touretzky,
M. Mozer, and M. Hasselmo, editors, *Advances in Neural Information
Processing Systems*, volume 8, pages 145–151. MIT Press, Cambridge,
MA, 1996.

[367] S. Makeig, S. Debener, J. Onton, and A. Delorme. Mining event-related
brain dynamics. *Trends in Cognitive Sciences (Regul Ed)*, 8:204–210,
May 2004.

[368] S. Makeig, T.P. Jung, A.J. Bell, D. Ghahremani, and T.J. Sejnowski.
Blind separation of auditory event-related brain responses into independent components. *Proc Natl Acad Sci USA*, 94:10979–10984, Sep 1997.

[369] A Maliani, A Pagani, F. Lombardi, and S. Cerutti. Cardiovascular neural
regulation explored in the frequency domain. *Circulation*, 84:482–492,
1991.

[370] M. Malik. Heart rate variability: standards of measurement, physiological interpretation and clinical use. Task Force of the European Society
of Cardiology and the North American Society of Pacing and Electrophysiology. *Circulation*, 93:1043–1065, Mar 1996.

[371] M. Malik and A.J. Camm. *Heart Rate Variability*. Futura Publishng,
Armonk, NY, 1995.

[372] U. Malinowska, P.J. Durka, K. Blinowska, W. Szelenberger, and A. Wakarow. Micro- and macrostructure of sleep EEG. *IEEE BME Magazine*, 25:26–31, 2006.

[373] U. Malinowska, P.J. Durka, J. Zygierewicz, W. Szelenberger, and A. Wakarow. Explicit parameterization of sleep EEG transients. *Computers in Biology and Medicine*, 37:534–541, 2007.

[374] U. Malinowska, H. Klekowicz, A. Wakarow, S. Niemcewicz, and P.J. Durka. Fully parametric sleep staging compatible with the classic criteria. *Neuroinformatics*, 7:245–253, 2009.

[375] S. Mallat. A theory for multiresolution signal decomposition: the wavelet representation. *IEEE Transactions on Pattern Analysis and Machine Intelligence*, 11(7):674–693, 1989.

[376] S. Mallat. *A Wavelet Tour of Signal Processing*. Academic Press, London, 1999.

[377] S.G. Mallat and Z. Zhang. Matching pursuits with time-frequency dictionaries. *IEEE Transactions on Signal Processing*, pages 3397–3415, 1993.

[378] J Malmivuo and R. Plonsey. *Bioelectromagnetism–Principles and Applications of Bioelectric and Biomagnetic Fields*. Oxford University Press, New York, 1995.

[379] V. Marozas, A. Janusauskas, A. Lukosevicius, and L. Sornmo. Multiscale detection of transient evoked otoacoustic emissions. *IEEE Transactions on Biomedical Engineering*, 53(8):1586–1593, Aug 2006.

[380] S. L. Jr Marple. *Digital Spectral Analysis with Applications*. Prentice-Hall Inc., Englewood Cliffs, NJ, 1987.

[381] S.M. Martens, C. Rabotti, M. Mischi, and R.J. Sluijter. A robust fetal ECG detection method for abdominal recordings. *Physiological Measurement*, 28:373–388, Apr 2007.

[382] W. Martin, L. Johnson, S.S. Viglione, and et al. Pattern recognition of EEG-EOG as a technique for all-night sleep stage scoring. *Electroencephalography and Clinical Neurophysiology*, 32(4):17–427, 1972.

[383] J. Martinerie, C. Adam, M. Le Van Quyen, M. Baulac, S. Clemenceau, B. Renault, and F.J. Varela. Epileptic seizures can be anticipated by non-linear analysis. *Nat Med*, 4:1173–1176, Oct 1998.

[384] M. Martinez, E. Soria, J. Calpe, J.F. Guerrero, and J.R. Magdalena. Application of the adaptive impulse correlated filter for recovering fetal electrocardiogram. In *Computers in Cardiology 1997*, pages 9–12. IEEE Comput. Soc. Press, 1997.

[385] E. Martínez-Montes, P. A Valdés-Sosa, F. Miwakeichi, R. I Goldman, and M. S Cohen. Concurrent EEG-fMRI analysis by multiway partial least squares. *Neuroimage*, 22(3):1023–1034, Jul 2004.

[386] E. Martínez-Valdés, F. Negro, C.M. Laine, D. Falla, F. Mayer, and D. Farina. Tracking motor units longitudinally across experimental sessions with high-density surface electromyography. *The Journal of Physiology*, 595:1479–1496, 2017.

[387] N. Marwan, N. Wessel, U. Meyerfeldt, A. Schirdewan, and J. Kurths. Recurrence-plot-based measures of complexity and their application to heart-rate-variability data. *Phys Rev E*, 66(2):026702, 2002.

[388] M. Matousek and I. Petersen. *Automation of clinical Electroencephalography*. Raven Press, New York, 1973.

[389] LK Matsuoka and SS Spencer. Seizure localization using subdural grid electrodes. *Epilepsia*, 34(Suppl.6):8, 1993.

[390] S. Mayer. *Studien zur Physiologie des Herzens und der Blutgefässe. Fünfte. Abh..: Über spontane Blutdruckschwankungen*. Sitzungsberichte Akademie der Wissenschaften in Wien, 1876.

[391] L. Mazurkiewicz and M. Piotrkiewicz. Computer system for identification and analysis of motor unit potential trains. *Biocybernetics and Biomedical Engineering*, 24(6):15–23, 2004.

[392] M.J. Mc Keown, C. Humphries, P. Achermann, and et al. A new method for detecting state changes in the EEG: exploratory application to sleep data. *Journal of Sleep Research*, 7:48–56, 1998.

[393] K.C. McGill, K.L. Cummins, and L.J. Dorfman. Automatic decomposition of the clinical electromyogram. *IEEE Transactions on Biomedical Engineering*, 32(7):470–477, Jul 1985.

[394] K.C. McGill, Z.C. Lateva, and H.R. Marateb. EMGLAB: an interactive EMG decomposition program. *Journal of Neuroscience Methods*, 149:121–133, Dec 2005.

[395] C.D. McGillem and J.I. Aunon. Measurements of signal components in single visually evoked brain potentials. *IEEE Trans Biomed Eng*, 24:232–241, May 1977.

[396] M. McKeown and T. Sejnowski. Independent component analysis of fMRI data: examining the assumptions. *Human Brain Mapping*, 6, 1998.

[397] P.E. McSharry, L.A. Smith, and L. Tarassenko. Prediction of epileptic seizures: are nonlinear methods relevant? *Nat Med*, 9:241–242, Mar 2003.

[398] J. Medvedev and O. Willoughby. Autoregressive modeling of the EEG in systemic kainic acid-induced epileptogenesis. *International Journal of Neuroscience*, 97:149–67, 1999.

[399] G.D. Meekins, Y. So, and D. Quan. Electrodiagnostic Medicine evidenced-based review: use of surface electromyography in the diagnosis and study of neuromuscular disorders. *Muscle Nerve*, 38:1219–1224, Oct 2008.

[400] V Menon, J M Ford, K O Lim, G H Glover, and A Pfefferbaum. Combined event-related fMRI and EEG evidence for temporal-parietal cortex activation during target detection. *Neuroreport*, 8(14):3029–3037, Sep 1997.

[401] R. Merletti, A. Holobar, and D. Farina. Analysis of motor units with high-density surface electromyography. *Journal of Electromyography & Kinesiology*, 18:879–890, Dec 2008.

[402] R. Merletti and S. Muceli. Tutorial. surface EMG detection in space and time: Best practices. *Journal of Electromyography and Kinesiology*, 49:102363, 2019.

[403] O. Meste, G. Blain, and S. Bermon. Analysis of the respiratory and cardiac systems coupling in pyramidal exercise using a time-varying model. In *Proc. Computers in Cardiology*, volume 29, pages 429–432. IEEE Press, 2002.

[404] B. Mijović, K. Vanderperren, N. Novitskiy, B. Vanrumste, P. Stiers, B. Van den Bergh, L. Lagae, S. Sunaert, J. Wagemans, S. Van Huffel, and M. De Vos. The "why" and "how" of jointica: results from a visual detection task. *NeuroImage*, 60(2):1171–1185, Apr 2012.

[405] K.J. Miller, D. Hermes, C.J. Honey, M. Sharma, R. P.N. Rao, M. den Nijs, E.E. Fetz, T.J. Sejnowski, A.O. Hebb, J.G. Ojemann, S. Makeig, and E.C. Leuthardt. Dynamic modulation of local population activity by rhythm phase in human occipital cortex during a visual search task. *Frontiers in Human Neuroscience*, 4(October):1–16, 2010.

[406] K. Minami, H. Nakajima, and T. Toyoshima. Real-time discrimination of ventricular tachyarrhythmia with Fourier-transform neural network. *IEEE Transactions on Biomedical Engineering*, 46:179–185, Feb 1999.

[407] P. Mitraszewski, K.J. Blinowska, P.J. Franaszczuk, and M. Kowalczyk. A study of electrocortical rhythm generators. *Biol Cybern*, 56:255–260, 1987.

[408] A. Moleti and R. Sisto. Objective estimates of cochlear tuning by otoacoustic emission analysis. *J Acoust Soc Am*, 113:423–429, Jan 2003.

[409] E. Moller, B. Schack, M. Arnold, and H. Witte. Instantaneous multivariate EEG coherence analysis by means of adaptive high-dimensional autoregressive models. *Journal of Neuroscience Methods*, 105:143–158, Feb 2001.

[410] A. Montalto, L. Faes, and D. Marinazzo. Mute: a MATLAB toolbox to compare established and novel estimators of the multivariate transfer entropy. *PLoS ONE*, 9:10:e109462, 2014.

[411] M. Monti. Statistical analysis of fMRI time-series: a critical review of the GLM approach. *Frontiers in Human Neuroscience*, 5, 2011.

[412] G.B. Moody and R.G. Mark. QRS morphology representation and noise estimation using the Karhunen-Loéve transform. In *Proc. Computers in Cardiology*, pages 269–272. IEEE Press, 1990.

[413] M. Moosmann, T. Eichele, H. Nordby, K. Hugdahl, and V.D. Calhoun. Joint independent component analysis for simultaneous EEG-fMRI: principle and simulation. *International Journal of Psychophysiology*, 67:212–221, Mar 2008.

[414] F. Mormann, T. Kreuz, C. Rieke, RG. Andrzejak, A. Kraskov, P. David, CE. Elger, and K. Lehnertz. On the predictability of epileptic seizures. *Clinical Neurophysiology*, 116(3):569–87, 2005.

[415] F. Mormann, R.G. Andrzejak, C.E. Elger, and K. Lehnertz. Seizure prediction: the long and winding road. *Brain*, 130(2):314–333, 2007.

[416] A. Mouraux and G.D. Iannetti. Across-trial averaging of event-related EEG responses and beyond. *Magnetic Resonance Imaging*, 26:1041–1054, Sep 2008.

[417] K. Mullinger and R. Bowtell. Combining EEG and fMRI. *Methods in Molecular Biology (Clifton, N.J.)*, 711:303–326, 2011.

[418] J. Munck, S. Gonçalves, L. Huijboom, J. Kuijer, P. Pouwels, R. Heethaar, and F. Silva. The hemodynamic response of the alpha rhythm: An EEG/fMRI study. *NeuroImage*, 35:1142–1151, 2007.

[419] M.M. Murray, D. Brunet, and C.M. Michel. Topographic ERP analyses: a step-by-step tutorial review. *Brain Topography*, 20:249–264, Jun 2008.

[420] H. Nakamura, M. Yoshida, M. Kotani, K. Akazawa, and T. Moritani. The application of independent component analysis to the multi-channel surface electromyographic signals for separation of motor unit action potential trains: part I-measuring techniques. *Journal of Electromyography & Kinesiology*, 14:423–432, Aug 2004.

[421] A. Nakhnikian, S. Ito, L.L. Dwiel, L.M. Grasse, G.V. Rebec, L.N. Lauridsen, and J.M. Beggs. A novel cross-frequency coupling detection method using the generalized Morse wavelets. *Journal of Neuroscience Methods*, 269:61–73, Aug 2016.

[422] S.M. Narayan. The pathophysiology guided assessment of t-wave alternans. In G.D. Clifford, F. Azuaje, and P.E. McSharry, editors, *Advanced Methods and Tools for ECG Data Analysis*, pages 197–214. Artech House Inc, Norwood, MA, 2006.

[423] N. Naseer and K.S. Hong. fNIRS-based brain-computer interfaces: a review. *Frontiers in Human Neuroscience*, 9:3, 2015.

[424] S.H. Nawab, S.S. Chang, and C.J. De Luca. High-yield decomposition of surface EMG signals. *Clinical Neurophysiology*, 121:1602–1615, Oct 2010.

[425] F. Negro, S. Muceli, A.M. Castronovo, A. Holobar, and D. Farina. Multichannel intramuscular and surface EMG decomposition by convolutive blind source separation. *Journal of Neural Engineering*, 13(2):026027, Apr 2016.

[426] J. Neonen, K. Pesola, H. Hanninen, K. Lauerma, P. Takala, T. Makela, M. Makijarvi, L. Knuuti, J. Toivonen, and T. Katila. Current-density estimation of exercise induced ischemia in patients with multivessel coronary artery disease. *Journal of Electrocardiology*, 34:37–42, 2001.

[427] T.I. Netoff, L. Caroll, L.M. Pecora, and S.J. Schiff. Detecting coupling in the presence of noise and nonlinearity. In J. Timmer B. Scheleter, M. Winterhalder, editors, *Handbook of Time Series Analysis*. Wiley VCH, 2006.

[428] M.E. Newman. Mixing patterns in networks. *Physical review. E*, 67(2 Pt 2):026126, Feb 2003.

[429] M.E. Newman and M. Girvan. Finding and evaluating community structure in networks. *Physical review. E*, 69(2 Pt 2):026113, Feb 2004.

[430] M.E.J. Newman. The structure and function of complex networks. *SIAM Rev*, 45:167–256, 2003.

[431] D.K. Nguyen, J. Tremblay, P. Pouliot, P. Vannasing, O. Florea, L. Carmant, F. Lepore, M. Sawan, F. Lesage, and M. Lassonde. Non-invasive continuous EEG-fNIRS recording of temporal lobe seizures. *Epilepsy Res*, 99(1–2):112–126, Mar 2012.

[432] R K Niazy, C F Beckmann, G D Iannetti, J M Brady, and S M Smith. Removal of fMRI environment artifacts from EEG data using optimal basis sets. *Neuroimage*, 28(3):720–737, Nov 2005.

[433] E. Niedermayer and F.H. Lopes da Silva. *Electroencephalography, Basic Principles, Clinical Applications, and Related Fields.* Williams & Wilkins, Baltimore, 5th edition, 2004.

[434] J. Niessing, B. Ebisch, K. E Schmidt, M. Niessing, W. Singer, and R. A W Galuske. Hemodynamic signals correlate tightly with synchronized gamma oscillations. *Science,* 309(5736):948–951, Aug 2005.

[435] S. Nieuwenhuis, G. Aston-Jones, and J.D. Cohen. Decision making, the P3, and the locus coeruleus-norepinephrine system. *Psychological Bulletin,* 131:510–532, Jul 2005.

[436] J. Ning, N. Atanasov, and T. Ning. Quantitative analysis of heart sounds and systolic heart murmurs using wavelet transform and AR modeling. *Conference proceedings - IEEE engineering in medicine and biology society,* 2009:958–961, 2009.

[437] G. Nolte, O. Bai, L. Wheaton, Z. Mari, S. Vorbach, and M. Hallett. Identifying true brain interaction from EEG data using the imaginary part of coherency. *Clinical Neurophysiology,* 115(10):2292–2307, Oct 2004.

[438] R.G. Noman, I. Pal, C. Steward, J.A. Walsleban, and D.M. Rapoport. Interobserver agreement among sleep scorers from different centers in a large dataset. *Sleep,* 23(7):901–908, 2000.

[439] G. Notaro, A.M. Al-Maamury, A. Moleti, and R. Sisto. Wavelet and matching pursuit estimates of the transient-evoked otoacoustic emission latency. *J Acoust Soc Am,* 122:3576–3585, Dec 2007.

[440] P.L. Nunez. *Electric Fields of the Brain.* Oxford University Press, New York, 1981.

[441] P.L. Nunez and K.L. Pilgreen. The spline-laplacian in clinical neurophysiology: a method to improve EEG spatial resolution. *Journal of Clinical Neurophysiology,* 8:397–413, 1991.

[442] P.L. Nunez and R. Sirnivasan. *Electric Fields of the Brain: The Neurophysics of EEG.* Oxford University Press, New York, 2nd edition, 2006.

[443] H. Obrig, C. Hirth, J.G. Junge-Hülsing, C. Döge, T. Wolf, U. Dirnagl, and A. Villringer. Cerebral oxygenation changes in response to motor stimulation. *J Appl Physiol,* 81(3):1174–1183, Sep 1996.

[444] H. Obrig, M. Neufang, R. Wenzel, M. Kohl, J. Steinbrink, K. Einhäupl, and A. Villringer. Spontaneous low frequency oscillations of cerebral hemodynamics and metabolism in human adults. *Neuroimage,* 12(6):623–639, Dec 2000.

[445] H. Obrig, H.R. Heekeren, J. Ruben, R. Wenzel, J.-P. Ndayisaba, U. Dirnagl, and A. Villringer M.D. Continuous spectrum near-infrared spectroscopy approach in functional activation studies in the human adult. In I.J. Bigio, W.S. Grundfest M.D., H. Schneckenburger, K. Svanberg M.D., and P. M. Viallet, editors, *Optical Biopsies and Microscopic Techniques*, volume 2926, pages 58–66. International Society for Optics and Photonics, SPIE, 1996.

[446] J. Onton and S. Makeig. Information-based modeling of event-related brain dynamics. *Progress in Brain Research*, 159:99–120, 2006.

[447] S. Orfanidis. *Optimum Signal Processing*. Macmillan, 1988. Chapter 5.

[448] M.R. Ortiz, E.R. Bojorges, S.D. Aguilar, J.C. Echeverría, R. Gonzalez-Camarena, S. Carrasco, M.J. Gaitan, and A. Martınez. Analysis of high frequency fetal heart rate variability using empirical mode decomposition. *Computers in Cardiology*, 32:675–678, 2005.

[449] S. Osowski, L.T. Hoai, and T. Markiewicz. Supervised learning methods for ECG classification/neural networks and SVM approaches. In G.D. Clifford, F. Azuaje, and P.E. McSharry, editors, *Advanced Methods and Tools for ECG Data Analysis*, pages 319–337. Artech House Inc, Norwood, MA, 2006.

[450] W. Ou, A. Nummenmaa, J. Ahveninen, J.W. Belliveau, M.S. Hämäläinen, and P. Golland. Multimodal functional imaging using fMRI-informed regional EEG/MEG source estimation. *NeuroImage*, 52(1):97 –108, 2010.

[451] O. Ozdamar, J. Zhang, T. Kalayci, and Y. Ulgen. Time-frequency distribution of evoked otoacoustic emissions. *British Journal of Audiology*, 31:461–471, Dec 1997.

[452] T.E. Özkurt and A. Schnitzler. A critical note on the definition of phase–amplitude cross-frequency coupling. *Journal of Neuroscience Methods*, 201(2):438–443, Oct 2011.

[453] O. Pahlm and L. Sornmo. Software QRS detection in ambulatory monitoring–a review. *Medical & Biological Engineering & Computing*, 22:289–297, Jul 1984.

[454] J. Pan and W.J. Tompkins. A real-time QRS detection algorithm. *IEEE Trans Biomed Eng*, 32:230–236, Mar 1985.

[455] A. Pantelopoulos and N. Bourbakis. Efficient single-lead ECG beat classification using matching pursuit based features and artificial neural network. In *Proc of Inter Confrence on Information technology and Application in Biomedicine*, 2010.

[456] D. Papo, M. Zanin, J.H. Martínez, and J.M. Buldú Beware of the Small-World Neuroscientist! *Frontiers in Human Neuroscience*, 10:96, 2016.

[457] J. Pardey, S.J. Roberts, L. Tarassnko, and J. Stradling. A new approach to the analysis of the human sleep-wakefulness continuum. *Journal of Sleep Research*, 5:201–210, 1996.

[458] T.W. Park and C.S. Burrus. *Digital Filter Design*. John Wiley & Sons, 1987.

[459] J. Parra, S.N. Kalitzin, J. Iriarte, W. Blanes, D.N. Velis, and F. H. Lopes da Silva. Gamma-band phase clustering and photosensitivity: is there an underlying mechanism common to photosensitive epilepsy and visual perception? *Brain*, 126:1164–1172, May 2003.

[460] R.D. Pascual-Marqui. Standardized low-resolution brain electromagnetic tomography (sLORETA): technical details. *Methods and Findings in Experimental and Clinical Pharmacology*, 24 Suppl D:5–12, 2002.

[461] D. Pascucci, M. Rubega, and G. Plomp. Modeling time-varying brain networks with a self-tuning optimized kalman filter. *PLOS Computational Biology*, 16(8):1–29, 08 2020.

[462] C.S. Pattichis and A.G. Elia. Autoregressive and cepstral analyses of motor unit action potentials. *Medical Engineering & Physics*, 21:405–419, 1999.

[463] C.S. Pattichis and M.S. Pattichis. Time-scale analysis of motor unit action potentials. *IEEE Transactions on Biomedical Engineering.*, 46:1320–1329, Nov 1999.

[464] A. Paulini, M. Fischer, S. Rampp, G. Scheler, R. Hopfengärtner, M. Kaltenhäuser, A. Dörfler, M. Buchfelder, and H. Stefan. Lobar localization information in epilepsy patients: MEG—a useful tool in routine presurgical diagnosis. *Epilepsy research*, 76(2):124–130, 2007.

[465] C.K. Peng, S. Havlin, H.E. Stanley, A.L. Goldberger, and A.L. Goldberger. Quantification of scaling exponents and crossover phenomena in nonstationary heartbeat time series. *Chaos*, 5:82–87, 1995.

[466] T. Penzel, H. Hirshkowitz, J. Harsh, R. Chervin, N. Butkov, M. Kryger, and et al. Digital analysis and technical specifications. *Journal of Clinical Sleep Medicine*, 3(2):109–120, 2007.

[467] E. Pereda, R.Q. Quiroga, and J. Bhattacharya. Nonlinear multivariate analysis of neurophysiological signals. *Progress in Neurobiology*, 77:1–37, 2005.

[468] E. Pereda, R. Rial, A. Gamundi, and J. Gonzalez. Assessment of changing interdependences between human electroencephalograms using nonlinear methods. *Physica D*, 148:147–158, 2001.

[469] J.S. Perkiomaki, T.H. Makikallio, and H.V. Huikuri. Fractal and complexity measures of heart rate variability. *Clin Exp Hypertens*, 27:149–158, 2005.

[470] K. Pesola, J. Nenonen, R. Fenici, J. Lotjonen, M. Makijarvi, P. Fenici, P. Korhonen, K. Lauerma, M. Valkonen, L. Toivonen, and T. Katila. Bioelectromagnetic localization of a pacing catheter in the heart. *Physics in Medicine & Biology*, 44:2565–2578, Oct 1999.

[471] G. Pfurtscheller. *Digital Biosignal Processing*, chapter Mapping procedures, pages 459–479. Elsevier, 1991.

[472] G. Pfurtscheller. Quantification of ERD and ERS in the time domain. In G. Pfurtscheller and F.H. Lopes da Silva, editors, *Event-related Desynchronization*, volume 6, pages 89–105. Elsevier, 1999.

[473] G. Pfurtscheller and A. Aranibar. Evaluation of event-related desynchronization (ERD) preceding and following voluntary self-paced movement. *Electroencephalography and Clinical Neurophysiology*, 46:138–146, Feb 1979.

[474] G. Pfurtscheller and R. Cooper. Selective averaging of the intracerebral click evoked responses in man: an improved method of measuring latencies and amplitudes. *Electroencephalogr Clin Neurophysiol*, 38:187–190, Feb 1975.

[475] G. Pfurtscheller and F.H. Lopes da Silva. Event-related EEG/MEG synchronization and desynchronization: basic principles. *Clinical Neurophysiology*, 110:1842–1857, Nov 1999.

[476] G. Pfurtscheller, A.R. Schwerdtfeger, B. Rassler, A. Andrade, G. Schwarz, and W. Klimesch. Verification of a central pacemaker in brain stem by phase-coupling analysis between hr interval- and bold-oscillations in the 0.10–0.15 hz frequency band. *Frontiers in Neuroscience*, 14:922, 2020.

[477] T. Picton, S. Bentin, P. Berg, E. Donchin, S. Hillyard, R. Johnson, G. Miller, W. Ritter, D. Ruchkin, M. Rugg, and M. Taylor. Guidelines for using human event-related potentials to study cognition: Recording standards and publication criteria. *Psychophysiology*, 37:127–152, 2000.

[478] J.F. Pieri, J.A. Crowe, B.R. Hayes-Gill, C.J. Spencer, K. Bhogal, and D.K. James. Compact long-term recorder for the transabdominal foetal and maternal electrocardiogram. *Medical & Biological Engineering & Computing*, 39:118–125, Jan 2001.

[479] J.P. Pijn, J. Van Neerven, A. Noest, and F.H. Lopes da Silva. Chaos or noise in EEG signals; dependence on state and brain site. *Electroencephalography and Clinical Neurophysiology*, 79:371–381, Nov 1991.

[480] J.P. Pijn, D.N. Velis, M.J. van der Heyden, J. DeGoede, C.W. van Veelen, and F.H. Lopes da Silva. Nonlinear dynamics of epileptic seizures on basis of intracranial EEG recordings. *Brain Topogr*, 9:249–270, 1997.

[481] S.M. Pincus. Approximate entropy as a measure of system complexity. *Proc Natl Acad Sci USA*, 88:2297–2301, Mar 1991.

[482] R. Pineas, R. Crow, and H. Blackburn. *The Minnesota Code Manual of Electrocardiographic Findings*. John Wright-PSG, Inc., Littleton, MA, 1982.

[483] J.A. Pineda. Are neurotransmitter systems of subcortical origin relevant to the electrogenesis of cortical ERPs? *Electroencephalogr Clin Neurophysiol Suppl*, 44:143–150, 1995.

[484] S Pinkus and B.H. Singer. Randomness and degress of irregularity. *Proc. Natl. Acad Sci USA*, 93:2083–2088, 1996.

[485] M. Pinsky. *Introduction to Fourier Analysis and Wavelets*. Brooks/Cole, ISBN 0-534-37660-6 edition, 2002.

[486] M. Piotrkiewicz. Perspectives and drawbacks of surface electromyography. In K. Kedzior and A. Morecki, editors, *Lecture notes of the 46th ICB seminar Biomechanics of musculoskeletal system. Medical robotics*, pages 219–227, 1999.

[487] C.S. Poon. The chaos about heart rate chaos. *J Cardiovasc Electrophysiol*, 11:235–236, Feb 2000.

[488] A. Porta, G. Baselli, F. Lombardi, N. Montano, A. Malliani, and S. Cerutti. Conditional entropy approach for the evaluation of the coupling strength. *Biological Cybernetics*, 81:119–129, Aug 1999.

[489] A. Porta, S. Guzzetti, N. Montano, M. Pagani, V. Somers, A. Malliani, G. Baselli, and S. Cerutti. Information domain analysis of cardiovascular variability signals: evaluation of regularity, synchronization and co-ordination. *Medical & Biological Engineering & Computing*, 38:180–188, 2000.

[490] S.A. Prahl. Tabulated molar extinction coefficient for hemoglobin in water. accessed September, 2020.

[491] M.B. Priestley. *Spectral Analysis and Time Series*. Academic Press Inc., London, 1981.

[492] R. Probst, B.L. Lonsbury-Martin, and G.K. Martin. A review of otoacoustic emissions. *J Acoust Soc Am*, 89:2027–2067, May 1991.

[493] R. Quian Quiroga and E.L. van Luijtelaar. Habituation and sensitization in rat auditory evoked potentials: a single-trial analysis with wavelet denoising. *International Journal of Psychophysiology*, 43:141–153, Feb 2002.

[494] R.Q. Quiroga, J. Arnhold, and P. Grassberger. Learning driver-response relationships from synchronization patterns. *Physical Review E*, 61:5142–5148, May 2000.

[495] R.Q. Quiroga, R. Kraskov, T. Kreuz, and P. Grassberger. Performance of different synchronization measures in real data: a case study on electroencephalographic signals. *Physical Review E*, 65:041903, 2002.

[496] L.R. Rabiner. A tutorial on Hidden Markov Models and selected applications in speech recognition. *Proceedings of the IEEE*, 77(2):257–286, 1989.

[497] I.J. Rampil. EEG signal analysis in anesthesia. In S. Tong and N. Thakor, editors, *Quantitative EEG Analysis Methods and Clinical Applications*. Artech House, Norwood, MA, 2009.

[498] S. Ray, N.E. Crone, E. Niebur, P.J. Franaszczuk, and S.S. Hsiao. Neural correlates of high-gamma oscillations (60-200 Hz) in macaque local field potentials and their potential implications in electrocorticography. *J Neurosci*, 28:11526–11536, Nov 2008.

[499] S. Ray, C.C. Jouny, N.E. Crone, D. Boatman, N.V. Thakor, and P. J. Franaszczuk. Human ECoG analysis during speech perception using matching pursuit: a comparison between stochastic and dyadic dictionaries. *IEEE Transactions on Biomedical Engineering*, 50:1371–1373, Dec 2003.

[500] S. Ray, W. L.ee, C.D. Morgan, and W. Airth-Kindree. Computer sleep stage scoring–an expert system approach. *International Journal of Biomedical Computing*, 19(1):43–61, 1986.

[501] Kline R.B. *Structural Equation Modeling*. Guilford, New York, 1998.

[502] A. Rechtschaffen and A. Kales, editors. *A Manual of Standardized Terminology, Techniques and Scoring System for Sleep Stages in Human Subjects*. US Government Printing Office, Washington DC, 1968. Number 204 in National Institutes of Health Publications.

[503] J.S. Richman and J.R. Moorman. Physiological time-series analysis using approximate entropy and sample entropy. *Am J Physiol Heart Circ Physiol*, 278:H2039–2049, Jun 2000.

[504] M. Richter, T. Schreiber, and D.T. Kaplan. Fetal ECG extraction with nonlinear state-space projections. *IEEE Trans Biomed Eng*, 45:133–137, Jan 1998.

[505] B. Rivet, A. Souloumiac, V. Attina, and G. Gibert. xDAWN Algorithm to Enhance Evoked Potentials: Application to Brain-Computer Interface. *IEEE Transactions on Biomedical Engineering*, 56(8):2035–2043, 2009.

[506] S. Roberts and L. Tarassenko. New method of automated sleep quantification. *Med Biol Eng Comput*, 30:509–517, Sep 1992.

[507] M.S. Robinette and T.J. Glattke. *Otoacoustic Emissions Clinical Applications*. Thieme, New York, Stuttgart, 2002.

[508] F. Roche, V. Pichot, E. Sforza, I. Court-Fortune, D. Duverney, F. Costes, M. Garet, and J.C. Barthelemy. Predicting sleep apnoea syndrome from heart period: a time-frequency wavelet analysis. *Eur Respir J*, 22(6):937–942, Dec 2003.

[509] Z. Rogowski, I. Gath, and E. Bental. On the prediction of epileptic seizures. *Biological Cybernetics*, 42:9–15, 1981.

[510] A.C. Rosa, L. Parrino, and M.R. Terzano. Automatic detection of cyclic alternating pattern (CAP) sequences in sleep preliminary results. *Clinical Neurophysiology*, 110:585–592, 1999.

[511] M J Rosa, J Daunizeau, and K J Friston. EEG-fMRI integration: a critical review of biophysical modeling and data analysis approaches. *Journal of Integrative Neuroscience*, 9(4):453–476, Dec 2010.

[512] V. Rosik, M . Tysler, S. Jurko, R. Raso, and M. Turzowa. Portable system for ECG mapping. *Measurment Science Review*, 1:27–30, 2001.

[513] N. Rulkov, M.M. Sushchik, L.S. Tsimring, and H. D.I. Abarbanel. Generalized synchronization of chaos in directionally coupled chaotic systems. *Phys Rev E*, 51:980– 994, 1995.

[514] M. Rupawala, H. Dehghani, S.J.E. Lucas, P. Tino, and D. Cruse. Shining a light on awareness: A review of functional near-infrared spectroscopy for prolonged disorders of consciousness. *Frontiers in Neurology*, 9:350, 2018.

[515] D. Ruppert, M.P. Wand, and R.J. Carroll. *Semiparametric Regression*. Cambridge University Press, Cambridge, 2003.

[516] L. Rutter, S.R. Nadar, T. Holroyd, F.W. Carver, J. Apud, D.R. Weinberger, and R. Coppola. Graph theoretical analysis of resting magnetoencephalographic functional connectivity networks. *Frontiers in Computational Neuroscience*, 7:93, 2013.

[517] S.A. Huettel, A.W. Song, and G. Mc Carthy *Functonal Magnetic Resonance*. Sinauer, 2009.

[518] M.E. Saab and J. Gotman. A system to detect the onset of epileptic seizures in scalp EEG. *Clinical Neurophysiology*, 116:427–442, Feb 2005.

[519] T. Sadoyama and H. Miyano. Frequency analysis of surface EMG to evaluation of muscle fatigue. *European Journal of Applied Physiology and Occupational Physiology*, 47(3):239–246, 1981.

[520] I. Sahin, N. Yilmazer, and M.A. Simaan. A method for subsample fetal heart rate estimation under noisy conditions. *IEEE Transactions on Biomedical Engineering*, 57:875–883, Apr 2010.

[521] Y. Salant, I. Gath, and O. Henriksen. Prediction of epileptic seizures from two-channel EEG. *Medical and Biological Engineering and Computing*, 36:549–556, 1998.

[522] R. Sameni, G.D. Clifford, C. Jutten, and M. B Shamsollahi. Multichannel ECG and noise modeling: Application to maternal and fetal ECG signals. *EURASIP Journal on Advances in Signal Processing*, 2007, 2007. Article ID 043407.

[523] S. Samiee and S. Baillet. Time-resolved phase-amplitude coupling in neural oscillations. *NeuroImage*, 159:270–279, Oct 2017.

[524] T.H. Sander, A. Liebert, B.M. Mackert, H. Wabnitz, S. Leistner, G. Curio, M. Burghoff, R. Macdonald, and L. Trahms. DC-magnetoencephalography and time-resolved near-infrared spectroscopy combined to study neuronal and vascular brain responses. *Physiological Measurement*, 28(6):651–664, Jun 2007.

[525] D.T. Sandwell. Biharmonic spline interpolation of geos-3 and seasat altimeter data. *Geophysical Research Letters*, 2:139–142, 1987.

[526] S Sanei and J Chambers. *EEG Signal Processing*. Willey & Sons, Chichester, 2009.

[527] S. Sasai, F. Homae, H. Watanabe, and G. Taga. Frequency-specific functional connectivity in the brain during resting state revealed by NIRS. *Neuroimage*, 56(1):252–257, May 2011.

[528] J.R. Sato, D.Y. Takahashi, S.M. Arcuri, K. Sameshima, P.A. Morettin, and L.A. Baccalá. Frequency domain connectivity identification: an application of partial directed coherence in fMRI. *Human Brain Mapping*, 30(2):452–461, Feb 2009.

[529] B.M. Sayers, H.A. Beagley, and W.R. Henshall. The mechanism of auditory evoked EEG responses. *Nature*, 247:481–483, 1974.

[530] R. Scheeringa, M. C.M. Bastiaansen, K. M. Petersson, R. Oostenveld, D. G. Norris, and P. Hagoort. Frontal theta EEG activity correlates negatively with the default mode network in resting state. *International Journal of Psychophysiology*, 67(3):242–251, 2008. Integration of EEG and fMRI.

[531] R. Scheeringa, P. Fries, K.-M. Petersson, R. Oostenveld, I. Grothe, D. G. Norris, P. Hagoort, and M. C M Bastiaansen. Neuronal dynamics underlying high- and low-frequency EEG oscillations contribute independently to the human bold signal. *Neuron*, 69(3):572–583, Feb 2011.

[532] R. Scheeringa, K. M. Petersson, A. Kleinschmidt, O. Jensen, and M. C M Bastiaansen. EEG α power modulation of fMRI resting-state connectivity. *Brain Connectivity*, 2(5):254–264, 2012.

[533] R. Scheeringa, K. M. Petersson, R. Oostenveld, D. G Norris, P. Hagoort, and M. C M Bastiaansen. Trial-by-trial coupling between EEG and bold identifies networks related to alpha and theta EEG power increases during working memory maintenance. *Neuroimage*, 44(3):1224–1238, Feb 2009.

[534] B. Schelter, J. Timmer, and M. Eichler. Assessing the strength of directed influences among neural signals using renormalized partial directed coherence. *Journal of Neuroscience Methods*, 179:121–130, Apr 2009.

[535] S.J. Schiff, T. Sauer, R. Kumar, and S.L. Weinstein. Neuronal spatiotemporal pattern discrimination: the dynamical evolution of seizures. *Neuroimage*, 28:1043–1055, Dec 2005.

[536] K. Schindler, H. Leung, C.E. Elger, and K. Lehnertz. Assessing seizure dynamics by analysing the correlation structure of multichannel intracranial EEG. *Brain*, 130:65–77, Jan 2007.

[537] A. Schlögl, P. Anderer, M.J. Barbanoj, G. Klösch, G. Gruber, J.L. Lorenzo, et al. Artefact processing of the sleep EEG in the "siesta"-project. *Proceedings EMBEC'99*, pages 1644–1645, 1999.

[538] T. Schreiber. Measuring information transfer. *Phys Rev Lett*, 85:461–464, 2000.

[539] C.E. Schroeder and P. Lakatos. Low-frequency neuronal oscillations as instruments of sensory selection. *Trends in Neurosciences*, 32(1), Jan 2009.

[540] M.L. Schroeter, M.M. Bücheler, K. Müller, K. Uludağ, H. Obrig, G. Lohmann, M. Tittgemeyer, A. Villringer, and D.Y. von Cramon. Towards a standard analysis for functional near-infrared imaging. *Neuroimage*, 21(1):283–90, Jan 2004.

[541] A.N. Sen, S.P. Gopinath, and C.S. Robertson. Clinical application of near-infrared spectroscopy in patients with traumatic brain injury: a review of the progress of the field. *Neurophotonics*, 3(3):031409, Jul 2016.

[542] L. Senhadji, G. Carrault, J.J. Bellanger, and G. Passariello. Comparing wavelet transforms for recognizing cardiac patterns. *IEEE Engineering in Medicine and Biology Magazine*, 14:167–173, 1995.

[543] C. Serio. Discriminating low-dimensional chaos from randomness: A parametric time series modelling approach. *Il Nuovo Cimento B (1971-1996)*, 107:681–701, 1992.

[544] C.E. Shannon. Communication in the presence of noise. *Proc. Institute of Radio Engineers*, 37(1):10–21, 1949. Reprint as classic paper in: Proc. IEEE, Vol. 86, No. 2, (Feb 1998).

[545] C.E. Shannon. A mathematical theory of communication. *Bell System Technical Journal*, 27:379–423, 1948.

[546] C.A. Shera and J.J. Guinan. Evoked otoacoustic emissions arise by two fundamentally different mechanisms: a taxonomy for mammalian OAEs. *J Acoust Soc Am*, 105:782–798, Feb 1999.

[547] A. Shoeb, H. Edwards, J. Connolly, B. Bourgeois, S.T. Treves, and J. Guttag. Patient-specific seizure onset detection. *Epilepsy & Behavior*, 5:483–498, Aug 2004.

[548] C. Sielużycki, R. König, A. Matysiak, R. Kuś, D. Ircha, and P. J. Durka. Single-trial evoked brain responses modeled by multivariate matching pursuit. *IEEE Transactions on Biomedical Engineering*, 56:74–82, Jan 2009.

[549] C. Sielużycki, R. Kuś, A. Matysiak, P.J. Durka, and R. König. Multivariate matching pursuit in the analysis of single-trial latency of the auditory M100 acquired with MEG. *International Journal of Bioelectromagnetism*, 11(4):155–160, 2009.

[550] D.F. Silva and G.E. A. P.A. Batista. *Proceedings of the 2016 SIAM International Conference on Data Mining*, chapter Speeding Up All-Pairwise Dynamic Time Warping Matrix Calculation, pages 837–845. SIAM, 2016.

[551] R. Simes. An improved Bonferroni procedure for multiple tests of significance. *Biometrika*, 73:751–754, 1986.

[552] A. Sinai, N.E. Crone, H.M. Wied, P.J. Franaszczuk, D. Miglioretti, and D. Boatman-Reich. Intracranial mapping of auditory perception: event-related responses and electrocortical stimulation. *Clinical Neurophysiology*, 120:140–149, Jan 2009.

[553] W. Singer. Synchronization of cortical activity and its putative role in information processing and learning. *Annu Rev Physiol*, 55:349–374, 1993.

[554] R. Sisto and A. Moleti. Modeling otoacoustic emissions by active nonlinear oscillators. *J Acoust Soc Am*, 106:1893–1906, Oct 1999.

[555] R. Sisto and A. Moleti. On the frequency dependence of the otoacoustic emission latency in hypoacoustic and normal ears. *J Acoust Soc Am*, 111:297–308, Jan 2002.

[556] W. Skrandies. Data reduction of multichannel fields: global field power and principal component analysis. *Brain Topography*, 2:73–80, 1989.

[557] W. Skrandies. Brain mapping of visual evoked activity–topographical and functional components. *Acta Neurologica Taiwanica*, 14:164–178, Dec 2005.

[558] J. Smith and I. Karacan. EEG sleep stage scoring by an automatic hybrid system. *Electroencephalography and Clinical Neurophysiology*, 31(3):231–237, 1971.

[559] S.M. Smith, M. Jenkinson, M.W. Woolrich, C.F. Beckmann, T. E.J. Behrens, H. Johansen-Berg, P.R. Bannister, M. De Luca, I. Drobnjak, D.E. Flitney, R.K. Niazy, J. Saunders, J. Vickers, Y. Zhang, N. De Stefano, J.M. Brady, and P.M. Matthews. Advances in functional and structural MR image analysis and implementation as FSL. *NeuroImage*, 23:S208–S219, 2004. Mathematics in Brain Imaging.

[560] D.E. Smylie, G.L. Clarke, and Ulrych T.J. Analysis of irregularities in the earth's rotation. *Comput Phys*, 13:391–430, 1973.

[561] E.N. Sokolov. Neuronal models and the orienting response. In M.A. Brazier, editor, *The Central Nervous System and Behavior III*. Macy Foundation, 1960.

[562] M. Sood, P. Besson, M. Muthalib, U. Jindal, S. Perrey, A. Dutta, and M. Hayashibe. NIRS-EEG joint imaging during transcranial direct current stimulation: online parameter estimation with an autoregressive model. *J Neurosci Methods*, 274:71–80, Dec 2016.

[563] E.P. Souza Neto, M.A. Custaud, J.C. Cejka, P. Abry, J. Frutoso, C. Gharib, and P. Flandrin. Assessment of cardiovascular autonomic control by the empirical mode decomposition. *Methods of Information in Medicine*, 43:60–65, 2004.

[564] H. Spekreijse, O. Esteves, and O. Reits. Visual evoked potentials and the physiological analysis of visual processes in man. In J.E. Desmedt, editor, *Visual Evoked Potentials in Man*, pages 3–15. Clarendon Press, 1976.

[565] C. Stam, J.P.M. Pijn, P. Suffczyński, and F.H. Lopes da Silva. Dynamics of the human alpha rhythm: evidence for non-linearity? *Clinical Neurophysiology*, 110:1801–1813, 1999.

[566] C.J. Stam, W. de Haan, A. Daffertshofer, B.F. Jones, I. Manshanden, A.M. van Cappellen van Walsum, T. Montez, J.P. Verbunt, J.C. de Munck, B.W. van Dijk, H.W. Berendse, and P. Scheltens. Graph theoretical analysis of magnetoencephalographic functional connectivity in Alzheimer's disease. *Brain*, 132(Pt 1):213–224, Jan 2009.

[567] C.J. Stam, B.F. Jones, G. Nolte, M. Breakspear, and P. Scheltens. Small-world networks and functional connectivity in Alzheimer's disease. *Cerebral Cortex*, 17(1):92–99, Jan 2007.

[568] C.J. Stam and B.W. Van Dijk. Synchronization likelihood: An unbiased measure of generalized synchronization in multivariate data sets. *Physica D: Nonlinear Phenomena*, 2002.

[569] D.W. Stashuk. Decomposition and quantitative analysis of clinical electromyographic signals. *Medical Engineering & Physics*, 21:389–404, 1999.

[570] P.D. Stein, H.N. Sabbah, J.B. Lakier, D.J. Magilligan, and D. Goldstein. Frequency of the first heart sound in the assessment of stiffening of mitral bioprosthetic valves. *Circulation*, 63:200–203, Jan 1981.

[571] G. Strang and T. Nguyen. *Wavelets and Filter Banks*. Wellesley-Cambridge Press, 1996.

[572] G. Stroink. Forty years of magnetocardiology. In S. Supek and A. Susac, editors, *Advances in Biomagnetism BIOMAG2010, IFMBE Proceedings*, volume 28, pages 431–435. Springer, 2010.

[573] G. Stroink, R.J. Meeder, P. Elliott, J. Lant, and M.J. Gardner. Arrhythmia vulnerability assessment using magnetic field maps and body surface potential maps. *Pacing and Clinical Electrophysiology*, 22:1718–1728, Dec 1999.

[574] S.C. Strother. Evaluating fMRI preprocessing pipelines. *IEEE Engineering in Medicine and Biology Magazine*, 25(2):27–41, 2006.

[575] S.M. Stufflebeam, N. Tanaka, and S.P. Ahlfors. Clinical applications of magnetoencephalography. *Human Brain Mapping*, 30:1813–1823, Jun 2009.

[576] P. Suffczynski, S. Kalitzin, and F.H. Lopes Da Silva. Dynamics of non-convulsive epileptic phenomena modeled by a bistable neuronal network. *Neuroscience*, 126:467–484, 2004.

[577] P. Suffczynski, S. Kalitzin, F. Lopes da Silva, J. Parra, D. Velis, and F. Wendling. Active paradigms of seizure anticipation: computer model evidence for necessity of stimulation. *Phys Rev E*, 78(5):051917, 2008.

[578] M.R. Symms, P.J. Allen, F.G. Woermann, G. Polizzi, K. Krakow, G.J. Barker, D.R. Fish, and J.S. Duncan. Reproducible localization of interictal epileptiform discharges using EEG-triggered fMRI. *Physics in Medicine & Biology*, 44:N161–168, Jul 1999.

[579] S.M. Szczepanski, N.E. Crone, R.A. Kuperman, K.I. Auguste, J. Parvizi, and R.T. Knight. Dynamic changes in phase-amplitude coupling facilitate spatial attention control in fronto-parietal cortex. *PLoS Biology*, 12(8), 2014.

[580] B. Taccardi. Distribution of heart potentials on dog's thoracic surface. *Circ Res*, 11(5):862–869, Nov 1962.

[581] B. Taccardi. Distribution of heart potentials on the thoracic surface of normal human subjects. *Circ Res*, 12(4):341–352, Apr 1963.

[582] A. Taddei, M. Emdin, M. Varanini, A. Macerate, P. Pisani, E. Santarcangelp, and C. Marchesi. An approach to cardiorespiratory activity monitoring through principal component analysis. *Journal of Ambulatory Monitoring*, 5(2–3):167–173, 1993.

[583] D.Y. Takahashi, L.A. Baccala, and K. Sameshima. Connectivity inference between neural structures via partial directed coherence. *Journal of Applied Statistics*, 34(10):1255–1269, 2007.

[584] F. Takens. Detecting strange attractors in turbulence. *Lecture Notes in Mathematics*, 898:366–381, 1981.

[585] M. Takeuchi, E. Hori, K. Takamoto, A.H. Tran, K. Satoru, A. Ishikawa, T. Ono, S. Endo, and H. Nishijo. Brain cortical mapping by simultaneous recording of functional near infrared spectroscopy and electroencephalograms from the whole brain during right median nerve stimulation. *Brain Topography*, 22(3):197–214, Nov 2009.

[586] C. Tallon-Baudry. Oscillatory synchrony and human visual cognition. *Journal of Physiology (Paris)*, 97:355–363, 2003.

[587] C. Tallon-Baudry, O. Bertrand, C. Delpuech, and J. Pernier. Stimulus specificity of phase-locked and non-phase-locked 40Hz visual responses in human. *Journal of Neuroscience*, 16(13):4240–4249, July 1996.

[588] C.L. Talmadge, A. Tubis, G.R. Long, and C. Tong. Modeling the combined effects of basilar membrane nonlinearity and roughness on stimulus frequency otoacoustic emission fine structure. *J Acoust Soc Am*, 108:2911–2932, Dec 2000.

[589] M.G. Tana, R. Sclocco, and A.M. Bianchi. GMAC: a MATLAB toolbox for spectral granger causality analysis of fMRI data. *Computers in Biology and Medicine*, 42(10):943–956, Oct 2012.

[590] H. Tang, T. Li, Y. Park, and T. Qiu. Separation of heart sound signal from noise in joint cycle frequency-time-frequency domains based on fuzzy detection. *IEEE Transactions on Biomedical Engineering*, 57(10):2438–2447, Oct 2010.

[591] D.D. Taralunga, W. Wolf, R. Strungaru, and G. Ungureanu. Abdominal fetal ECG enhancement by event synchronous canceller. *Conf Proc IEEE Eng Med Biol Soc*, 2008:5402–5405, 2008.

[592] M. Ten Hoopen. Variance in average response computation: regular versus irregular stimulation. In A. Remond, editor, *Handbook of Electroencephalography and Clinical Neurophysiology. Part A*, volume 8, pages 151–158. Elsevier, Amsterdam, 1975.

[593] N. Thakor, X. Jia, and R.G. Geocardin. Monitoring neurological injury by qEEG. In S. Tong and N. Thakor, editors, *Quantitative EEG Analysis Methods and Clinical Applications*. Artech House, Norwood, MA, 2009.

[594] J. Theiler, S. Eubank, A. Longtin, B. Galdrikian, and J.D. Farmer. Testing for nonlinearity in time series: the method of surrogate data. *Physica D*, 58((1-4)):77–94, 1992.

[595] J. Theiler. Spurious dimension from correlation algorithms applied to limited time-series data. *Phys Rev A*, 34:2427–2432, Sep 1986.

[596] R.F. Thompson and W.A. Spencer. Habituation: a model phenomenon for the study of neuronal substrates of behavior. *Psychol Rev*, 73:16–43, 1966.

[597] D.J. Thomson. Spectrum estimation and harmonic analysis. *Proceedings of the IEEE*, 70:1055–1096, 1982.

[598] S. Thurner, M.C. Feurstein, and M.C. Teich. Multiresolution wavelet analysis of heartbeat intervals discriminates healthy patients from those with cardiac pathology. *Phys Rev Lett*, 80(7):1544–1547, Feb 1998.

[599] G. Tognola, F. Grandori, and P. Ravazzani. Time-frequency distributions of click-evoked otoacoustic emissions. *Hear Res*, 106:112–122, Apr 1997.

[600] G. Tognola, M. Parazzini, P. de Jager, P. Brienesse, P. Ravazzani, and F. Grandori. Cochlear maturation and otoacoustic emissions in preterm infants: a time-frequency approach. *Hear Res*, 199:71–80, Jan 2005.

[601] E. Toledo, O. Gurevitz, H. Hod, M. Eldar, and S. Akselrod. Wavelet

analysis of instantaneous heart rate: a study of autonomic control during thrombolysis. *Am J Physiol Regul Integr Comp Physiol*, 284(4):1079–1091, Apr 2003.

[602] S. Tong, B. Hong, L. Vigderman, and N.V. Thakor. Subband EEG complexity after global hypoxic-ischemic brain injury. *Conf Proc IEEE Eng Med Biol Soc*, 1:562–565, 2004.

[603] V. Toronov, A. Webb, J.H. Choi, M. Wolf, A. Michalos, E. Gratton, and D. Hueber. Investigation of human brain hemodynamics by simultaneous near-infrared spectroscopy and functional magnetic resonance imaging. *Medical Physics*, 28(4):521–527, Apr 2001.

[604] A.B.L. Tort, M.A. Kramer, C. Thorn, Daniel J. Gibson, Yasuo Kubota, Ann M. Graybiel, and Nancy J. Kopell. Dynamic cross-frequency couplings of local field potential oscillations in rat striatum and hippocampus during performance of a T-maze task. *Proceedings of the National Academy of Sciences of the United States of America*, 105(51):20517–20522, 2008.

[605] A.B.L. Tort, R. Komorowski, H. Eichenbaum, and N. Kopell. Measuring phase-amplitude coupling between neuronal oscillations of different frequencies. *Journal of Neurophysiology*, 104(2):1195–1210, May 2010.

[606] A.B.L. Tort, R.W. Komorowski, J.R. Manns, N.J. Kopell, and H. Eichenbaum. Theta-gamma coupling increases during the learning of item-context associations. *Proceedings of the National Academy of Sciences of the United States of America*, 106(49):20942–20947, December 2009.

[607] G. Tropini, J. Chiang, Z. Wang, and M.J. McKeown. Partial directed coherence-based information flow in Parkinson's disease patients performing a visually-guided motor task. *Annu Int Conf IEEE Eng Med Biol Soc*, 2009:1873–1878, 2009.

[608] G.L. Turin. An introduction to matched filters. *IRE Transactions on Information Theory*, 6(3):311–329, 1960.

[609] T.J. Ulrych and T.N. Bishop. Maximum entropy spectral analysis and autoregressive decomposition. *Rev Geophys Space Phys*, 13:183–200, 1975.

[610] A. Urbano, C. Babiloni, P. Onorati, F. Carducci, A. Ambrosini, L. Fattorini, and F. Babiloni. Responses of human primary sensorimotor and supplementary motor areas to internally triggered unilateral and simultaneous bilateral one-digit movements. A high-resolution EEG study. *Eur J Neurosci*, 10:765–770, Feb 1998.

[611] K. Urbanowicz, J. Zebrowski, R. Baranowski, and J.A. Holyst. How random is your heart beat? *Physica A: Statistical Mechanics and its Applications*, 384(2):439–447, 2007.

[612] Pedro A Valdés-Sosa, Jose M Sánchez-Bornot, Agustín Lage-Castellanos, Mayrim Vega-Hernández, Jorge Bosch-Bayard, Lester Melie-García, and Erick Canales-Rodríguez. Estimating brain functional connectivity with sparse multivariate autoregression. *Philosophical Transactions of the Royal Society of London. Series B, Biological Sciences*, 360(1457):969–981, May 2005.

[613] M.P. van den Heuvel, C.J. Stam, M. Boersma, and H.E. Hulshoff Pol. Small-world and scale-free organization of voxel-based resting-state functional connectivity in the human brain. *Neuroimage*, 43(3):528–539, Nov 2008.

[614] M.P van den Heuvel and H.E Hulshoff Pol. Exploring the brain network: a review on resting-state fMRI functional connectivity. *European Neuropsychopharmacology*, 20(8):519–534, Aug 2010.

[615] J. van Driel, R. Cox, and M. X Cohen. Phase-clustering bias in phase-amplitude cross-frequency coupling and its removal. *Journal of Neuroscience Methods*, 254:60–72, Oct 2015.

[616] B.C.M. van Wijk, M. Beudel, A. Jha, A. Oswal, T. Foltynie, M.I. Hariz, P. Limousin, L. Zrinzo, T.Z. Aziz, A.L. Green, P. Brown, and V. Litvak. Subthalamic nucleus phase-amplitude coupling correlates with motor impairment in Parkinson's disease. *Clinical Neurophysiology*, 127(4):2010–2019, 2016.

[617] S Vanni, J Warnking, M Dojat, C Delon-Martin, J Bullier, and C Segebarth. Sequence of pattern onset responses in the human visual areas: an fMRI constrained vep source analysis. *Neuroimage*, 21(3):801–817, Mar 2004.

[618] A. Villringer and B. Chance. Non-invasive optical spectroscopy and imaging of human brain function. *Trends in Neurosciences*, 20(10):435–442, Oct 1997.

[619] A.J. Viterbi. Error bounds for convolutional codes and an asymptotically optimum decoding algorithm. *IEEE Transactions on Information Theory*, 13(2):260–269, 1967.

[620] M. von Spreckelsen and B. Bromm. Estimation of single-evoked cerebral potentials by means of parametric modeling and Kalman filtering. *IEEE Trans Biomed Eng*, 35:691–700, Sep 1988.

[621] B. Voytek, R.T. Canolty, A. Shestyuk, N.E. Crone, J. Parvizi, and R.T. Knight. Shifts in gamma phase-amplitude coupling frequency from theta to alpha over posterior cortex during visual tasks. *Frontiers in Human Neuroscience*, 4(October):1–9, 2010.

[622] B. Voytek, M. D'Esposito, N. Crone, and R.T. Knight. A method for event-related phase/amplitude coupling. *NeuroImage*, 64:416–424, Jan 2013.

[623] S. Waldert, M. Bensch, M. Bogdan, W. Rosenstiel, B. Scholkopf, C.L. Lowery, H. Eswaran, and H. Preissl. Real-time fetal heart monitoring in biomagnetic measurements using adaptive real-time ICA. *IEEE Trans Biomed Eng*, 54:1867–1874, Oct 2007.

[624] F. Wallois, M. Mahmoudzadeh, A. Patil, and R. Grebe. Usefulness of simultaneous EEG-NIRS recording in language studies. *Brain and Language*, 121(2):110–123, May 2012.

[625] D.O. Walter. A posteriori Wiener filtering of average evoked response. *Electroenceph Clin Neurophysiol*, 27:61–70, 1969.

[626] D.O. Walter and W.R. Adey. Spectral analysis of EEG recorded during learning in cat before and after subthalamic lesions. *Experimental Neurology*, 7:481, 1963.

[627] H. Wang and F. Azuaje. An introduction to unsupervised learning for ECG classification. In G.D. Clifford, F. Azuaje, and P.E. McSharry, editors, *Advanced Methods and Tools for ECG Data Analysis*, pages 339–366. Artech House Inc, Norwood, MA, 2006.

[628] W. Wang, Z. Guo, J. Yang, Y. Zhang, L.G. Durand, and M. Loew. Analysis of the first heart sound using the matching pursuit method. *Medical & Biological Engineering & Computing*, 39(6):644–648, Nov 2001.

[629] K.R. Ward, R.R. Ivatury, R.W. Barbee, J. Terner, R. Pittman, I.P. Filho, and B. Spiess. Near infrared spectroscopy for evaluation of the trauma patient: a technology review. *Resuscitation*, 68(1):27–44, Jan 2006.

[630] D.J. Watts and S.H. Strogatz. Collective dynamics of 'small-world' networks. *Nature*, 393(6684):440–442, Jun 1998.

[631] N,R. Waytowich, V.J. Lawhern, A.W. Bohannon, K.R. Ball, and B.J. Lance. Spectral transfer learning using information geometry for a user-independent brain-computer interface. *Frontiers in Neuroscience*, 10(SEP), 2016.

[632] C.S. Weaver. Digital filtering with application to electrocardiogram processing. *IEEE Trans. on Audio and Electroacoustics*, 16(3):350–391, 1968.

[633] B. Welch. The significance of the difference between two means when the population variances are unequal. *Biometrika*, 29:350–362, 1938.

[634] B.L. Welch. The generalization of "Student's" problem when several different population variances are involved. *Biometrika*, 34(1–2):28–35, 1947.

[635] G. Welch, G. Bishop, and R.E. Kalman. A new approach to linear filtering and prediction problems. *Transaction of the ASME – Journal of Basic Engineering*, pages 35–45, 1960.

[636] F. Wendling, K. Ansari-Asl, F. Bartolomei, and L. Senhadji. From EEG signals to brain connectivity: a model-based evaluation of interdependence measures. *Journal of Neuroscience Methods*, 183:9–18, Sep 2009.

[637] F. Wendling, F. Bartolomei, J.J. Bellanger, and P. Chauvel. Epileptic fast activity can be explained by a model of impaired GABAergic dendritic inhibition. *Eur J Neurosci*, 15:1499–1508, May 2002.

[638] J.W. Wheless, L.J. Willmore, J.I. Breier, M. Kataki, J.R. Smith, D. W. King, K.J. Meador, Y.D. Park, D.W. Loring, G.L. Clifton, J. Baumgartner, A.B. Thomas, J.E. Constantinou, and A.C. Papanicolaou. A comparison of magnetoencephalography, MRI, and V-EEG in patients evaluated for epilepsy surgery. *Epilepsia*, 40:931–941, 1999.

[639] J.C Whitman, L.M Ward, and T.S Woodward. Patterns of cortical oscillations organize neural activity into whole-brain functional networks evident in the fMRI bold signal. *Frontiers in Human Neuroscience*, 7:80, 2013.

[640] M. Wibral, C. Bledowski, A. Kohler, W. Singer, and L. Muckli. The timing of feedback to early visual cortex in the perception of long-range apparent motion. *Cerebral Cortex*, 19(7):1567–1582, Jul 2009.

[641] M. Wibral, C. Bledowski, and G. Turi. Integration of separately recorded EEG/MEG and fMRI data. In Markus Ullsperger and Stefan Debener, editors, *Simultaneous EEG and fMRI: Recording, Analysis, and Application*. Oxford Scholarship Online, 2010. Chapter 3.7.

[642] N Wiener. *Extrapolation, Interpolation, and Smoothing of Stationary Time Series*. Wiley, New York, 1949.

[643] C. Wilke, L. Ding, and B. He. Estimation of time-varying connectivity patterns through the use of an adaptive directed transfer function. *IEEE Trans Biomed Eng*, 55(11):2557–2564, Nov 2008.

[644] J.D. Wilson, R.B. Govindan, J.O. Hatton, C.L. Lowery, and H. Preissl. Integrated approach for fetal QRS detection. *IEEE Trans Biomed Eng*, 55:2190–2197, Sep 2008.

[645] I. Winkler, S. Haufe, and M. Tangermann. Automatic Classification of Artifactual ICA-Components for Artifact Removal in EEG Signals. *Behavioral and Brain Functions*, 7(1):30, Aug 2011.

[646] J. Winslow, M. Dididze, and C.K. Thomas. Automatic classification of motor unit potentials in surface EMG recorded from thenar muscles paralyzed by spinal cord injury. *Journal of Neuroscience Methods*, 185:165–177, Dec 2009.

[647] M. Winterhalder, T. Maiwald, H.U. Voss, R. Aschenbrenner-Scheibe, J. Timmer, and A. Schulze-Bonhage. The seizure prediction characteristic: a general framework to assess and compare seizure prediction methods. *Epilepsy & Behavior*, 4:318–325, Jun 2003.

[648] M. Winterhalder, B. Schelter, W. Hesse, K. Schwab, L. Leistritz, D. Klan, R. Bauer, J. Timmer, and H. Witte. Comparison of linear signal processing techniques to infer directed interactions in multivariate neural systems. *Signal Processing*, 85:2137–2160, 2005.

[649] H.P. Wit, P. van Dijk, and P. Avan. Wavelet analysis of real ear and synthesized click evoked otoacoustic emissions. *Hearing Research*, 73:141–147, Mar 1994.

[650] C. Woody. Characterization of an adaptive filter for the analysis of variable latency neuroelectric signals. *Medical and Biological Engineering and Computing*, 5:539–554, 1967.

[651] S.C. Wriessnegger, J. Kurzmann, and C. Neuper. Spatio-temporal differences in brain oxygenation between movement execution and imagery: a multichannel near-infrared spectroscopy study. *International Journal of Psychophysiology*, 67(1):54–63, Jan 2008.

[652] J.J. Wright. *Brain Dynamics*, chapter Linearity and nonlinearity in electrocortical waves and their elementary statistical dynamics, pages 201–213. Springer Series in Brain Dynamics 2. Springer Verlag, 1989.

[653] J.J. Wright, P.D. Bourke, and C.L. Chapman. Synchronous oscillations in the cerebral cortex and object coherence: Simulation of basic electrophysiological findings. *Biol Cybern*, 83:341–353, 2000.

[654] M. Wyczesany, M.A. Ferdek, and S.J. Grzybowski. Cortical functional connectivity is associated with the valence of affective states. *Brain and Cognition*, 90:109–115, Oct 2014.

[655] S. Yamada and I. Yamaguchi. Magnetocardiograms in clinical medicine: unique information on cardiac ischemia, arrhythmias, and fetal diagnosis. *Internal Medicine*, 44:1–19, Jan 2005.

[656] F. Yger, M. Berar, and F. Lotte. Riemannian approaches in brain-computer interfaces: A review. *IEEE Transactions on Neural Systems and Rehabilitation Engineering*, 25(10):1753–1762, 2017.

[657] S.N. Yu and Y.H. Chen. Electrocardiogram beat classification based on wavelet transformation and probabilistic neural network. *Pattern Recognition Letters*, 28:1142–1150, 2007.

[658] H. Yuan, T. Liu, R. Szarkowski, C. Rios, J. Ashe, and B. He. Negative covariation between task-related responses in alpha/beta-band activity and bold in human sensorimotor cortex: an EEG and fMRI study of motor imagery and movements. *Neuroimage*, 49(3):2596–2606, Feb 2010.

[659] M.A. Yücel, J. Selb, C.M. Aasted, P.Y. Lin, D. Borsook, L. Becerra, and D.A. Boas. Mayer waves reduce the accuracy of estimated hemodynamic response functions in functional near-infrared spectroscopy. *Biomedical Optics Express*, 7(8):3078–3088, Aug 2016.

[660] P. Zaramella, F. Freato, A. Amigoni, S. Salvadori, P. Marangoni, A. Suppiej, A. Suppjei, B. Schiavo, and L. Chiandetti. Brain auditory activation measured by near-infrared spectroscopy (NIRS) in neonates. *Pediatric Research*, 49(2):213–219, Feb 2001.

[661] J.J. Żebrowski, W. Popławska, and R. Baranowski. Entropy, pattern entropy, and related methods for the analysis of data on the time intervals between heartbeats from 24-h electrocardiograms. *Physical Review E*, 50(5):4187–4205, 1994.

[662] X. Zeng, Y. Dong, and X. Wang. Flexible electrode by hydrographic printing for surface electromyography monitoring. *Materials*, 13(10), 2020.

[663] D. Zennaro, P. Wellig, V.M. Koch, G.S. Moschytz, and T. Laubli. A software package for the decomposition of long-term multichannel EMG signals using wavelet coefficients. *IEEE Trans Biomed Eng*, 50:58–69, Jan 2003.

[664] X. Zhang, L.G. Durand, L. Senhadji, H.C. Lee, and J.L. Coatrieux. Analysis-synthesis of the phonocardiogram based on the matching pursuit method. *IEEE Trans Biomed Eng*, 45(8):962–971, Aug 1998.

[665] X. Zhang, W. Zhong, Jurij Brankačk, Sascha W. Weyer, Ulrike C. Müller, Adriano B.L. Tort, and Andreas Draguhn. Impaired theta-gamma coupling in APP-deficient mice. *Scientific Reports*, 6(1), Apr 2016.

[666] Z.G. Zhang, V.W. Zhang, S.C. Chan, B. McPherson, and Y. Hu. Time-frequency analysis of click-evoked otoacoustic emissions by means of a minimum variance spectral estimation-based method. *Hear Res*, 243:18–27, Sep 2008.

[667] H. Zhao, J.F. Strasburger, B.F. Cuneo, and R.T. Wakai. Fetal cardiac repolarization abnormalities. *Am J Cardiol*, 98:491–496, Aug 2006.

[668] H. Zhao and R.T. Wakai. Simultaneity of foetal heart rate acceleration and foetal trunk movement determined by foetal magnetocardiogram actocardiography. *Physics in Medicine & Biology*, 47:839–846, Mar 2002.

[669] M. Zijlmans, W. Zweiphenning, and N. van Klink. Changing concepts in presurgical assessment for epilepsy surgery. *Nature Reviews Neurology*, 15, Jul 2019.

[670] A.M. Zoubir and B. Boashash. The bootstrap and its application in signal processing. *IEEE Signal Processing Magazine*, 15:56–76, 1988.

[671] M.J. Zwarts, G. Drost, and D.F. Stegeman. Recent progress in the diagnostic use of surface EMG for neurological diseases. *Journal of Electromyography & Kinesiology*, 10:287–291, Oct 2000.

[672] G. Zweig and C.A. Shera. The origin of periodicity in the spectrum of evoked otoacoustic emissions. *J Acoust Soc Am*, 98:2018–2047, Oct 1995.

[673] J. Zygierewicz, K.J. Blinowska, PJ. Durka, W. Szelenbeger, S. Niemcewicz, and W. Androsiuk. High resolution study of sleep spindles. *Clinical Neurophysiology*, 110(12):2136–2147, 1999.

[674] J. Zygierewicz, P.J. Durka, H. Klekowicz, P.J. Franaszczuk, and N.E. Crone. Computationally efficient approaches to calculating significant ERD/ERS changes in the time-frequency plane. *Journal of Neuroscience Methods*, 145:267–276, Jun 2005.

[675] Q. Zhang, A. I. Manriquez, C. Medigue, Y. Papelier, and M. Sorine. An algorithm for robust and efficient location of T-wave ends in electrocardiograms. *IEEE Trans Biomed Eng*, 53:2544-2552, 2006.

Index

For Product Safety Concerns and Information please contact our EU
representative GPSR@taylorandfrancis.com
Taylor & Francis Verlag GmbH, Kaufingerstraße 24, 80331 München, Germany